Solar Energy Dictionary

Solar Energy Dictionary

V. DANIEL HUNT

Director, The Energy Institute

INDUSTRIAL PRESS INC.

200 Madison Ave.

New York, N.Y. 10157

Notice

This book was prepared as an account of work sponsored by the Industrial Press. Neither the Industrial Press nor The Energy Institute, nor any of their employees, nor any of their contractors, subcontractors, consultants, or their employees, make any warranty, expressed or implied, or assumes any legal liability or responsibility for the accuracy, completeness or usefulness of any information, apparatus, product or process disclosed, or represents that its use would not infringe on privately owned rights.

The views, opinions and conclusions in this book are those of the author and do not necessarily represent those of the United States Government or the United States Department of Energy.

Public domain information and those documents abstracted or used in full, edited or otherwise used are noted in the acknowledgments or on specific pages or illustrations.

Library of Congress Cataloging in Publication Data

Hunt, V. Daniel.
Solar energy dictionary.

Bibliography: p.
1. Solar energy—Dictionaries. I. Title.
TJ810.H86 621.47'03'21 81-13393
ISBN 0-8311-1139-9 AACR2

FIRST PRINTING
SOLAR ENERGY DICTIONARY

Acknowledgments

I am indebted to the support, assistance, and kindness received from the following organizations for their cooperation in providing research materials utilized in the writing of this book.

- The Solar Energy Research Institute
- The Department of Energy
- The Electric Power Research Institute
- The National Solar Heating and Cooling Information Center
- The Environmental Protection Agency
- The U.S. Department of Agriculture
- The Library of Congress

Credit for the photographs and illustrations is indicated where appropriate throughout this dictionary. The Department of Energy provided the majority of the photographs.

I acknowledge and wish to specifically credit all of the sources reviewed for the selection of the terms and definitions used in this book:

A Dictionary of Mining, Mineral, and Related Terms, Paul W. Thrush and Staff of Bureau of Mines (eds.)
Annual Review of Solar Energy, Solar Energy Research Institute
An Overview of the Impact of Carbon Dioxide on Climate, The MITRE Corporation
A Survey of Passive Solar Buildings, U.S. HUD and DOE
Biological Paths to Self-Reliance, Russell E. Anderson
Building a Sustainable Future, Solar Energy Research Institute
Build Your Own Solar Water Heater, Stu Campbell with Doug Taff
Buying Solar, Federal Energy Administration/DOE
Common Environmental Terms, A Glossary, U.S. Department of Commerce
Direct Use of the Sun's Energy, Farrington Daniels
DOE Facilities Solar Design Handbook, U.S. DOE and HUD
Earth, Water, Wind and Sun, Dan S. Halacy
Energy Alternatives, A Comparative Analysis, U.S. Council on Environmental Quality
Energy Conservation Standards, Dubin and Long
Energy for Survival—The Alternative to Extinction, Wilson Clark
Energy from the Waves, David Ross
Energy Future, Daniel Yergin *et al.* (eds)
Energy Policy, Congressional Quarterly Inc.
Energy Storage, William W. Eaton

Glossary of Electric Utility Terms, Edison Electric Institute
Handbook of Conservation and Solar Energy—Trends and Perspectives, V. Daniel Hunt
Heating Your Home with Wood, Neil Soderstrom
Home Energy How-to, A. J. Hand
Hot Water from the Sun, U.S. HUD and DOE
How to Build a Solar Heater, Ted Lucas
How to Buy Solar Heating—Without Getting Burnt, Malcolm Wells and Irwin Spetgang
How to Make Your Own Alcohol Fuels, Larry W. Carley
Installation Guidelines for Solar DHW Systems, U.S. HUD and DOE
Making and Using Electricity from the Sun, Technical Staff of Solarex Corp.
Methane: Planning a Digester, Peter-John Meynell
Morsø Wood Heat Handbook, Southport Stoves, Inc.
National Program Plan for Passive and Hybrid Solar Heating and Cooling, DOE/CS-0089, DOE
Natural Solar Architecture and Passive Primer, David Wright
Oceans of Energy, Augusta Goldin
Photovoltaics—Sunlight to Electricity in One Step, Paul D. Maycock and Edward N. Stirewalt
Principles of Solar Engineering, Frank Kreith and Jan F. Kreider
Protecting Solar Access for Residential Development, U.S. HUD and DOE
Put the Sun to Work Today, DOE
Residential Solar Design Review, U.S. HUD and DOE
Site Planning for Solar Access, U.S. HUD and DOE
Solar Energy, William W. Eaton
Solar Energy for Space Heating and Hot Water, DOE, Division of Solar Energy
Solar Energy: Fundamentals in Building Design, Bruce Anderson
Solar Energy in America, William D. Metz and Allen L. Hammond
Solar Energy Information Locator, SERI
Solar Energy In Review, draft, SERI
Solar Energy Systems—Copper Brass Bronze Design Handbook, Copper Development Association Inc.
Solar Energy Systems and Resources, The MITRE Corporation
Solar Energy—Technology and Applications, J. Richard Williams
Solar Energy Technology Handbook, William C. Dickinson and Paul N. Cheremisinoff
Solar Energy Thermal Process, John A. Duffie and William A. Beckman
Solar Energy Utilization, Tim Michels
Solar Greenhouse, Bill Yanda and Rick Fisher
Solar Heating and Cooling, Jackson Hand
Solar Heating and Cooling, Jan F. Kreider and Frank Kreith
Solar Heating Design—By The F-Chart Method, William A. Beckman
Solar Houses, Louis Gropp
Successful Solar Energy Solutions, Spruille Braden III and Kathleen Steiner
Sun Language, Solar Energy Institute of America
Sun Up to Sun Down, Shawn Buckley
The Coming Age of Solar Energy, Dan S. Halacy
The First Passive Solar Home Awards, U.S. HUD and DOE
The Gasohol Handbook, V. Daniel Hunt
The Passive Solar Energy Book, Edward Mazria
The Salton Sea Solar Pond, DOE
Thermal Shutters and Shades, William A. Shurcliff
Windpower—A Handbook on Wind Energy Conversion Systems, V. Daniel Hunt
Wood Energy, Mary Twitchell

Preface

Solar energy research, development, and demonstration have rapidly expanded over the past decade; as a result, new terminology is appearing at a phenomenal rate. This dictionary provides an introduction to the various solar technologies and definitions of more than 3000 solar energy terms, processes, or products. Extensive support material is provided by delineating points of contact for additional information, acronyms, a detailed bibliography, and other reference data.

The definitions encompass processes, products, and renewable solar energy technologies including:

Passive solar
Active solar
Photovoltaics
Biomass
Alcohol fuels (Gasohol)
Wind energy conversion
Solar thermal energy
Ocean energy
Satellite power systems
Industrial process heat
Waste energy conversion

In addition to the specific renewable solar energy technologies, this book provides definitions of terms in fields that are closely related such as energy conservation, environment, heating, ventilation and cooling, and solar installation plumbing.

Presenting definitions as simply and clearly as possible, this dictionary also gives you more than 200 additional aids—charts, graphs, process diagrams, and photographs—that further clarify many of the definitions.

Solar Energy Dictionary is based on information acquired from numerous experts; it includes information from the Solar Energy Research Institute, the Department of Energy, the U.S. Department of Agriculture, the Environmental Protection Agency, NASA, the Conservation and Renewable Energy Information Network, and many other organizations.

Contents

Acknowledgments v
Preface vii
Rediscovering the Sun 1
Entries 25
Appendices
  Metric Conversion Tables 351
  For More Information 355
  Bibliography 369
  Acronyms and Abbreviations 405

REDISCOVERING THE SUN

For millions of years nature processed, developed, and stored energy resources in the earth. Until quite recently these resources went untouched, an unexploited legacy of prehistoric time. By the American Revolution, most of the energy the nation consumed came from the muscle power of humans and animals, and from renewable energy supplied by the sun through the action of water, wind, and the burning of wood.

Then, at steadily escalating rates, industrialized society started drawing down its capital energy assets. First coal, then oil, gas, and nuclear power were used to produce steam and electricity for factories and homes. Ever larger amounts of energy were used by trains, trolleys, automobiles, buses, and airliners. By the mid-twentieth century, commerce, industry, and personal life-styles were based on a cornucopia of cheap and abundant energy.

THE PRESENT

Today, most of the energy used throughout the world still is drawn from the earth's capital assets. Each year the world consumes as much oil as it took nature 2 million years to create. A vast and complicated distribution system has been built to make petroleum and other finite energy sources serve people. Meanwhile the United States, once self-sufficient in petroleum, is now dependent on imported oil.

THE ENERGY FUTURE

Energy trends indicate that world oil production will peak and start a long downward slide around the 1990s. Replacement with alternative energy sources will be difficult, time-consuming, and expensive. Nuclear power, unlike coal and oil, produces only electricity for consumer use. Although there is an abundance of coal in the United States, its increased use and application in the economic mainstream may not be accomplished swiftly.

With the continued depletion of the world's nonreplaceable fuels, it is important that more attention be paid to energy conservation, the use of alternative resources, and renewable energy.

Coal must be substituted for oil and gas wherever possible, especially in electric power plants and other industries. Municipal and agricultural wastes must be converted into liquid and gaseous fuels.

We must better harness the wind with machines based on sound designs, and use sunlight to heat buildings and to provide power for a number of industrial and agricultural processes. There are other energy applications, like hydrogen and fusion, now being developed, but their practical uses are decades away. Meanwhile, by using energy more efficiently in transportation, buildings, and industrial processes, much of the flagrant waste that has prevailed can be avoided.

SOLAR HEATING SYSTEMS

Using heat from the sun to warm buildings and produce hot water represents the simplest and most practical application of solar technology. The basic material that makes this solar heating possible is ordinary window glass.

Sunlight passes easily through glass and strikes a black or darkened surface. Part of the sunlight striking this surface changes from visible light to heat. Light can be reflected back out again through the glass, but heat is absorbed or trapped by the glass.

This phenomenon, termed the greenhouse effect, is familiar to anyone who has left his car exposed to full sunlight, with the windows tightly closed. The interior of the car is much warmer than the outside air, even on a cold winter day.

Whether you realize it or not, your own home probably incorporates some solar features now. Something as simple as a south-facing window, for example, admits direct sunlight during most of the day. The window glass traps the solar heat, adding warmth to the room. If the window is covered with a storm sash or is double glazed, the warming effect is enhanced: the tighter window seal means that less warm air is lost to the outside; less cold air infiltrates the house, and less solar heat escapes through the double-glass panes.

DIRECT PASSIVE SOLAR HEATING

Solar designers and architects recognize a number of building features that can take advantage of solar energy. These direct solar systems, often called passive systems, are usually an integral part of the building and are, therefore, most suitable for new construction. They require few, if any, mechanical or working parts.

Ideally, the designer of a direct solar system begins by positioning or orienting the building so that walls and roof receive maximum solar radiation in winter. This means that, with a rectangular structure, the long axis runs east-west, allowing a southerly exposure for the long side of the building. In practice, it is often best to orient the long side facing slightly to the southeast; this permits the building to receive more sunlight earlier in the day, when it needs to recover warmth that has been lost overnight. By the same token, this orientation reduces solar heat gain in the late afternoon, an obvious advantage in the hot summer months.

Such simple features as large window openings on the south, and small windows, or none at all, on the north, serve to maximize solar heat gain and retention of heat in winter. Movable shutters and awnings, or fixed overhangs, can be used in summer to ward off excess solar radiation.

Heavy construction materials, such as brick, concrete, stone, or adobe, are favored by solar designers because they provide *thermal mass.* These materials are capable of storing heat overnight or for much longer periods of time. During the day, for example, walls with good thermal mass properties absorb heat. They release the heat slowly during the night, thus helping to moderate the day-night temperature differences.

Direct Solar Building Designs Although there are many building plan variations that use direct solar heating principles, most designs fall into one of several basic types.

In the direct heat gain plan, which is the simplest, the building's south wall features a large expanse of double-pane glass. In winter the low-altitude sun enters through the window, and heat is stored in the building's masonry or concrete floor and solid northern wall. This design makes the whole building a live-in solar collector.

On cold days and nights, heat stored in the floors and walls radiates throughout the living space. Movable insulating shutters are used at night to cover the south walls to prevent heat loss.

Excessive solar heat gain is prevented during the summer by a roof overhang or awnings which shield the southern glass wall from the high-altitude sun.

Another basic design features a thermal storage wall which faces south. The wall is usually painted black or another dark heat-absorbing color, and is faced with glass. A few inches of air space separate the glass from the wall to trap the solar heat. Sunshine enters through the glass, and heat is immediately stored in the wall. The stored heat radiates throughout the building at night and on cloudy days.

The thermal storage wall is usually made of massive, 2-foot-thick (0.6-meter-thick) concrete, although other materials are possible. For example, since water is an excellent heat storage medium, walls made of stacks of water-filled oil drums have been successfully used.

A variation of the thermal storage wall is the trombe wall, named for its designer. Vents are located at the top and bottom of a glass-faced thermal storage wall. Cool room air entering the lower vents rises as it is heated by sunlight and reenters the room through the upper vents. This natural process, termed thermosiphoning, is based on the simple and familiar principle that air rises as it absorbs heat. An advantage of the trombe wall is that it provides gentle circulation and a continual replenishment of warm air to the room.

Space does not permit the listing of all systems that are either available or are on the drawing boards. Many inventive people are at work designing solar application units.

Direct Solar-Heating Potential Although direct solar-heating systems are not now widely used in the United States, their future potential could be significant. It should be noted that some direct solar-heated homes will require a conventional heating system to serve as a back-up unit during periods of cloudy weather or extreme cold.

Studies by the Department of Energy indicate that direct solar-heating systems could provide practically all of the heat needed for a home in Los Angeles and a substantial fraction of home heating needs in other areas. For example, direct solar designs could supply 60 percent of the heat needed for a home in New York, 57 percent in Seattle, and 42 percent in Madison, Wisconsin.

For most existing buildings of conventional architectural style, direct solar systems are not a cost-effective retrofit. These systems are such an integral part of building design that they do not lend themselves to an economical retrofit.

INDIRECT ACTIVE SOLAR-HEATING SYSTEMS

Homeowners do not need to build new houses to take advantage of solar energy. Many existing structures can be easily retrofitted with indirect solar-heating systems. Indeed, indirect systems are the most common types of solar units being installed in homes today.

The indirect solar-heating system, often called an active solar system, is simple, consisting of six main parts, each with an essential function:

A collector, to capture the solar heat;

A transport system, to move the heat from the collector to storage;

A storage unit, to retain heat for use at night or on cloudy days;

A distribution system, to move the heat from storage to rooms where it is needed;

An auxiliary system, or conventional heating unit, for use in emergencies, or to provide any warmth that cannot be supplied by the solar system alone; and

A system of controls, to regulate the operation of the total system.

Collectors Unlike the direct solar collector, which is an integral part of a building (usually a wall), the indirect solar collector is a separate unit. It can be mounted on the roof, on the south wall of a building, or on its own supporting structure away from the house.

The typical collector is a shallow rectangular box, painted a heat-absorbent black inside, and covered with glass to retain the solar energy.

These flat-plate collectors are of two general types, depending upon whether they use air or a liquid medium to absorb the heat and transport it to where it can be stored or used.

In the air-type unit, air is blown over the blackened surface of the collector. This surface has corrugations or other irregularities that create turbulence in the air flow. The turbulent air absorbs heat from the collector, then carries it to a storage unit.

Water or another liquid can also be used to absorb solar heat. In the simplest liquid-type collector, cool water trickles down the length of a vertically corrugated, blackened absorber surface. It gathers as hot water in a trough at the bottom edge of the collector, from which it is moved to storage.

In other units, the liquid transport medium is confined to tubes running on, under, or embedded in, the blackened absorber surface. In yet another variation, the absorber is made of two pieces of sheet metal, with space for the liquid transport medium to flow in between.

Air and liquid collectors each have their own advantages and disadvantages. Air systems are simpler in design and generally easier to install. Freezing and corrosion problems are of less concern than they are in liquid units. Also, unlike the liquid unit, there is no risk of damage to the building or household effects if a leak develops in the air-distribution system. On the other hand, heat-storage units used with liquid collector systems can be smaller than those used with air systems. This may be a considerable advantage in homes where space is at a premium. Liquid systems are much more practical in situations where solar energy is to be used for domestic hot water.

Transport and Distribution Systems There are two transport loops or subsystems in any indirect solar-heating system: one to carry heat from the collector to storage and another to distribute it throughout the house or wherever it is needed. It is possible, by manipulating controls on the system, to distribute solar heat directly from the collectors to rooms where it is needed, bypassing the storage unit.

Some points to consider when choosing between air and liquid collector/distribution systems:

- Air systems, which use ductwork, are generally cheaper to install than liquid systems, which rely on more complex plumbing.
- In cold climates, it is necessary to use antifreeze or automatic drain-down devices in liquid systems, or run a risk of frozen pipes and costly damage.
- Air systems can be more expensive to operate than liquid systems. This is so because more electric power is typically needed to run the air-system fans than the pumps that are used in liquid systems.
- In cities, or especially dusty areas, it may be necessary to install filters to clean and purify the air for use in air systems.

Storage Systems Two types of storage units are in common use with indirect solar heating systems.

With air systems the storage unit usually consists of a bin or container filled with rocks. The rocks absorb heat from air blown in from the solar collectors. They can remain hot for several days if necessary, thus storing the sun's energy.

A tank of water serves much the same purpose for liquid systems. Water, like rock, is inexpensive and has the ability to retain heat for long periods of time. Hot water from the storage tank can be circulated directly to room radiators. In some designs, a heat exchanger (usually a simple copper coil or finned tube) is located in the storage tank. Fluid circulating through the heat exchanger picks up the stored heat and carries it to rooms where it is needed.

Most storage units, because of their size and weight, are located in basements or crawl spaces. In general, it is best to locate the storage unit as close as possible to the collector, in order to minimize heat loss through pipes and ducts.

Auxiliary Heating Systems Most solar-heating systems are designed to provide 50 to 90 percent of a building's heating requirements. Therefore, auxiliary heating systems are needed in nearly all solar buildings.

These may be as costly as conventional oil furnaces or as simple and inexpensive as wood-burning stoves. In areas where the climate is not severe, several electric space heaters may be adequate.

Controls Nearly all indirect solar systems use electrical controls to regulate system operation. A differential thermostat is frequently used to automatically compare the temperatures in different parts of the system.

If the temperature is higher in the collector than in the storage unit, for example, the thermostat will turn on pumps in the transport system to carry collector heat to storage. If the room temperature is too low, and the heat in storage is insufficient to raise it to the desired level, the thermostat will turn on the auxiliary system.

SOLAR WATER HEATING

Even if homeowners decide that solar space heating is too expensive, the chances are still good that a solar water-heating system would be economical.

A number of solar water-heater designs are possible, including both direct and indirect systems.

In the simplest direct systems water is heated in a black-painted glass-enclosed vessel. This collector vessel may also double as the hot-water storage tank. In some cases the heated water in the collector flows of its own accord, by thermosiphoning through a pipe to a storage tank.

Indirect hot-water systems are similar to indirect liquid-type space heating systems. Water is pumped from a flat-plate collector to a storage unit.

The solar-heated water can be used directly for domestic purposes, or it can just be used as the transport medium to absorb heat, then transfer it to the domestic water supply via a heat exchanger. The latter design is often used in cold climates where the antifreeze is added to water in the collector to prevent ice formation and damage to plumbing. When antifreeze is used, a heat exchanger transfers heat to the domestic water supply without contaminating it.

In most hot-water systems a conventional auxiliary hot-water heater is needed.

Heating water for swimming pools is an excellent application of solar technology. In some states it may be the only option, because the use of fossil fuels for such "nonessential" purposes is limited.

In principle, solar swimming-pool heating systems are similar to other indirect solar-heating systems, though they are simpler and can be less expensive. Very simple collectors can be used because required temperatures are low, no antifreeze is required, and the pool itself serves as the storage tank. The same pump needed to run the pool-filter system can also be used to circulate water through the collector and back again.

BUYING SOLAR

Now you know something about how solar-heating systems work. Perhaps at this point you're really interested in building or retrofitting your own solar home. But before you plunge ahead, there are things you should find out about your home and about your own motives for wanting to go solar.

ENERGY CONSERVATION FIRST

If you are thinking about retrofitting an existing home, it is *essential* that the building first be well-insulated and as energy-efficient as possible. It makes neither common sense nor economic sense to spend money on systems that capture solar heat if this same heat is allowed to escape through drafty walls, attics, and window/door openings.

If you are installing a solar hot-water system, do not overlook some of the simple things you can do to avoid unnecessary loss of hot water. For example, you should install flow restrictors in shower heads and insulate all exposed hot-water pipes. There are many good books that can help you evaluate the energy efficiency of your home and that describe the most cost-effective conservation measures.

SOLAR IS INVENTIVE

When most people are asked why they are considering installing solar energy systems, they typically respond, "To save money on utility bills, of course." And indeed you *can* save money on utility bills. But there's much more to the economics of solar than that—it appeals to inventive people. As a result, a number of people are attracted to solar energy for other than purely economic reasons:

- It can provide a measure of self-sufficiency.
- Using renewable solar energy helps conserve our fossil fuel resources.
- Direct solar architecture can be aesthetically pleasing, simple and elegant in design.

The weight attached to such noneconomic arguments for solar is personal. Only you can decide how much value they hold for you.

THE ECONOMICS OF SOLAR

Sunlight may be free, but the equipment needed to capture and use solar energy is not.

The key to understanding solar economics is life-cycle cost. This means the cost of purchasing, financing, and maintaining a solar unit over the period of its useful life—which may be 20 to 30 years, or even more. At some point, the end of the payback period, the amount of money saved in utility bills will equal the extra cost of installing a solar system.

The length of this payback period determines just how economic solar is for you. As the cost of fuel rises, the payback period will be shortened, and the economics of solar will become more favorable. Since the future price of fuels cannot be known precisely, payback periods for solar equipment must be estimates, at best.

The Department of Energy has found that, in most parts of the country, solar heating makes good economic sense when it can be financed as part of a new home mortgage:

- If the homeowner's only other choice is electric resistance heating, and
- If the solar unit can be installed at a total cost equal to about $20 per square foot of collector area.

Many factors in addition to initial purchase price must be considered in any comprehensive analysis of solar economics. For example, prevailing interest rates, or the cost of financing a solar system, will affect the length of the payback period.

BUILDING CODES, ZONING, AND MASTER PLANS

Before proceeding to build or install a solar system, it is prudent to check your local building code and the master plan for the area in which you live.

Building codes regulate the materials and construction techniques that may be used in buildings. Master plans and zoning ordinances establish control over land use, including such aspects as location of structures on building lots, height of structures, and the aesthetics or architectural styles that are permitted. Both codes and zoning ordinances can affect how, and even whether, you can build a solar system.

If you engage an architect or contractor, he normally is responsible for seeing that the design and installation meet local building codes. But even if you do hire professionals, it is good to become personally acquainted with the master plan and local zoning ordinances.

It is especially important to find out whether they would permit construction of a building, on a neighboring lot, high enough to cast a shadow on your solar collector. Even a tree planted by a neighbor could limit the sunlight falling on your collector. Local ordinances vary; many areas do not yet have legislation to provide protection against such acts or to deal with sun rights.

EVALUATING SOLAR EQUIPMENT

Prior to purchasing any solar equipment survey manufacturers' data. This information is summarized in publications such as *The Solar-Industry Index; The Com-*

prehensive Guide to Manufacturers and Service Organizations, Solar Age magazine, and information available from the Conservation and Renewable Energy Network.

Any product you are considering should meet standards for solar equipment cited in the Department of Housing and Urban Development's Minimum Property Standards, used for evaluating Federal Housing Administration and Veterans Administration loans.

As a further precaution, any equipment you buy should be backed up by a warranty. Some warranties are "limited" and cover only the cost of faulty parts. Others are "full" warranties and promise payment for both parts and labor to correct problems or defects. Read all warranties carefully and be sure you fully understand the terms that are offered. The Latin term of caution, caveat emptor, "Let the buyer beware," applies to this new and growing industry; not all dealers are solar experts with lengthy experience.

USING SOLAR ENERGY TO PRODUCE ELECTRICITY AND FUELS

The introductory solar technology discussed to this point has involved ways to capture, store, and use heat received from the sun. But energy from the sun can also be used to generate electricity and to produce liquid or gaseous fuels.

There are many ways to produce electricity and fuels using solar power. Some techniques are commonly used today. Others are still experimental, or they are presently uneconomic except for specialized purposes.

For the most part, electricity and liquid and gaseous fuels are now produced from nonrenewable fossil energy sources. The federal government has supported research and demonstrations of solar-electric and solar-fuels technologies in order to reduce our dependence on fossil fuels and to place more reliance on renewable energy from the sun.

PASSIVE SOLAR ENERGY

Among the proven, and even ancient, methods of capturing solar energy are direct passive solar systems. Properly designed, located, and constructed buildings can take full advantage of natural heating and cooling from the sun. A structure can be built to absorb the winter sun's heat and to circulate warm air in and cool air out. The same structure can deflect the summer sun while absorbing cooler air from a breeze.

Passive solar energy refers to the process requiring no mechanical devices to collect and distribute solar-gained heat or solar-generated coolness. However, in cases where passive principles can be used but not relied on totally for sufficient comfort, mechanical equipment supplements the passive system. This combination is called a hybrid system. Certain architectural and building principles are the cornerstone of both passive and hybrid systems.

For thousands of years, people have used the sun to heat their shelters. Cities such as Olynthus in ancient Greece and Priene in Asia Minor utilized passive solar principles in urban planning and in building design. For example, the designers of these cities laid out streets in north-south and east-west directions, giving houses a direct southern exposure. Similar urban plans existed in ancient China. In the Roman Empire, the courts established solar access laws to guarantee a person's right to unobstructed sunlight.

In North America, before European settlers left for the New World, Pueblo Indians in the Southwest constructed passive solar communities. During the eleventh and twelfth centuries, the Anasazi people built community shelters in south-facing cliffs. Mesa Verde and Pueblo Bonito are examples of these terraced units which allowed sun absorption during the winter.

A passive solar building's effectiveness depends on how efficiently it can collect and use solar radiation. Elements such as seasonal solar angles, location and size of vegetation, and building orientation can affect the efficiency of a solar-heated or cooled building. Buildings are designed in consideration of these variables and are constructed with heat-absorbent materials such as masonry or stone.

Insulation plays a vital role in improving the efficiency of a solar house. Insulation can range in type from fiberglass, urethane, or styrene to the earth itself.

Air circulation is also an important part of any space-heating or cooling system. Interior floor and ceiling vents, exterior vents, wall angles, and air ducts leading from one level to another all enhance the passive heating or cooling modes.

Passive solar buildings can be constructed in hillsides called earth berms, or underground, taking advantage of the earth's thermal "lag time." Underground buildings remain cool in summer months and warm in winter months, unlike buildings on the surface.

Passive solar homes rely on the direct gain of solar radiation, either in the form of direct sunlight or daylight. This means solar energy, as heat, falls on thermal mass walls, half-walls, or floors where the energy can be stored. In the evening, the stored heat naturally passes from the thermal mass back to the air, thus heating the building. Primary absorption areas are oriented to the south to permit the maximum amount of winter solar radiation to enter, while preventing the summer sun from overheating the rooms.

A less conventional yet effective method of extra solar collection is a rock bin, wall, or room below the first floor of a building. The sun hits this exposed area just as it does the thermal mass floors and walls above. The rocks collect and store heat which then naturally rises into the living areas. Circulation vents also allow cooler room air to fall into the rocks below where it can be reheated, becoming part of a heating and distribution cycle.

A greenhouse can function both as a solar collector and a storage unit. To do so, it must be well-insulated, airtight, and sufficiently glazed, and must contain a thermal storage system. The glass walls of a greenhouse permit solar radiation to fall on thermal storage masses where it is absorbed. Specially designed walls prevent much of the radiation from reflecting into the atmosphere and being lost. Shades or movable insulation prevent heat from seeping out at night. In the summer, deciduous trees provide cover to shade the greenhouse. During the winter, solar radiation can fall unimpeded onto the greenhouse.

The various layers of insulation built into an attached greenhouse, and the extra insulation it provides for the solid wall to which it is attached, together can reduce heat loss by 9 to 37 percent for a building.

The federal government has focused research and development in five major sectors: residential buildings, commercial buildings, products, agricultural applications, and solar cities and towns.

Specific projects include collection of winter data on passive solar buildings; examining methods of applying solar principles in mass-produced, factory-built buildings; development of a passive system costing technique; and development of selective coatings for windows to enhance their passive solar properties.

ACTIVE SOLAR ENERGY

The sun can be utilized in many ways to provide heat, light, and energy for our modern needs and comforts. Some methods of capturing solar energy are old, but others have been developed recently through the applications of industrial technology. Active solar heating and cooling systems rely on various types of mechanical equipment to collect and distribute solar heat.

A major feature of the active solar system is its adaptability to existing buildings. Collector panels can be attached to flat and most slanted roofs, and connected to the existing distribution system. Active systems can also be incorporated into new structures designed in modern or traditional architectural styles.

In eighteenth century Europe the forerunners of solar collectors, solar hot boxes, were constructed of glass to absorb heat for cooking. In the 1880s, Edward S. Morse of Salem, Massachusetts, designed a sophisticated hot box made from glass and slate, which resembled a large window. The Massachusetts Institute of Technology (MIT) began two decades of solar-collector research in 1938. During this period, MIT built four homes for research and development, discovering the importance of proper panel angle and construction materials. Also in 1938, Arthur Brown built the first solar-heated and -cooled school in Tucson, Arizona. He was able to keep costs down by eliminating a storage system which was not needed during school hours, by using metal for the roof which served as the collector, and by utilizing the existing hot-air distribution system. In summer, a second roof with extended overhangs was built over the ordinary roof to reduce heat absorption while ducts used to circulate hot air in winter were opened to blow warm air outside.

In California during the 1890s, a solar water heater industry thrived. Conventional fuels of the time (coal, wood, and manufactured gas) were expensive as well as inconvenient and dangerous to use. Yet as the solar industry dwindled in the West, presumably due to the discovery of natural gas, it prospered in the South. An estimated 40,000 solar water-heating units were sold in Florida alone by the start of World War II. In 1939, Savannah, Georgia, was the site of a 480-unit public housing project which employed solar water-heaters.

Active solar systems are currently available on the consumer market. Active systems are characterized by the common flat-plate collector which has the ability to absorb heat up to approximately 140°F (60°C). They can be constructed as part of the roof on a new building, attached to the roof of an existing building, or even mounted on the ground. Normally, the collectors should face south and be tilted at an angle which is equal to the site's latitude plus 10°. A change in this angle within a range of plus or minus an additional 15° will still allow the collector to gain virtually all available solar energy. Retrofit often relies on this flexibility factor to accommodate the slant of an existing roof. Angle problems can be easily remedied when collectors are located on the ground.

A significant component of an active system is the solar collector, also referred to as a solar panel. Each unit in the system is composed of glass or plastic to increase absorption and reduce heat loss to the air. Tubes located under the glass transfer a heated fluid or air. Collectors are attached to a roof section facing south. Tubes inside the collectors are connected to tubing at the base of the panels along the roof. The heated transfer fluid or air passes through this tubing into the building for use by conventional distribution methods or for storage in a tank for later use. The cycle is completed with tubing that carries the transfer fluid back to the collectors for reheating.

Fans and pumps are a key mechanical component of an active solar system.

They are used to move the warmed air into the building space, and to recirculate the cooler room air through the collector system. Controller units which are also part of the mechanical system function as regulators to shut down and reactivate the collector and distribution systems.

Heat pumps can also be components of an active solar system. A heat pump works according to the principles of a basic refrigeration cycle. The electric motor in a heat pump provides the energy source to take heat out of cool outside air and transfer it to a warm, interior space. By doing work the heat pump reverses the natural thermodynamic law that says energy is transferred from a hot area to a cool area. In this way a heat pump follows the process of an air conditioner or refrigerator, in which warm air is removed from the interior of an ice box or a house, leaving it cool. A heat pump captures heat from the outside atmosphere and transfers it inside a building. Outside air possesses heat, even in the winter; for example, air at 0°F (−18°C) contains 89 percent of the heat in air at 100°F (38°C).

Supported, where necessary, by a backup conventional system, active solar systems can provide reliable and efficient sources of heating and cooling.

Active solar water-heating systems function on the same principle as active space-heating units. Liquid or air from the collectors heats the water in the tank. The transfer fluid is then returned, often by a pump, to the collectors through a closed system of tubes. When an open system is used, the water warmed by the collectors is used directly from a tank which simply stores any reserve quantities.

Low installation costs relative to more expensive solar space-heating systems account in part for the current popularity of solar hot-water systems. They require fewer collectors than needed for space heating, and they can be adapted to many buildings' needs, thereby significantly reducing conventional fuel costs. Today, homes as well as businesses such as laundries, restaurants, and motels use active solar hot-water systems.

Despite its seeming improbability, rooms can be cooled with solar heat. Solar collectors provide the heated air or liquid necessary to perform the energy function in a cooling system. There are three basic types of solar cooling systems:

- The absorption cooling system transfers a heated liquid from the solar collectors to run a generator or boiler which activates a refrigeration loop. This loop cools a storage reservoir from which cool air is drawn for space distribution.
- The Rankine steam turbine engine runs a vapor compressor air conditioner or water chiller. This cooling system has the advantage of a wider temperature range over the absorption system and, with electricity as an auxiliary power source, can be more practical than systems which must use oil or gas.
- The solar desiccant system produces cooled air by drying, extracting heat from, and rehumidifying room air. After air passes through the drying wheel covered in a desiccant material (silica gel or lithium chloride), it is heated by a solar coil before it goes into a heat exchanger. The resulting cooler air is rehumidified by an evaporative cooler.

Research and tests are still being conducted to increase the cost effectiveness of these technologies.

Active heating and cooling systems appear in homes as well as public and commercial buildings across the country. In Shenandoah, Georgia, a new community center utilizes 58,000 square feet (5390 square meters) of collectors to heat and cool its earth-berm building. In Keystone, South Dakota, the Mount Rushmore National Memorial Visitor Center has a heating and cooling system which includes a 3000-gallon (11,360-liter) water storage tank, heat exchangers, and 2000 square feet (186 square meters) of flat-plate collectors. This solar system provides approximately 53 percent of total heating and 41 percent of cooling needs of the observation room.

At the present time, the solar industry is working toward application development and commercialization of solar-heating and -cooling technologies.

PHOTOVOLTAIC ENERGY CONVERSION

Photovoltaic energy conversion is the direct transformation of light into electrical energy. Most methods of electric power generation convert the thermal energy of fossil and nuclear fuels, or the kinetic energy of the wind or falling water, into mechanical energy to produce electric power. Photovoltaic energy systems, however, are silent and environmentally clean because they contain no moving parts.

In 1931, Dr. Bruno Land demonstrated a solar system in which a battery exposed to sunlight was able to power a small electric motor almost indefinitely. This system used copper oxide, silver selenide, and a third unknown ingredient.

The primary component of photovoltaic energy systems is the photovoltaic array, which is composed of photovoltaic cells, sometimes called solar cells. These cells consist of semiconductor material, such as silicon or cadmium sulfide, and are wired to draw off the current created by the photovoltaic effect.

The photovoltaic effect is the creation of an electrical charge as a result of light photons striking atoms in a semiconductor material. When a light photon of a sufficient energy level strikes an atom, it causes an electron (negative charge) to be dislodged from its fixed position in the atom and move freely in the surrounding material. If the vacant electron "hole" (which acts as a positive charge) and the electron are separated by an intrinsic voltage in the cell material, an electric current will be created.

This intrinsic voltage in the semiconductor material can be created by three methods. One involves adding small amounts of impurities or "dopants" to the pure material to produce a homojunction cell. Another method is to join two dissimilar semiconductor materials such as cadmium sulfide and copper sulfide, creating a heterojunction cell. A third method involves joining the semiconductor material directly to a metal, thus creating a "Schottky" barrier.

Two factors limit the performance efficiencies of these photovoltaic cells:

- Light energy reaching the cell sometimes lacks sufficient energy to dislodge an electron from its position in the semiconductor material.
- Energy given to the electrons which exceeds the level needed to dislodge the electron cannot be recovered as useful electric current.

Photovoltaic cells can be arranged in one of two configurations: flat-plate or concentrating. The more commonly used flat-plate arrays are flat and nonconcentrating. Crystalline silicon is the most commonly utilized semiconductor material in flat-plate cells, although gallium arsenide is also being used.

Photogenerated charges can be lost because of imperfections in the silicon crystal structure. Losses can be minimized if perfect crystals are used; but producing such crystals is expensive. Therefore, recent attempts to reduce manufacturing costs of silicon cells have focused on using small crystal grains to lessen the distance that electrons must travel in the material in order to be separated from their "holes" by the intrinsic voltage.

Concentrating arrays increase the amount of light energy which is available to dislodge electrons from their positions in the semiconductor material, thereby improving the conversion efficiency of cells to as much as 25 to 30 percent.

Several types of concentrators are being studied for use in photovoltaic systems, including Fresnel lenses, mirrors, and compound parabolic concentrators. Depending on how the collectors are arranged, concentrating systems require either a single-axis or double-axis tracking system to "follow" the sun in its path through the sky. Cells need to be constructed carefully for use in concentrating systems and they require more expensive materials. Both of these factors increase the cost of arrays per unit area, but this is expected to be offset by increased cell efficiency.

Photovoltaic array prices are presently about $7 per peak watt (Wp). This is a substantial decrease from 4 years ago when the figure was $28 per Wp.

The federal government has an extensive research and development program designed to commercialize photovoltaic energy by the late 1980s. Current efforts identify promising photovoltaic applications, develop test facilities to evaluate performance, establish performance standards, analyze potential markets, and identify potential incentive programs for accelerating the technology.

The overall objective of the federal program is to ensure that photovoltaic conversion systems play a significant role in supplying energy to the nation by the year 2000, providing approximately 50 gigawatts electric of generating capacity. To meet this objective, several time-phased goals have been established:

- Near-term: Photovoltaic array prices of $2.40 per peak watt (Wp) at an annual production rate of 20 peak megawatts (MWp) in 1982. Installed system prices will be $6 to $13 per Wp.
- Midterm: Photovoltaic array prices of $0.70 per Wp at an annual production rate of 500 MW in 1986. Installed prices will be $1.60 to $2.20 per Wp.
- Far-term: Photovoltaic array prices of $0.15 to $0.40 per Wp in 1990 with installed system prices of $1.10 to $1.30 per Wp and a production rate of 50 peak gigawatts (GWp) in 2000.

If the mid- and long-term goals of $0.70 per Wp in 1986 and $0.15 to $0.40 per Wp in 1990 are met, the cost of photovoltaic energy would be equivalent to or less than that of conventional utility-generated power.

Current Projects Photovoltaic systems are being operated for a variety of uses. Their initial applications came in the U.S. space program when photovoltaic arrays were used to provide power for spacecraft. A prime example of this utilization was the recent Skylab satellite, whose photovoltaic arrays performed beyond design expectations.

Recent terrestrial applications have focused on remote electric power needs. The applications vary from electric power for automated weather stations and fire lookout towers to electric power for entire villages. One 25 kilowatt (kW) photovoltaic system located near Mead, NB, has been powering an 80-acre irrigation system for more than 3 years under the auspices of the University of Nebraska. Another system is providing 3.5 kW of peak power for the Schuchuli Indian Village in AZ. This system, utilizing storage batteries, provides electric energy for 15 refrigerators, 1 washing machine, 1 sewing machine, an 1100-gallons per hour (gph) water pump, and light for 19 buildings.

Other current applications include providing electrical power for remote radio stations, seismic sensor stations, and radar stations. A 100-kW system for the Natural Bridges Monument project in UT, and an AM radio station in Ohio are presently in operation. Large load applications include a 300-kW photovoltaic energy system at the Mississippi County Community College in AR. The SOLARES Project developed jointly by the United States and Saudia Arabia has demonstrated the use of photovoltaic systems to power a village in the Kingdom of Saudia Arabia.

FUELS FROM BIOMASS

Organic matter, such as crop and forestry residues, marine plant life, and municipal and animal wastes, has great potential as an energy source. This organic matter is collectively known as biomass.

Biomass is usually considered a solar-related energy source, because all organic matter depends upon sunlight and the photosynthesis process. It is through photosynthesis, of course, that green plants grow and store energy within their tissues.

The federal government has sponsored a number of projects to develop and refine the various biomass technologies. For example, a comprehensive systems study has recently been conducted on biomass production by silviculture, or tree-farming. Other experiments indicate that plants, such as algae, that grow rapidly in water can also serve as a practical source of biomass.

In some areas of the United States, municipal waste or garbage is being converted to electricity in limited amounts. Thus, some of the discards of our consumer society can be reclaimed to provide power.

Burning, or direct combustion, is an efficient way to obtain heat energy from biomass. Other techniques to produce fuels from biomass include: gasification and liquefaction to make natural gas- and petroleumlike products; anaerobic digestion, or treating biomass by certain bacteria to produce a synthetic gas; and fermentation to produce gasoline substitutes.

Biomass has important energy supply potential because of its extensive and well-located resources in the United States, the potential of conversion technologies, and the availability of a strong commercial market. Biomass is already providing approximately 2 percent of our energy needs, primarily by the direct combustion of wood. The forest products and pulp and paper industries are major biomass consumers. They are working toward 100 percent energy self-sufficiency in the near future by using manufacturing and forest residues for production processes.

Annual biomass production goals for the year 2000 are about 7 quads of energy. One quad, or quadrillion (10^{15}) Btu's, is eqivalent to approximately 494,000 barrels of oil per day for one year, or enough energy to heat 500,000 homes for 20 years.

Biomass is the oldest source of energy known. For thousands of years, people have burned wood for both heat and protection. Until the mid-nineteenth century, wood accounted for about 90 percent of the energy supply in the United States. In 1940, 20 percent of U.S. homes still used wood for space heating.

Although fossil fuels replaced wood as the prime energy source in the twentieth century, interest in biomass never completely faded. During the 1930s, several American studies investigated the feasibility of converting grain to ethyl alcohol as a gasoline substitute. During World War II, many nations worked to develop the fuel potential of trees; Sweden, for example, cultivated large forest areas specifically for energy consumption and research. Today, private industry and government in the United States have renewed interest in biomass by researching conversion methods and encouraging the application of available technologies.

Because biomass is 50 to 90 percent water, the most cost-effective way to utilize the organic matter is at or near its source. If converted into liquids or gases, however, it can be transported easily and economically. Because biomass is available in wooded areas, on farm land, and in water, conversion sites are located throughout the country.

Wood, of course, is a major biomass resource. It serves as a fuel in the form of logs or residues. Residues consist of excess forest growth, insect-infested and diseased trees, and mill remains. To improve wood fuel production, researchers are developing new tree varieties such as hybrid poplars which grow rapidly and are totally consumable. Silvichemicals such as turpentine and resin, and carbon-based compounds made from lignin and cellulose, are other energy-related wood derivatives. These materials serve as feedstocks for the production of a variety of chemicals such as alcohol, aldehydes, ketones, ethylene, acids, and ammonia which can be used as petrochemical substitutes.

Nonwoody plants, such as herbs and grasses, have a potential as cost-effective energy feedstocks. These plants have a high yield capacity and can grow in arid or marginal land with minimal management.

Crops are another popular biomass resource. Sugar beets, sugarcane, sweet sorghum, and grains such as corn and wheat are processed for their carbohydrate content to make ethanol (ethyl alcohol). Alcohol development is receiving special emphasis under the new DOE Office of Alcohol Fuels. Gasohol, a combination of ethanol and gasoline, is our only near-term synthetic fuel. Animal manure is an excellent source of methane gas, which can be manufactured by a process called anaerobic digestion. This fuel supplies farmers with the energy necessary to grow crops and raise livestock. It can be viewed as completing a biomass energy cycle whose goal is energy self-sufficiency on farms.

Kelp and algae are aquatic sources of biomass. Through a process called biophotolysis, blue-green algae produce hydrogen as a waste product. Kelp produces hydrogen gas through photoelectrolysis, a photochemical process which uses a catalyst to separate hydrogen and oxygen.

Biomass conversion techniques are divided into two categories including thermochemical and biochemical processes. The direct combustion of wood is the most common thermochemical biomass conversion method. The number of woodburning stoves in homes has increased dramatically in the past few years, indicating a resurgence in the popularity of wood as a residential heating fuel. It is estimated, for example, that 40 percent of the homes in Maine use wood stoves. Wood can also supplement oil, natural gas, or solar energy heating systems. Nationwide, approximately 5 million American homes already have a working wood stove, and the practical potential exists for another 13 to 18 million homes.

Direct combustion of biomass residues has industrial applications for the production of process heat or electricity. Currently, direct burning is economically competitive with fossil fuels for some industries, especially those having easy access to biomass resources. A sugar company in Hawaii meets 75 percent of its mill and irrigation pump electricity needs from the burning of bagasse, sugarcane residue.

Utility plants in Vermont and Oregon also produce electricity from direct combustion. Two 10-megawatt (MW) units in Burlington, VT are 100 percent wood burning. They serve 40,000 people and sell an additional 15 percent of their output to surrounding utilities. In Eugene, OR, a 33.8 MW utility plant has been producing steam and electric power since the 1930s by burning mill wastes supplied by local sawmills. In addition to proving very economical, this plant reduces air pollution and helps solve wood residue waste-disposal problems.

Other direct combustion efforts include brick and textile industries in the South which burn sawdust and wood, respectively, to provide process heat. The federal government is working with industry to evaluate wood conversion processes and to retrofit large conventionally fueled process heat boilers for direct firing of green wood.

Gasification and liquefaction are two technologies which convert biomass into gas or liquid fuels. Gasification is the reaction of biomass with steam and oxygen at high temperature and pressure. The products include synthetic natural gas (methane), low and medium Btu gas, and hydrogen.

A large-scale, experimental liquefaction facility in Albany, OR, uses wood chips to produce oil. Biomass reacts with carbon monoxide in the presence of an alkaline catalyst under moderate temperature and high pressure. The resulting oil product has a consistency similar to heavy heating oil and a heating value (15,000 Btu/lb) similar to bunker fuel oil. The Albany facility has proven that liquefaction is an energy-efficient process which produces nonpolluting water, flue gases, and ash.

Pyrolysis is a thermochemical conversion process which functions without oxygen and at a lower temperature than gasification. Residue materials are gasified, ignited, and completely burned in a thermal reactor to produce gases. The biomass resources used in pyrolysis include manure, agricultural and wood wastes, and municipal solid wastes. Oil, charcoal, and synthetic gas as well as steam are the products.

The biochemical process, referred to as anaerobic digestion, is the controlled decomposition of organic matter. In the absence of oxygen, bacteria decays the material, producing methane. The feasibility of using animal manure in this process has been demonstrated at a full-scale facility in Bartow, FL. Located in the state's largest slaughtering and packing plant, this facility can handle 25 tons of manure daily. By processing the wastes from 10,000 cattle, the operation can produce enough methane to meet nearly all of the packing plant's energy needs, the equivalent of 35 barrels of oil per day. A byproduct of this process, protein-rich digester solids, will be concentrated to serve as a dietary supplement for 6000 cattle. Anaerobic digestion also plays a role in the production of energy from aquatic plants.

In addition to liquid and gaseous fuels, biochemical and thermochemical conversion systems produce petrochemical substitutes such as aldehydes, ketones, ethylene, acids, and ammonia. These chemical compounds are used in a wide variety of manufacturing processes.

Certain biomass conversion technologies, most notably direct combustion, anaerobic digestion, and gasification, are sufficiently advanced for near-term energy development. Direct combustion of wood is providing energy to the residential, industrial, and utility sectors. Anaerobic digestion of manure and agricultural wastes on farms is a first step in the attempt to make farmers energy self-sufficient. Low Btu gasifiers using wood as a feedstock are already in use by industry.

WIND-ENERGY CONVERSION SYSTEMS

Unequal heating of the earth's surface by the sun produces air masses of differing density and heat content, causing atmospheric movements from gentle breezes to tornadoes. Wind power is therefore a form of solar energy that can be converted by windmills to mechanical energy and to electric power.

People in Western Europe have used wind energy for hundreds of years to grind grain or pump water. Almost everyone has heard of the windmills of Holland and those of Don Quixote's Spain.

In America, wind supplied a significant amount of the energy consumed in rural areas until the mid-1930s. At that time the Rural Electrification Administration introduced electric cooperatives and, with centrally generated cheap electricity available, the need for wind-generated energy disappeared.

However, with the dramatic increases in the cost of conventional energy sources, wind energy is becoming an economically important alternative, and there is a resurgence of interest in this ancient form of solar energy. Many farmers are installing commercially available windmills to power mechanical pumps or to turn an electrical generator. Wind energy can be especially practical in remote areas where large quantities of power are not essential, and where electric transmission costs from distant generators are high. The Federal Wind Energy program is developing three major types of wind systems:

1. Advanced small-scale wind systems which will be used by farms, rural residences, and remote communities. These small-scale wind systems yield up to 100 kilowatts (kW) in electrical output.
2. An intermediate class of wind-energy conversion systems is being developed for irrigation and small utilities. These systems will yield between 100 kW and 1 megawatt (MW).
3. Large-scale wind systems will have a capacity of at least 1MW. With rotor diameters of up to 300 feet (91 meters), these wind-energy conversion systems are being designed for utility and industrial use.

Structural limits prevent even the largest systems from exceeding 4 to 5 MW. Since typical coal-fired generators produce 500 to 1000 MW, power to replace utility plants will probably be provided by multiunit wind-energy conversion system clusters or "wind farms."

Efforts are on-going to develop wind-powered methods of pumping water for deep well irrigation, direct heating, and similar applications. U.S. Department of Agriculture and the Rocky Flats Test Center are field testing these newly developed systems.

The technical and economic aspects of small-scale wind-energy conversion systems are being evaluated at sites throughout the country. To date, machines have been installed in New Hampshire, Hawaii, Wyoming, Alaska, Nebraska, and Kansas.

The goal of the federal small-scale wind-energy program is to accelerate development of reliable systems and to ensure their earliest possible commercialization. Meeting this objective requires technological advances and a strong industrial applications base as well as technical and nontechnical marketing of the systems. By the mid-1980s, available systems are expected to generate electricity at or below \$0.04 to \$0.06 per kW hour.

Improvements in these wind systems and climbing energy costs will make wind energy economically more attractive. At the same time, demonstrations should promote acceptance and spur manufacturing and installation capacities.

Utilities such as Southern California Edison Company, Bonneville Power Administration, and the Bureau of Reclamation are showing more interest in the potential of intermediate- and large-scale wind-energy conversion systems. Technical and cost complications, however, will probably prevent wind systems from replacing more than 20 to 30 percent of any existing utility's generating capacity.

Two first-generation systems have been connected to electric utility systems. One is the MOD-OA, a 200-kW system with a 125-foot (37.5 meter) rotor diameter. The other is the MOD-1, with an output of 2 MW and a rotor diameter of 200 feet (61 meters).

Four MOD-OA systems are now in place in Clayton, NM; Culebra, PR; Block Island, RI; and Oahu, HI. These systems are providing valuable experience in the utilities' day-to-day operation of wind-energy conversion systems.

The Clayton unit is producing 2 to 3 percent of the local utility's power. That figure increases to as much as 15 percent at night when the load is reduced and if the wind system is operating at maximum power.

During 1979, MOD-1, the first operational large-scale system, was installed at Boone, NC. The Blue Ridge Electrical Cooperative is now testing it.

The 2.5-MW MOD-2, second-generation successor to MOD-1, with a 300-foot (91 meter) diameter rotor, began operation in early 1981. Three units erected at Goodnoe Hills in Washington are being tested by the Bonneville Power Administration. This will represent the first test of multiple units interconnected to a utility system. The MOD-2 is particularly significant, and it could be cost-effective in some high-wind, high-fuel-cost utility markets.

Most wind turbines operate from a horizontal-axis and rely on the same principles used by windmills or airplanes. Innovative vertical-axis designs are being tested with the hope of improving wind system performance and reducing costs.

The Darrieus is a vertical-axis wind turbine (VAWT). The egg-beaterlike shape is called a troposkien from the Greek meaning "turning rope." The 56-foot (17-meter) Darrieus being tested at Sandia Labratories has two struts on each of three blades to enhance its structural strength. Its starter motor is activated by a 10 mph wind and disengages once the turbine is under wind power. Its rated output of 60 kW is attained with a 32 mph wind speed. The turbine shuts down with wind speeds greater than 60 mph.

The VAWT has several advantages over horizontal-axis turbines. Its simplicity reduces potential construction and maintenance difficulties. The transmission/generator equipment can be located on the ground. The VAWT's overall weight is likely to be substantially less than that of conventional turbines. Reduced construction and other costs are also a significant advantage of the VAWT. In addition to the Darrieus, the federal government is developing other small-scale system designs such as the gyromill, cycloturbine, and self-twisting blade concepts.

SOLAR THERMAL ENERGY

Heat from the sun can be converted to electricity by solar thermal systems. The system may consist of a large field of mirrors focusing on a boiler mounted on a tower. Fluid in the boiler is vaporized to run a turbine coupled to an electric generator. In an alternate arrangement, solar heat is absorbed by a transport medium in a solar collector. This heated fluid is pumped to a central plant where the heat is used to vaporize another fluid to run a turbine electric generator.

Solar thermal energy conversion systems can provide a renewable energy source for all four major energy consumption sectors in the United States: industrial, utility, residential/commercial, and transportation.

Solar thermal technology utilizing concentrating solar collectors produces high-quality thermal energy, or heat. This heat can then be used directly for industrial process heat applications or advanced solar heating and cooling. It can also be converted into electrical energy for utility or onsite electricity generation, or mechanical energy for applications such as irrigation pumping.

Three classes of systems are under development, each oriented toward specific applications. These are: linear concentrators (parabolic troughs and hemispherical bowls); parabolic dishes; and central heliostat/receiver systems. Linear systems applications include solar heating and cooling, irrigation pumping, and midtemperature industrial process heat. Applications for the parabolic dish range from small community and remote electric systems to small-scale high-temperature process heat. The heliostat/central receiver concept is focused on on-site cogeneration, repowering and stand-alone bulk electric power generating systems, and solar-driven fuels production processes.

The technical feasibility of solar thermal energy was established a century ago. A parabolic dish-powered solar printing press was demonstrated at the 1878 Paris Exposition and a parabolic trough solar irrigation pump was operating in Egypt in 1913. There are still basic impediments to the commercialization of these systems.

The barriers to the widespread utilization of solar thermal systems include cost, reliability, durability, and performance. Studies give strong support to the belief that these barriers can be overcome.

The federal program emphasizes advanced research and development to improve performance, laboratory subsystem testing to verify design adequacy, field tests to demonstrate systems-level reliability and durability, and the establishment of a manufacturing industry to produce low-cost concentrators and other subsystems. The program's objective is to establish technical readiness for distributed receiver (less than 10 megawatts of electricity [MWe]) and centralized receiver (more than 10 MWe) energy production applications within a competitive price range.

For parabolic trough and hemispheric bowl technology, subsystem testing is underway at the Sandia Laboratories test facility. Field experiments include:

1. Shallow well 50-horsepower (hp) (37-kW) irrigation pumping experiment using parabolic troughs operating at Willard, NM.
2. Deep-well 150 kWe irrigation pumping experiment using parabolic troughs operating at Coolidge, AZ.
3. Four industrial process steam applications projects in the construction phase using parabolic troughs.
4. Two enhanced oil recovery applications projects using parabolic troughs.
5. International Energy Agency distributed receiver experiment in Spain using parabolic troughs.

Subsystem testing of parabolic dish technology is taking place at the Jet Propulsion Lab Parabolic Dish Test Site at Edwards AFB, CA, at the Sandia Laboratory, and at the Solar Energy Research Institute in Colorado.

1. Initiated parabolic dish module experiments for remote applications.
2. Industrial total energy system field test using parabolic dishes at a knitwear factory in Shenandoah, GA.
3. Small community 1 MWe experiment using parabolic dishes.

Heliostats/central receivers subsystem development and pilot plants include:

1. Solar One 10 MW central receiver pilot plant at Barstow, CA.
2. Utility repowering/industrial retrofit project.
3. International Energy Agency central receiver experiment in Spain for electric power generation.
4. DOE/Electric Power Research Institute central receiver/gas turbine solar field test.

Performance improvements under investigation include the development of high-temperature [2000° to 2500°F (1094° to 1371°C)] central receivers, advanced thermal storage concepts and advanced materials. Studies and development activity may also prove the technical feasibility of solar thermal high-temperature processes for the production of transportable fuels from renewable resources.

Cost reductions have already been achieved in concentrator development programs. Costs of troughs, dishes, and heliostats have been reduced to within two to five times the program's 1990 cost goals which were based on competitive economics with fossil-based conventional systems.

OCEAN ENERGY

The oceans store vast amounts of solar energy, enough to make sizable contributions toward the energy needs of the United States and other coastal nations. The seas continually collect this energy in the form of wind-driven waves, global ocean currents, heated surface layers in the tropics, tides, and salt concentrated in the oceans by evaporation.

With the natural storage capacity of the oceans, these forms of energy are available continuously, day and night, year-round.

Four of the ocean-energy technologies offer significant promise for the United States: Ocean Thermal Energy Conversion (OTEC), wave power, ocean currents, and salinity gradients. At present, OTEC is the most advanced and is the likeliest to realize commercial success. The OTEC Research, Development, and Demonstration Act of 1980 established a national goal of 10,000 megawatts (MW) of electrical capacity or energy product equivalent from OTEC systems by the year 1999.

For centuries man has tried to extract useful energy from the oceans. The Chinese used wave machines to crush iron ore in the thirteenth century, and Leonardo da Vinci designed several wave-powered devices. In colonial Massachusetts a flour mill was operated by the tides for many years.

Beginning around the turn of the century, experimentation with ocean power burgeoned. There are many patents on record for such devices, most exploiting wave power, dating back to the early 1900s. OTEC was first theorized in 1881 by Jacques d'Arsonval. George Claude built the first OTEC plant in 1930, but his technical success was marginal and OTEC was far from competitive with the abundant cheap energy then available.

Experimentation on salinity gradients for power production began in earnest only during the past two decades. Ocean current utilization, while theorized for centuries, has only been seriously pursued for the past 5 years.

Following the 1973 oil embargo, the federal government launched several projects to provide energy alternatives. The Ocean Energy Program was born out of those projects, first under the National Science Foundation, then under the Energy Research and Development Administration, and now under DOE.

Surface waters of the tropical oceans have a temperature of about 83°F (28°C). Nearer the bottom, however, the water is much colder, ranging down to 35°F (2°C). This temperature differential is the basis for ocean thermal energy conversion.

If a vessel containing liquid ammonia or Freon is placed in the warm water of a tropical sea, that liquid will boil and vaporize. The vapor can drive a low-pressure turbine to generate electricity.

Unlike other solar-energy systems which lose their magic when the sun goes down, ocean thermal power plants could operate continuously without storage facilities; this is because surface waters in the Gulf Stream, the Caribbean, and generally throughout the tropics and subtropics, remain warm continuously. The ocean acts as a giant collector and storage facility for solar heat.

Ocean thermal plants located sufficiently close to shore could provide baseload electricity directly to land areas by submarine cable. But electricity as an end use is only one likely application. More remote harnessing of thermal energy could be undertaken in any tropical waters by producing an intermediate fuel or chemical product on-site with electricity generated at sea. Hydrogen, methanol, and ammonia are frequently mentioned as products that could be manufactured in quantity at sea. These energy-intensive products could be shipped ashore to provide the fuel for either piston-engine or turbine autos, helping to free the world's automobile-dominated economies from their dependence on gasoline.

Whenever two solutions of different salt concentrations exist, there also exists an energy potential between them referred to as salinity gradients. Techniques are being studied which utilize the salinity difference, or gradient, of water-based solutions to produce energy. The key element is a semipermeable membrane that must separate the two solutions.

In one scheme, fresh water will migrate through a semipermeable membrane into the saltier solution, increasing pressure. The pressurized flow is discharged through a hydraulic turbine, transforming into electricity the energy resulting from mixing fresh and salt water.

Alternatively, cells of varying salinity can be separated by membranes which permit selective passage of the positive or negative ions of the salt solution. The ion passage creates an electrical current in electrodes immersed in the cells.

There are numerous wave-power machines in existence and on the drawing board that can extract energy from ordinary ocean waves. Generally they rely on three types of motion:

1. Relative motion between two or more bodies, e.g., hinged barges, "flip-flop" floats on a shaft.
2. Motion of the water within the device, e.g., hydraulic effect on a turbine of the water column rising and falling in a chamber.
3. Concentrating waves, rising and spilling into a chamber, and then running out through a turbine. Variations of these three basic schemes are in operation today in small-scale applications.

Within all seas there are ocean currents that constitute a global circulation pattern. One such current is the Gulf Stream. Proposals have been made to harness some of the power in these ocean currents by anchoring various devices within them.

Recent research and modeling indicate that it should be possible to install giant, low-speed water-turbines resembling windmills in some of the ocean currents near the United States. The electricity they generate would then be cabled to shore.

By 1982 DOE expects to complete the conceptual designs and small-scale at-sea experiments for wave power, ocean currents, and salinity gradients. At that time information should be available to determine which alternatives possess sufficient technical and economic feasibility.

"Mini-OTEC," a floating energy platform, is now stationed off the island of Hawaii. The product of a consortium of private firms and the State of Hawaii, it produced 50 kW of electric power for an experimental period. "Mini-OTEC" uses off-the-shelf components to demonstrate the production of electricity at sea and will provide technical data for future development.

OTEC-1 is a floating heat-exchanger test platform corresponding to a 1-MW electrical plant. In mid-1980 the testing of full-scale components in actual at-sea conditions began aboard OTEC-1, a modified Navy tanker. If funded, OTEC-1 will lead to a pilot plant which will produce 10 to 40 MW of electricity by 1986. This electricity will be used directly by a U.S. commercial power grid or onboard a large floating industrial plant. The floating plant would make products such as aluminum in processes that normally rely heavily on electrical power and fossil fuels.

HYDROELECTRIC POWER

Falling water in streams and rivers can provide an infinitely renewable energy resource. Warmth from the sun evaporates water, which rises to the upper atmosphere. There it condenses and falls again to the earth as rain. Water that collects in rivers can be made to pass through turbines which spin to produce electricity.

Hydroelectric power has been established as a source of energy in the United States for almost a century. In 1882, the first hydro power plant went into operation at Appleton, WI. It generated 200 watts (W) and transmitted electric power 1 mile (1.6 km) at 110 volts (V). Since then, large hydroelectric power dams on major rivers have been developed as one of the cleanest and safest energy sources in the nation. Presently, about 15 percent of the nation's electricity is produced by hydroelectric generators.

A recent study identified approximately 48,500 dams throughout the United States which generate no electricity. It has been estimated that a total of 54,100 megawatts (MW) of generating capacity could be developed at these dams representing a potential increase of about 80 percent over present hydroelectric capacity.

The federal government is sponsoring research and demonstrations of new and innovative methods to take advantage of hydroelectric power. This work includes evaluation of new equipment, standardization of power-plant designs, feasibility studies for installation of dams at specific sites throughout the country.

Today, after years of neglect, small hydro power is becoming an attractive energy source. Costs are calculated to be comparable to new coal or nuclear capacity, and, furthermore, the use of existing dams involves no new environmental impact. Hydro power has been overlooked for many years because of higher priorities for the use of other sources of energy. Now, the position of hydro power has been altered dramatically by changing economic patterns and the renewed interest in renewable resource energy technologies. Such dams would help offset the use of imported petroleum. The main barriers to utilizing hydro power include uncertainties about the resource potential and financial and institutional constraints.

Hydro power plants offer several advantages. They are virtually nonpolluting. They utilize a renewable resource—flowing water. The facilities are small and can be aesthetically acceptable. In addition, the effects upon stream ecology are minor compared to larger facilities. Dams may in fact enhance the streams by maintaining water depths sufficient to support aquatic life. Finally, technology for hydro power plants is well established. The required components are available throughout private industry. This could speed formation of an industrial base with potential for mass production and capacity to develop substantial small hydro resources.

The generating machinery for lower heads, which often are characteristic of hydro, can be divided into three groups: vertical-shaft turbines, "axial" turbines, and cross-flow turbines.

Of the vertical-shaft types, the adjustable blade propeller turbine is often preferred because of its greater operational flexibility. The axial turbine has a horizontal or inclined shaft and fixed or adjustable runner blades. The cross-flow turbine is a radial, impulse-type turbine; it is a low-speed machine with its generator located outside the water passage.

The major expenditure for hydro power development is the initial capital investment, including costs for the main structures and generating machinery. Operational and maintenance costs for hydro facilities have proven very low. And importantly, once completed, the plants costs are largely unaffected by inflation.

The federal government's work with the public and private sectors is aimed at further reduction in the cost of the structures, and reduction in turbine cost through mass production and standardization, or by simplified turbine design. New developments in automatic control systems promise to make hydro more efficient and economical.

A FUTURE FOR SOLAR ENERGY

The down-to-earth applications of solar hot water, biomass fuels (gasohol), and passive solar renewable technology are economic, if not yet fashionable. With predictable increases in the cost of energy due to oil decontrol, other solar-renewable sources should become increasingly reliable and economically acceptable, while also relieving some of the dependence on imported fossil fuels.

The United States and the rest of the world are in an energy transition period as the competition increases for finite supplies of oil and gas. Applications of conservation, solar, and renewable energy techniques are within the reach of every citizen. You can make an important contribution toward our energy independence by learning about and utilizing some of these solar energy options.

A

abatement. The method of reducing the degree or intensity of pollution; also the use of such a method.

absolute alcohol. Completely dehydrated ethyl alcohol of the highest proof obtainable (200 proof); also "neat" alcohol. See **anhydrous ethanol.**

absolute humidity. Weight of water vapor per unit volume of air-steam mixture. Usually expressed in grains per cubic foot. (Not in common use.)

absolute pressure. Pressure measured with respect to a vacuum (zero pressure); absolute pressure can be zero only in a perfect vacuum.

absolute temperature. Temperature measured with respect to absolute zero on a thermodynamic or Kelvin scale. The zero of the Kelvin scale is −273.16°C.

absolute zero. The temperature at which a perfect gas kept at constant volume would exhibit no pressure. It is equal to −273.16°C.

absorbent. The less volatile of the two working fluids in an absorption cooling device.

absorber plate. The surface in a flat-plate collector upon which incident solar radiation is absorbed. A black surface that absorbs solar radiation and converts it to heat; a component of a flat-plate collector.

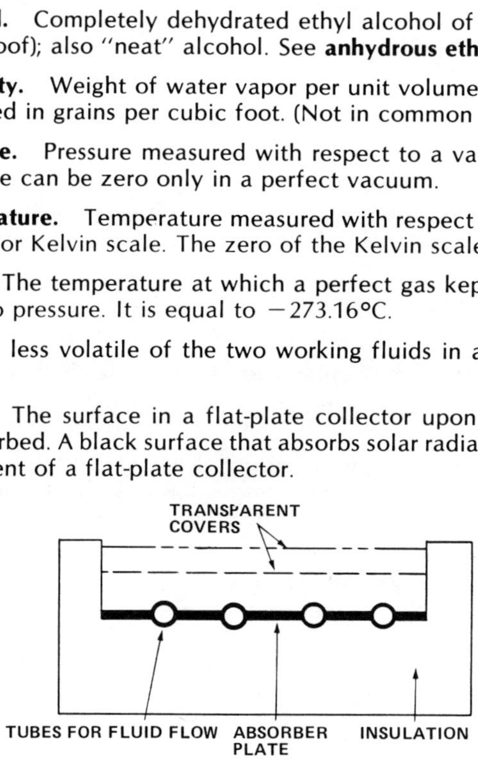

Courtesy SERI

Absorber plate on cross section of typical flat plate solar collector.

absorptance. The dimensionless ratio of solar energy absorbed by a surface to the incident solar energy striking it; the frequency spectrum of the ratio must be specified.

absorption. The process by which radiation is captured within a material. Also used to describe a cooling technology in which one material is first evaporated with heat input and later reabsorbed with heat release displacing the need for a compressor.

absorption chiller. Air-conditioning device which uses heat at 190°F (88°C) or higher to generate cooling; it may be powered by solar-heated water.

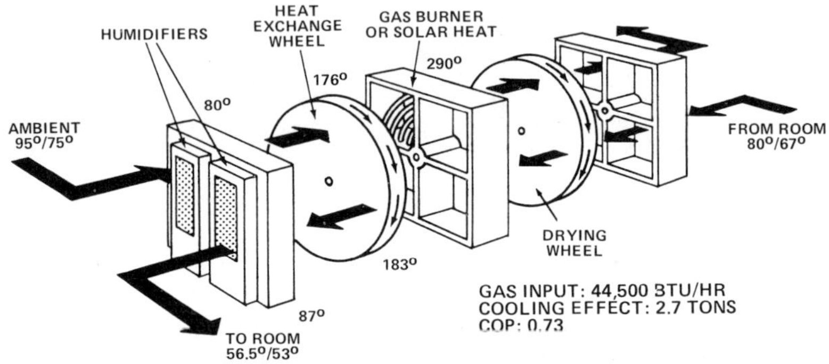

Source: Based on a drawing by W. F. Bush, statement relative to H.R. 10952. Institute of Gas Technology, Illinois Institute of Technology, Chicago, Ill.

Pictorial diagram of absorption chiller system.

absorption cooling. Cooling provided by evaporating a volatile fluid which is then (1) absorbed in a strong solution, (2) desorbed under pressure by outside heat, and (3) recondensed at a temperature high enough that the heat of condensation can be rejected to the outside.

absorption curve. A plot of the relationship between thickness of absorbing material and intensity of transmitted radiation.

absorption dehydration. A new process for dehydrating alcohol fuels. A material such as cornstarch, ground corn, avicel, or cellulose through which the fuel is passed. The material absorbs the water and not the ethanol.

absorption loss. The loss of water occurring during initial filling of a reservoir in wetting rocks and soil; that part of the transmission loss which is due to the dissipation or the conversion of energy into some other form of energy, usually heat.

absorption refrigeration. A system in which a secondary fluid absorbs the refrigerant, thus giving up heat, and then releases the refrigerant and allows the heat to be reabsorbed. The absorption cycle consists of the expansion of a vapor, absorbing heat; the absorption of that vapor by a liquid or solid; the desorption of the vapor from the liquid or solid by the application of heat sometimes from a solar collector, a flame, or steam; the condensation of the vapor, giving off heat; the expansion that starts the cycle again. Ammonia or water is used as the vapor in commercial absorption cycle refrigerators and water or lithium bromide is the absorber.

absorptive coating. An absorber coating which improves the absorptance of the absorber to sunlight.

Material Performance Characteristics of Several Absorptive Coatings				
	Property			
Materials	Absorptance	Emittance	Breakdown temperature(°F)	Comments
Black chrome	0.87-0.93	0.1		
Alkyd enamel	0.9	0.9		Durability limited at high temperature.
Black acrylic paint	0.92-0.97	0.84-0.90		
Black inorganic paint	0.89-0.96	0.86-0.93		
Black silicone paint	0.86-0.94	0.83-0.89		Silicone binder.
PbS/silicone paint	0.94	0.4	662	Has a high emittance for thicknesses 10 mm.
Flat black paint	0.95-0.98	0.89-0.97		
Ceramic enamel	0.9	0.5		Stable at high temperatures.
Black zinc	0.9	0.1		
Copper oxide over aluminum	0.93	0.11	392	
Black copper over copper	0.85-0.90	0.08-0.12	842	Pastinates with moisture.
Black chrome over nickel	0.92-0.94	0.07-0.12	842	Stable at high temperatures.
Black nickel over nickel	0.93	0.06	842	May be influenced by moisture at elevated temperatures.
Ni-Zn-S over nickel	0.96	0.97	536	
Black iron over steel	0.90	0.10		

absorptivity. The ratio of the radiant energy number absorbed by a black body to that falling upon it. It is equal to the emissivity for radiation of the same wavelength.

ABS plastic pipe. Black plastic pipe with excellent thermal and abrasion-resistant properties. It is available in many standard diameters from 1 inch to 6 inches or larger, along with such fittings as elbows, tees, and couplings. It is widely used by plumbers and heating/ventilating/air-conditioning contractors.

acceleration. The rate of change of velocity, measured as a change in velocity in unit time. It is expressed as feet per square second (ft/sec^2).

acceptor. A dopant material such as boron which has fewer outer shell electrons than required in an otherwise balanced crystal structure, providing a hole which can accept a free electron.

acceptor level. Energy level formed within the energy gap by a deficiency of electrons.

acclimation. The physiological and behavioral adjustments of an organism to changes in its immediate environment. The acclimation or adaptation of a particular species over several generations to a marked change in the environment.

acid. A compound that dissociates in a water solution to furnish hydrogen ions, having acid-forming constituents present in excess of the proportion required to form a neutral or normal compound.

acid gas. Gaseous mixture of carbon dioxide (CO_2) and hydrogen sulfide (H_2S).

acid hydrolysis. Decomposition or alteration of a chemical substance by water in the presence of acid.

acidity. The measure of how many hydrogen ions a solution contains per unit volume; may be expressed in terms of pH.

acid rain. Abnormally acidic rainfall, most often dilute sulfuric or nitric acid.

acid soil. A soil deficient in available bases, particularly calcium; a soil having a preponderance of hydrogen over hydroxyl ions in the soil solution; generally, a soil that is acid throughout most or all of the parts of it that plant roots occupy.

acid spoil. Spoil material which contains sufficient pyrite so that weathering produces acid water.

acre-foot. A quantity of water covering 1 acre of surface area to a depth of 1 foot. One acre-foot of water can satisfy the municipal and industrial energy demands of four people for 1 year. An acre-foot equals 326,000 gallons (1.23×10^6 liters).

activated carbon. A highly adsorbent form of carbon, used to remove odors and toxic substances from gaseous emissions. In advanced waste treatment, activated carbon is used to remove dissolved organic matter from waste water.

activated sludge. Sludge that has been aerated and subjected to bacterial action; used to remove organic matter from sewage. A process of waste treatment used to biologically degrade organic matter in a dilute water suspension. High-rate air diffusion through the liquor promotes the growth of aerobic bacteria and other organisms, which, acting on the organic matter, produce a sludge.

activated sludge process. The process of using biologically active sewage sludge to hasten breakdown of organic matter in raw sewage during secondary waste treatment.

activation. The processing of charcoal or carbon or other substance to give it a high degree of porosity in pores so small that they will adsorb molecules.

active solar heating system. A solar heating system that is combined with the active primary heating system in the house. (See diagram, p. 29.)

active solar system. An assembly of collectors, a thermal storage device, and transfer fluid which converts solar energy into thermal energy, and in which energy in addition to solar is used to accomplish the transfer of thermal energy. (See p. 29.)

active system. A solar heating or cooling system that requires external power to operate the system.

adaptation. A change in structure or habit of an organism that produces better adjustment to the environment.

additive. Any materials incorporated in finished petroleum products for the purpose of improving their performance in existing applications or for broadening the areas of their utility.

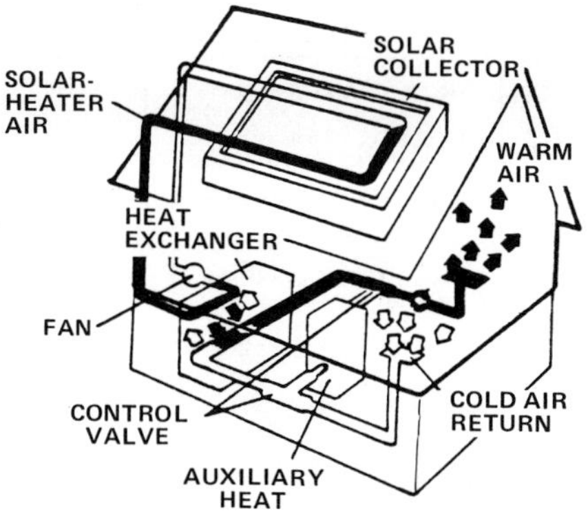

Courtesy DOE

Active solar space heating system contains several major subsystems.

add-on systems. Storage components that may be added optionally to basic space-conditioning systems.

adiabatic compression. Compression in which no heat is added to or subtracted from the air, and the internal energy of the air is increased by an amount equivalent to the external work done by the air. The increase in temperature of the air during adiabatic compression tends to increase the pressure on account of the decrease in volume alone; therefore, the pressure during adiabatic compression rises faster than the volume diminishes.

adiabatic lapse rate. The rate of decrease of temperature with height of a parcel of air lifted adiabatically through an atmosphere in hydrostatic equilibrium. For dry air this is equal to g/C_p, where g is the acceleration of gravity and C_p is the specific heat of air.

adobe. A sun-dried, unburned brick of clay (earth) and straw used in construction. Within the United States, adobe is used primarily in the Southwest. (See table, p. 30.)

adsorption. The adhesion in an extremely thin layer of molecules (as of gases, solutions, or liquids) to the surface of solid bodies or liquids with which they are in

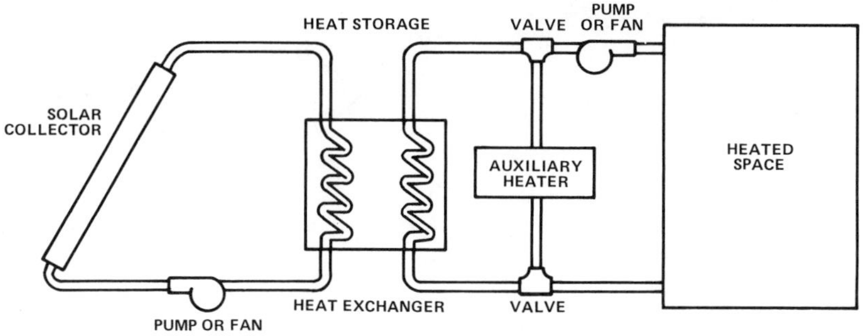

Courtesy DOE

Block diagram of active solar system.

CHARACTERISTICS OF ADOBE			
SPECIFIC HEAT BTU/LB. °F	DENSITY LB./FT.^{3}	HEAT TRANSMISSION BTU/FT.^{2} °F HR. FOR A ONE FT. THICKNESS	R-VALUE FOR A ONE FT. THICKNESS
0.24	106	0.3 (DIVIDE THIS NUMBER BY THICKNESS IN FEET)	3.3

Courtesy DOE

Insulation characteristics of adobe building materials.

contact. Used also to describe a form of solar cooling employing this principle.

advanced battery. Large battery storage systems for harnessing solar or wind energy or storing excess electricity during low-demand periods for use during periods of higher demand.

advanced waste treatment. Wastewater treatment beyond the secondary or biological stage that includes removal of nutrients such as phosphorus and nitrogen and a high percentage of suspended solids. Advanced waste treatment, known as tertiary treatment, is the "polishing stage" of wastewater treatment and produces a high-quality effluent.

adventitious buds. Buds of tissues that develop abnormal positions, e.g., roots growing from stems.

adverse hydro. Water conditions that limit the production of hydroelectric power.

aeration. The process of being supplied or impregnated with air. Aeration is used in wastewater treatment to foster biological and chemical purification.

aerobes. Bacteria which require the presence of free oxygen for their metabolic processes. Oxygen in chemical combination will not support aerobic organisms.

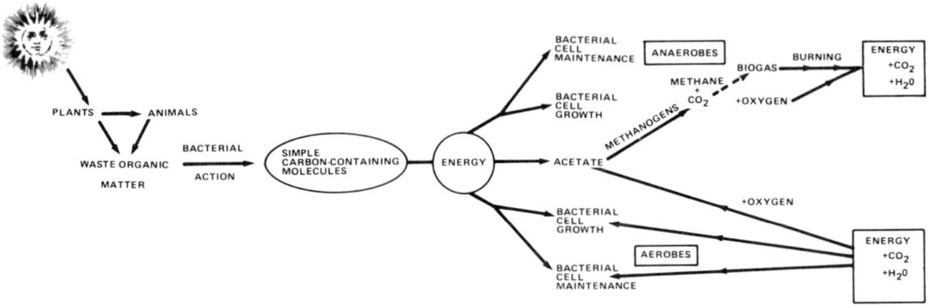

Reprinted with permission from Russell E. Anderson, Biological Paths to Self-Reliance, *Van Nostrand Reinhold*

Flow diagram shows differences in aerobic and anaerobic bacterial breakdown of organic matter.

aerobic. Bacteria or other microorganisms that require free oxygen, or air, to live in a medium are aerobic. Some organisms can live both in an aerobic and in an anaerobic (airless) state. Yeast is one of these. When yeast lives in an aerobic state, it metabolizes any sugars present to produce new material for its growth and reproduction. The wastes discarded by yeast working in an aerobic state are water and carbon dioxide.

aerodynamics. The branch of dynamics that deals with the motion of air and other gases, and the forces acting upon bodies passing through them.

aerosol. A suspension of liquid or fine solid particles in the air.

aerosphere. The atmosphere considered as a spherical shell of gases surrounding the earth.

afforestation. Forest crops established artificially by planting or sowing on land which has not previously grown tree crops.

aflatoxin. A highly toxic carcinogen produced by the fungus *Aspergillus flavus*. High-moisture grains, such as corn, are susceptible to this fungus. Grains that are contaminated by aflatoxin should not be used for animal or human food. The starch in such grains can be converted to alcohol, but the residue (distiller's feeds) must be properly disposed of (burned or buried), since the toxin survives the distillation process. It would also be advisable to thoroughly clean any machinery and the distillery that has processed aflatoxin-infected grain before processing good grain.

afterburner. An air-pollution abatement device that removes undesirable organic gases through incineration.

aggregate. Sand, gravel, or any material in a bedded iron ore, sometimes so abundant as to make it resemble a puddingstone; uncrushed or crushed gravel, crushed stone or rock, sand, or artificially produced inorganic materials which form the major part of concrete. The term also means to bring together, or collect, or to gather into a mass.

agricultural applications. Solar energy can be used in agriculture for heating structures such as chicken houses, pig farrowing and brooding facilities, milking parlors, and greenhouses. In addition, solar energy may be substituted for fossil fuels in several agricultural processes. Solar dryers may be used to prepare such products as fruits, hay, grains, tobacco, peanuts, and wood for the marketplace. These systems can be purchased or site-built. The U.S. Department of Energy has funded a number of agricultural solar projects, which are managed by the Science Education Administration of the U.S. Department of Agriculture.

agricultural and industrial process heat. Both an end-use application for thermal energy and the name applied to that portion of the national solar program developing solar technologies for this end-use sector.

agricultural pollution. The liquid and solid wastes from all types of farming, including runoff from pesticides, fertilizers, and feedlots; erosion and dust from plowing; animal manure and carcasses; and crop residues and debris. It has been estimated that agricultural pollution in the U.S. has amounted to more than 2½ billion (10^9) tons per year.

air. The mixture of gases enveloping the earth and forming its atmosphere. It is composed, by volume, of 21 percent oxygen and 78 percent nitrogen; by weight, about 23 percent oxygen and 77 percent nitrogen. It also contains about 0.03 percent carbon dioxide, some aqueous vapor, and some argon.

air-based solar-heating system. System in which air is heated in a solar collector and is used to transfer heat to the rest of the system. (See diagram, p. 32.)

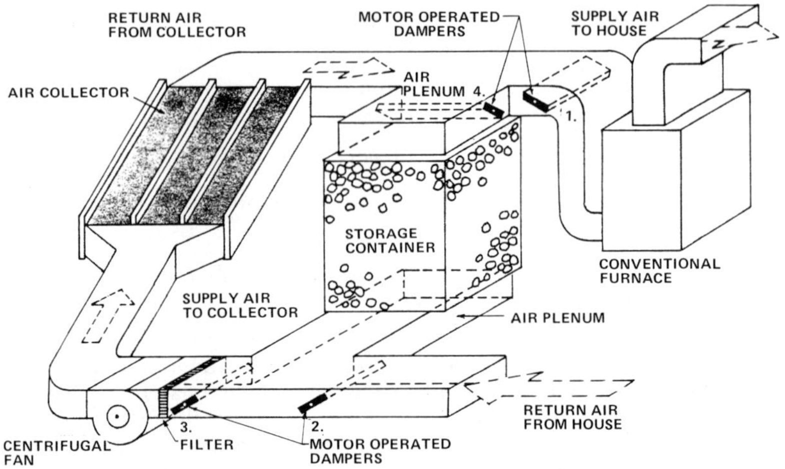

Courtesy National Copper Association

Air collection system with solid material storage for air distribution solar space heating system.

air battery. Advanced battery system which uses air for one electrode; there are iron-air and lithium-air systems.

airborne pollutants. Include aerosols, such as dusts, smoke, and fumes; and gases or organic vapors, such as sulfur dioxide, chlorine, oxides of nitrogen, aldehydes, and carbon monoxide.

air chamber. Closed section of pipe or a container filled with air entrapped at atmospheric pressure which, when mounted in a water supply line, absorbs the pressure surges caused by the rapid opening and closing of valves.

air change. The quantity of air infiltration and/or ventilation in cubic feet per hour or per minute, divided by the volume of the room, gives the number of air changes during that interval of time.

air conditioning. The control of the quality, quantity, and temperature-humidity of the air in a room or building within prescribed limits.

air diffuser. A device, such as a louver, which allows mixing of the air, delivered by it into a room, with the room air.

air dry. Wood containing 20 percent moisture is considered air dry.

air filter. A filter for purifying air by removing either particulate matter or gases or vapors which can be removed by chemical action.

airfoil. A curved surface designed to create lift as air flows over its surface.

air-gap separation. The unobstructed vertical distance through the free atmosphere measured between the lowest opening from any pipe or faucet supplying water to a tank, plumbing fixture, or other device and the flood-level rim of the receptacle.

air gasification. A partial combustion process in which coal is burned with roughly 50 percent of the air needed for complete combustion. The process results in a nonfuel gas, composed of carbon monoxide, a fuel species, and nitrogen (has a heating value one-sixth that of a unit volume of natural gas).

air-heating system. A solar-heating system in which air is heated in the solar collector and used as the energy transfer medium to the rest of the system.

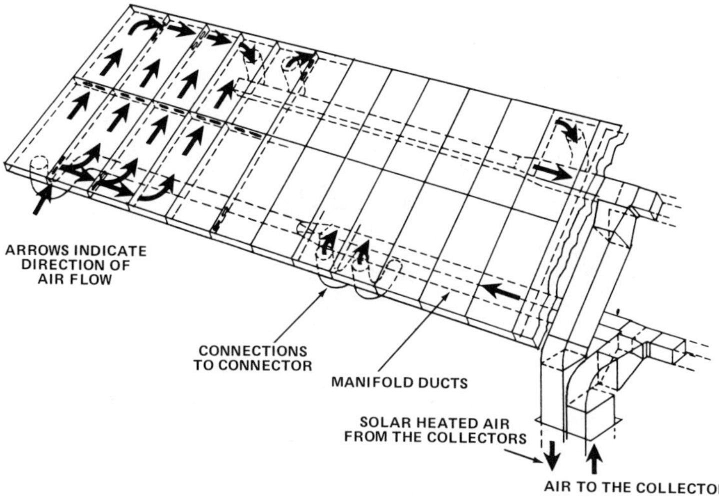

Courtesy Solaron Inc.

Typical air collector installation for air heating system.

air horsepower. The rate at which energy is used in horsepower units, in moving air between two points. Assuming 100 percent efficiency, it is the horsepower required to move a given volume of air against a given pressure.

airlock entry. A vestibule enclosed with two airtight doors; it reduces heat loss by limiting the movement of heated air.

air mass. The length of the path through the earth's atmosphere traversed by the direct solar radiation, expressed as a multiple of the path length with the sun at zenith.

air-mass-one (AM-1). Air mass 1 is the shortest path through the atmosphere and occurs when there is a 90 degree angle between the sun's radiation and the earth's surface. (See diagram, p. 34.)

air pollution. The presence of contaminants in the air in concentrations that prevent the normal dispersive ability of the air and that interfere directly or indirectly with people's health, safety, or comfort or with the full use and enjoyment of their property.

air-pollution episode. The occurrence of abnormally high concentrations of air pollutants usually due to low winds and temperature inversion and accompanied by an increase in illness and death. See **inversion.**

air-quality control region. An area designated by the federal government where two or more communities—either in the same or different states—share a common air-pollution problem.

air-quality criteria. The criteria for establishing the basis for air-quality standards,

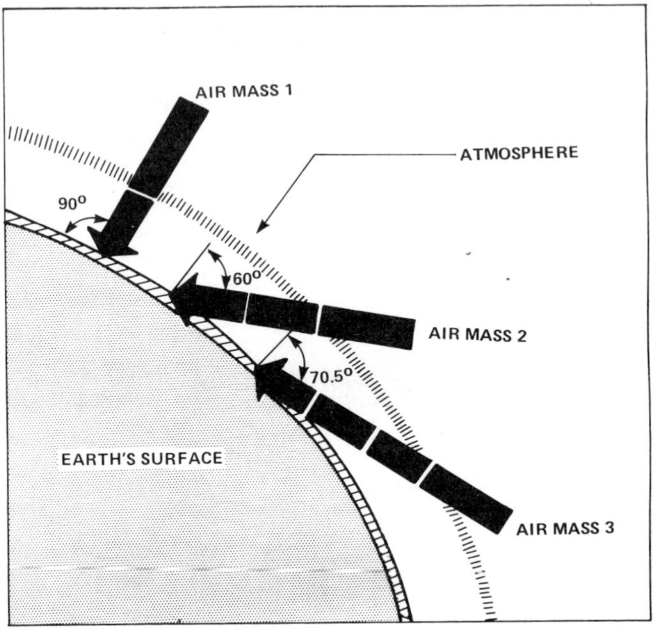

Courtesy National Copper Association

Relationship of air mass 1, 2, and 3 and declination is shown.

as determined from scientific compendiums documenting the relationship between various concentrations of air pollutants and their adverse effects.

air-quality standards. The prescribed level of pollutants in the outside air that cannot be exceeded legally during a specified time in a specified geographical area.

air sampling. The collection and analysis of samples of air to measure its radioactivity or to detect the presence of radioactive substances or chemical pollutants.

air shutter. Device for varying the size of air inlet(s).

air-source heat pump. Heat pump that transfers heat from outdoor air to indoor air during the heating season; the most common type of heat pump installed in homes.

air-tight stove. One with all seams filled and a very tight door. No air can enter the stove except at the draft control. Gives the operator full control over burning rate.

air-to-air heat pumps. Air-to-air heat pumps use outdoor air as a heat sink or heat source with the heat being delivered to or removed from the indoor air through the use of a direct refrigerant. Atmospheric air is the most universally available heat source and heat sink; hence, it is the one most commonly utilized.

The air-to-air heat pump is generally the smaller-tonnage unit in integral packaged form for residential and commercial application up to 30 tons. However, built-up systems of any size may be constructed. The models that are presently being marketed are either self-contained in one unit or split into two sections. Both types offer definite advantages, depending mainly on the type of application.

air-to-rock solar heating system. A solar-heating system of flat-plate collectors which circulates sun-heated air into a bin of rocks where its heat is transferred to the rocks to be stored for later use.

air-to-water heat pumps. A heat pump which uses the outdoor air as a heat source or sink, and which delivers the heat to the indoor area to be conditioned through the use of a secondary medium—in this case, water.

This type is marketed in sizes of 3 tons and up. No upper limit is anticipated with plants being designed to compete with conventional heating-cooling systems of all sizes.

air-type system. Active system that uses air as the medium to collect and carry solar heat; not commonly used for water heating alone because of the greater cost and poorer efficiency of air-to-water heat exchangers.

air washer. A device through which air is passed and washed with water from sprays for the purpose of humidifying, dehumidifying, or cleaning.

albedo. The reflecting power expressed as the ratio of light reflected from an object to the total amount falling on it.

alcohol. The family name of a group of organic chemical compounds composed of carbon, hydrogen, and oxygen; a series of molecules that vary in chain length and are composed of a hydrocarbon plus a hydroxyl group: $CH_3(CH_2)_n$-OH; includes methanol, ethanol, isopropyl alcohol, and others. See **ethanol**.

alcohol fuel plant. Under Bureau of Alcohol, Tobacco, and Firearms (BATF) regulations, a distilled spirits plant established solely for producing, processing, and using or distributing distilled spirits to be used extensively for fuel use. See **Bureau of Alcohol, Tobacco, and Firearms (BATF).**

alcohol fuel producer's permits. The document issued by BATF pursuant to the Crude Oil Windfall Profit Tax Act (26 USC 5181, P.L. 96-223) authorizing the person named to engage in business as an alcohol fuel production facility.

alcohol gasoline. The use of alcohol as a blending agent in gasoline.

aldehydes. Any of a class of highly reactive organic chemical compounds obtained by controlled oxidation of primary alcohols, characterized by the common group CHO, and used in the manufacture of resins, dyes, and organic acids.

algae. Primitive plants, one or many-celled, usually aquatic, and capable of synthesizing their own foodstuffs by photosynthesis.

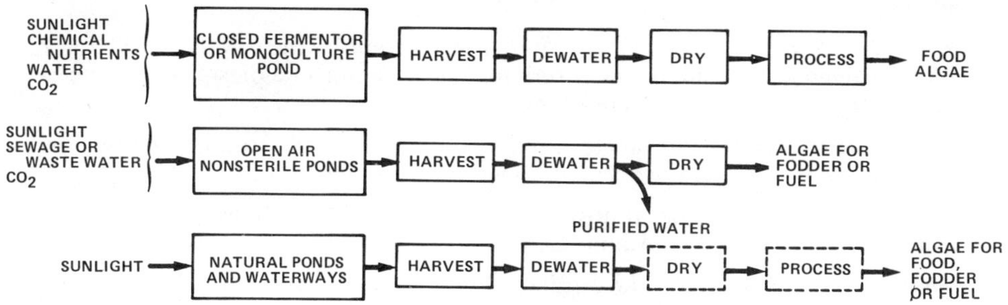

Reprinted with permission from Russell E. Anderson, Biological Paths to Self-Reliance, *Van Nostrand Reinhold*

A variety of alternative methods of algae production are shown in this block diagram.

algal bloom. A proliferation of living algae on the surface of lakes, streams, or ponds. Algal blooms are stimulated by phosphate enrichment.

alicyclic. Group of organic compounds visualized as an arrangement of carbon atoms in closed-ring structures. The three main subgroups are cycloparaffins (saturated), cycloolefins, and cycloacetylenes (cyclynes).

aliphatic. Group of organic compounds characterized by straight-chain structures of the constituent carbon atoms. The three main subgroups are paraffins (alkanes), olefins (alkenes or alkadienes), and acetylenes (alkynes). In complex structures, their chains can be branched or cross-linked.

alkadiene. Olefinic compound, which is composed of two double bonds that are either adjacent or conjugated.

alkali. Soluble mineral salts, usually of sodium, potassium, magnesium, or calcium, which combine with acids to form neutral salts. They are often used in water and wastewater treatment plants to control acidity. Salt of a low-density, low-melting point, highly reactive metal; characteristically "basic" in nature.

alkalinity. A quantitative measure of the capacity of liquids to neutralize strong acids. Alkalinity results from the presence of bicarbonates, carbonates, hydroxides, volatile acids, and salts.

alkene. Aliphatic hydrocarbon compound, consisting of one double bond, e.g., ethylene.

alkylation. The process of introducing one or more alkyl groups into the structure of hydrocarbons to form high-octane fuels.

alloy pipe. A steel pipe with one or more elements other than carbon which give it greater resistance to corrosion and more strength than carbon steel pipe.

alluvial. Particular types of, or minerals found associated with, deposits made by flowing water.

alpha-amylase. Enzyme which liquefies starch by conversion to dextrins.

alternating current (AC). Electrical current in which the flow of electrons reverses at regular intervals, usually 120 reversals per second or 60 cycles per second.

alternative combustion engines. Alternatives to the current spark ignition and diesel combustion engines, emphasizing the gas turbine and Stirling engines (q.v.) which can burn a variety of liquid fuels, including kerosene, diesel fuels, methanol, and coal-derived liquids.

alternator. A generator producing alternating current by the rotation of its rotor, which is driven by a steam or water turbine. A gas turbine or a diesel engine can be used as a prime mover in certain cases.

altitude angle. One of two angles used to specify the sun's position at any given time; altitude is the angle of the sun above the horizontal.

ambient. The surrounding atmosphere; encompassing on all sides; the environment surrounding a body but undisturbed or unaffected by it.

ambient air. Ambient air is the outdoor air in the vicinity of the solar collector being tested.

ambient temperature. The natural temperature of the atmosphere surrounding a particular location; it usually refers to outdoor temperature.

amino acids. The naturally occurring, nitrogen-containing building blocks of protein.

ammonia (NH_3). A "fixed" nitrogen form suitable as fertilizer for plant uptake and conversion to protein.

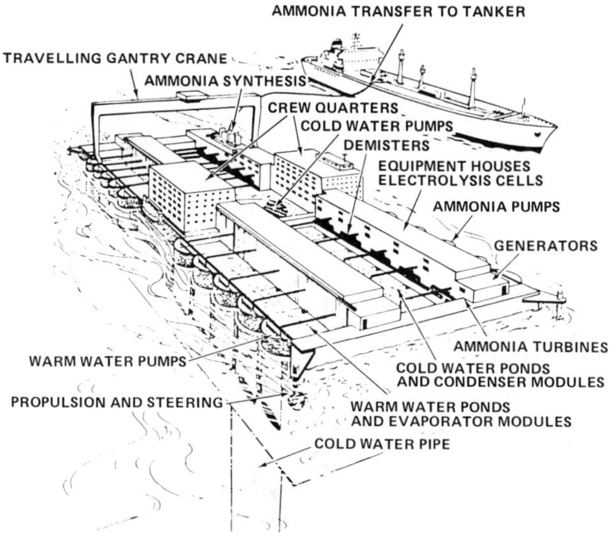

Courtesy DOE

An ammonia plant ship designed by the Johns Hopkins University/Applied Physics Laboratory.

ammonia machine. A mechanical refrigeration machine that uses ammonia as the refrigerant.

amorphous. The condition of a solid in which the atoms are not arranged in an orderly pattern; not crystalline.

amorphous silicon. A noncrystalline semiconductor material having only short-range order in its structure. Amorphous semiconductors hold promise for low-cost photovoltaics.

ampere. A unit of measure for an electric current; the current produced by 1 volt acting through a resistance of 1 ohm (Ω).

amp-hours. Calculated by multiplying current flow in amperes by number of hours the current flows.

amplitude. The maximum displacement from the mean position in connection with vibration or movement of a wave.

amylaceous feedstocks. Materials, such as cereal grains and potatoes, that are composed of saccharides in the form of starches.

amylase. A class of enzymes that hydrolyzes starches into sugars. All plants that store their sugars in the form of starches produce amylase enzymes. Many fungi and bacteria can produce various amylases and the commercially prepared amylase enzymes come from such sources. See **enzymes**.

amylodextrin. See **dextrin**.

amyloglucosidase. An enzyme, produced by microorganisms, which completely hydrolyzes starch to glucose. See **enzymes**.

amylopectin. This is the outer, virtually insoluble portion of the starch granule. The D-glucose units are joined together in a manner resembling a highly branched structure.

amylose. The inner, soluble portion of the starch granule. Amylose consists of a long chain of D-glucose units (about 200 to 500) joined together.

anabatic wind. An upslope wind; usually applied only when the wind is blowing up a hill or mountain as the result of local surface heating, and apart from the effects of the larger-scale circulation.

anadromous. Type of fish that ascends rivers from the sea to spawn.

anaerobic. Bacteria or other microorganisms that can live without free oxygen, or air, in a medium are anaerobic. Yeast can live in both an aerobic and anaerobic state. When yeast uses up all the free oxygen in a medium, it changes its metabolism and begins to produce ethyl alcohol and carbon dioxide; living, active, or occurring in the absence of free oxygen.

anaerobic digestion. The process by which organic matter is decomposed by bacteria which work in the absence of free oxygen, with the release of a burnable mixture of gases.

ancillary energy. A measure of the external energy required for an energy process, such as energy for process heat, electricity for pumps, and fuel for truck, train, or barge transportation.

anemometer. An instrument for measuring air velocity.

angle of incidence. The angle at which direct sunlight strikes a surface. The angle of incidence affects the amount of energy absorbed by a solar collector. Sunlight with an incident angle close to 90° (perpendicular to the surface) tends to be absorbed, while lower angles tend to be reflected. (See diagram, p. 39.)

angle valve. A valve, usually of the globe type, in which the axes of the inlet and outlet are at right angles.

anhydrous. Devoid of water; refers to a compound that does not contain water either absorbed on its surface or as water of crystallization.

anhydrous ethanol. 100 percent alcohol, neat alcohol; 197 to 200 proof water-free alcohol.

animal waste conversion. The process for obtaining oil from animal wastes. A government experiment has obtained 80 gallons of oil per ton from cow manure. In comparison, average oil shale yields 25 gallons of oil per ton of ore.

annual capital fixed charge rate. The annual repayment of an initial investment plus the annual discount rate.

Annual Cycle Energy System. The Annual Cycle Energy System (ACES) heating cycle consists of a methanol–water loop, a refrigerant loop, and a loop for domestic hot water. Heat is removed from a large tank of water for winter heating; this operation turns the water into ice, which is stored for cooling the house in summer.

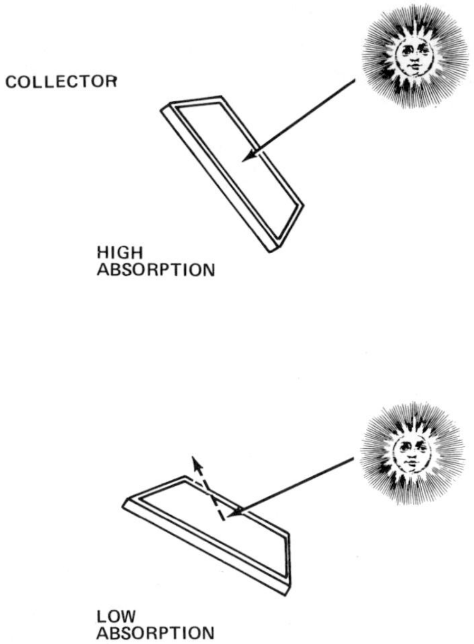

Courtesy HUD

Illustration of angle of incidence on a flat-plate collector.

annual energy displacement. Quantity of energy replaced by energy conversion systems and/or energy storage discharge.

annual load fraction. A fraction of the annual heating needs supplied by solar energy.

annual maximum demand. The greatest of all demands of the electrical load which occurred during a prescribed interval in a calendar year.

annual plant. A plant that completes its life cycle in 1 year or less.

anode. The positive terminal of an electrolytic cell; the negative terminal of a primary cell or of a storage battery that is delivering current; the electrode at which oxidation or corrosion occurs, or from which the current is transmitted to the electrolyte.

anoxic. Refers to an oxygen deficiency.

anthracite. A dense hard coal with a very high percentage of fixed carbon, usually about 90 percent.

anthraxylon. Vitreous-appearing components of coal, derived from the woody tissues of plants, e.g., stems, limbs, twigs, roots. The woody tissues remain as distinct units as coal.

antidegradation clause. A provision in air-quality and water-quality laws that prohibits deterioration of air or water quality in areas where the pollution levels are presently below those allowed.

antitrade winds. The prevailing westerly winds frequently present above the trade winds.

aperture. Opening or projected area of a solar collector, through which the solar radiation is admitted and directed to the absorber.

apparatus dew point. A term used interchangeably with average coil surface temperature. It is the temperature at which the sensible heat ratio line intersects the saturation curve on a psychrometric chart (q.v.).

apparent day. Solar day; interval between successive transits of the sun's center across the observer's meridian. The time thus measured is not uniform or clock time.

apparent proof. The proof indicated by a hydrometer after correction for temperature but without correction of the obscuration caused by the presence of solids.

application range. The range of operating volumes and pressures, determined by the manufacturer, at which a fan will operate satisfactorily.

aquaculture. Cultivation of fish, algae, water plants, and other waterborne organisms.

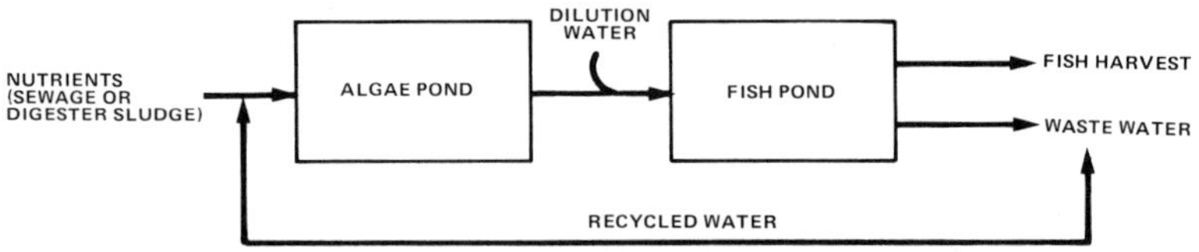

Reprinted with permission from Russell E. Anderson, Biological Paths to Self-Reliance, *Van Nostrand Reinhold*

Block diagram of two-stage aquaculture system.

aquaculture project. A controlled discharge of pollutants to enhance growth or propagation of harvestable freshwater, estuarine, or marine life plant or animal species.

aquatic biota. All living organisms of any designated aquatic area.

aquatic plants. Plants that grow in water either floating on the surface, growing up from the bottom of the body of water, or growing under the surface of the water.

aquifer. A water-bearing stratum of permeable rock, sand, or gravel used for storage of low-temperature thermal energy.

area drain. A drain that collects rain or surface water from an open space.

area source. In air pollution, any small individual fuel combustion source, including any transportation sources. See **point source.**

aromatics. Cyclic hydrocarbons and their derivatives; characterized by the presence of at least one benzene ring. The principal aromatics are benzene, toluene, and xylene (commonly referred to as the BTX group).

array. As in thermal or electrical solar array. A number of individual solar collection devices arranged in a suitable pattern to collect solar energy effectively.

Courtesy DOE

View of roof-mounted solar collectors at Disney World. Solar energy powers air conditioning, space heating, and hot-water supply.

artesian. Groundwater under sufficient pressure to cause the water level in a drilled hole to rise above the top of the rock unit.

artificial satellites. Man-made vehicles designed to orbit the earth, moon, or any other heavenly body. Satellites are being considered as platforms for solar power stations.

asbestos. A mineral fiber with countless industrial uses; a hazardous air pollutant when inhaled. Its effects as a water pollutant are under scrutiny.

aspect ratio. In a ventilation duct, the depth of elbow along (parallel to) the axis of bend, divided by the width in the plane of the bend.

assimilation. Conversion or incorporation of absorbed nutrients into protoplasm. Also refers to the ability of a body of water to purify itself of organic pollution.

asynchronous. An electric generator designed to produce an alternating current that matches an existing power source (e.g., utility mains) so the two sources can be combined to power one load (e.g., your home). The generator does not have to turn at a precise rpm to remain at correct frequency or phase. See **synchronous generator**.

atmosphere. The gaseous envelope surrounding the earth; the outdoor air in general. A mixture of gases within any specific chamber such as a heat-treating furnace.

atmospheric pollution. Degradation of the quality of the atmosphere due to combustion products from industrial plants, power plants, or vehicular engines.

atmospheric pressure. The pressure of the air at sea level. The standard atmospheric pressure will support a column of mercury 760 millimeters (mm) high, or 29.9 inches (in.) high, in a barometer.

atmospheric water. Water which exists in the atmosphere in a gaseous, liquid, or solid state.

atom. A particle of matter whose nucleus is indivisible by chemical means. It is the fundamental building block of the chemical elements.

atrium. A closed interior court to which other rooms open; it is often used for passive solar collection.

attenuation loss. Loss of solar power by absorption and scattering due to atmospheric conditions.

attic fan. A fan so mounted in the attic that it can draw air from the house at night, causing cool outside air to flow into the house, and exhausting the warm air to the outside.

auger kiln. Biomass gasifier which uses a steam gasification process. Biomass feedstock is carried through the stationary, inclined reactor by the motion of an internal auger.

autohydrolysis. Self-hydrolysis; in the case of biomass, the term refers to the conversion of cellulosic matter to sugars under elevated temperatures and pressure, and in the absence of an external source of acid.

automatic damper. A device which cuts off the flow of hot or cold air to or from a room when the thermostat indicates that the room is warm enough or cool enough.

automatic heating. A central heating system operated without manual attention. Usually means oil-fired, gas-fired, or stoker-fired furnaces and boilers.

autotrophic. Self-nourishing; denoting those organisms capable of constructing organic matter from inorganic substances.

auxiliary. The component physically connected as part of a solar system and that provides less than one-half of the annual energy supplied by that solar system.

auxiliary energy. Auxiliary heat plus the energy required to operate pumps, blowers, or other devices.

auxiliary energy subsystem. Equipment using conventional energy sources to supplement the output provided by the solar energy system. It may be integrated directly into the solar energy system, or it may be completely separate from it and contain its own means for delivery of heating, cooling, and/or hot water to the building.

auxiliary system. A supplementary heating unit to provide heat to a space when its primary unit cannot do so. This usually occurs during periods of cloudiness or intense cold, when a solar heating system cannot provide enough heat to meet the needs of the space. (See diagram, p. 43.)

availability factor. The ratio of time a wind-energy conversion system is producing power (or is producing some specified amount of power) to the total time interval under consideration. This excludes reliability factors such as maintenance.

available carbon. Free carbon not chemically bound to oxygen and, therefore, available for combustion.

available energy. That part of the total energy which can be usefully employed; in a perfect engine, that part which is converted to work.

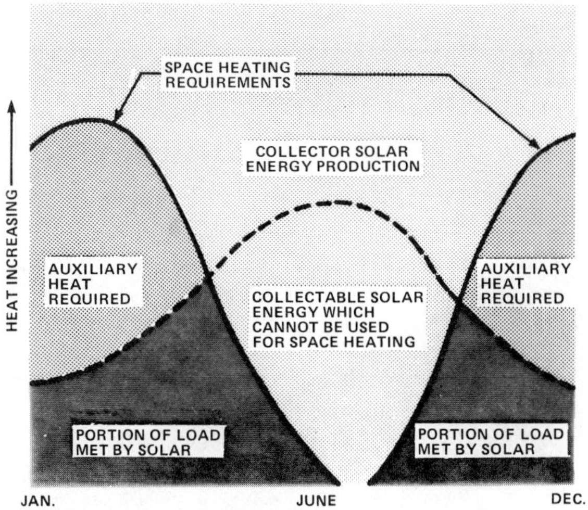

Courtesy National Copper Association

Illustration of space heating demand versus solar collector energy production illustrates periods requiring auxiliary systems.

available heat. The amount of energy that may be converted into useful heat from the chemical energy in a unit of fuel.

available power loss. The available power loss of a transducer connecting an energy source and an energy load is the transmission loss measured by the ratio of the source power to the output power of the transducer.

average demand. The demand on a system or any of its parts over an interval of time, determined by dividing the total energy supplied by the number of units of time in the interval.

average wind speed. The mean wind speed over a specified period of time. (See diagram, p. 44.)

axial fans. Axial fans are divided into three groups—propeller, tubeaxial, and vaneaxial. The propeller fan is used for high volumes of air at little or no static pressure differential. Tubeaxial and vaneaxial fans are propeller fans mounted in a cylinder and are similar except for vane-type straighteners on the vaneaxial. These vanes remove much of the swirl from the air and improve efficiency. Thus, a vaneaxial fan is more efficient than a tubeaxial and can reach higher pressures. Note that, with axial fans, the brake horsepower (BHP) is maximum at block-tight static pressure. With centrifugal fans, the BHP is minimum at block-tight static pressure.

axial flow compressor. A compressor in which air is compressed in a series of stages as it flows axially through a decreasing tubular area.

axial flow fan. A fan with a disc or air-foil-shaped blade mounted on a shaft, and which moves the air in the general direction of the shaft axis.

azeotrope. The chemical term for two or more liquids that, at a certain concentration, boil as though they are a single substance; alcohol and water cannot be

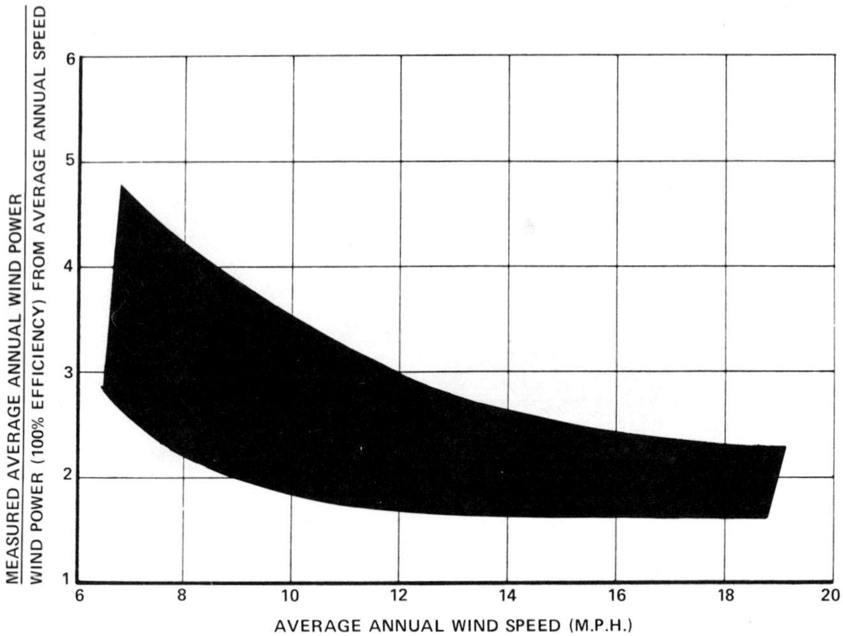

Courtesy DOE

Illustration of average wind speed range.

separated further than 194.4 proof because at this concentration, alcohol and water form an azeotrope and vaporize together.

azeotropic distillation. In ethanol distillation, an azeotrope forms and prevents a pure product from being produced. To obtain a pure product, additional stills must be added to the fractionating column(s) and an entrainer is introduced to produce a ternary azeotrope. A typical entrainer is benzene, although others, such as trichloroethylene, can be used. The benzene, combined in a certain ratio with water and ethanol, forms a ternary azeotrope and is taken off the top of the azeotropic column, while a pure, or anhydrous, alcohol is withdrawn off the bottom. The ternary azeotropic mixture has the lowest boiling point and will come off the top as the distillate.

azeotropic mixture. When a liquid mixture of two or more substances behaves like a single substance—i.e., the vapor produced by the partial evaporation of the liquid has the same composition as the liquid—it is an azeotropic mixture. For example, when the composition of an ethyl alcohol/water ratio is 95.57/4.43 percent, an azeotrope forms. In simple distillation, the vapors will remain at that proportion, making a pure product of 200 proof impossible to obtain. To get 200 proof ethanol, other types of distillation must be used. See **azeotropic distillation; vacuum distillation.**

azimuth (solar). One of two angles used to specify the sun's position at any given time; azimuth is the angle between south and the point on the horizon directly below the sun. South is 0° and angles to the east and west are described as 0° to 180°E or 0° to 180°W.

B

backfill. The material used to refill a ditch or other excavation, or the process of doing so.

backflow. Unintentional reversal of flow in a potable-water distribution system, which may result in the transport of foreign materials or substances into the other branches of the distribution system. Backflow may occur under two conditions: pressure greater than atmospheric and pressure that is subatmospheric. See **back pressure.**

backflow preventer. A device to prevent backflow into a water supply line from the connections at its outlet.

background level. With respect to air pollution, amounts of pollutants present in the ambient air due to natural sources.

back pressure. Pressure created, by any means, in the water distribution system of a building, which by being greater than the pressure in the water supply main could cause backflow.

back set (also called set back). The liquid portion of stillage recycled as part of the process liquid in alcohol mash preparation.

back siphonage. Backflow of possibly contaminated water into the potable water supply system as a result of the pressure in the potable water system becoming unintentionally less than the atmospheric pressure in the plumbing fixtures, pools, tanks, or vats that may be connected to the potable water distribution piping.

backslopping. The adjustment of the pH of alcohol mash by adding 10 to 20 percent of mash stillage residue prior to fermentation.

backup energy system. A backup energy system using conventional fuels should be provided for heating and domestic hot water. This system should be capable of providing all of the energy demand during any period when the solar energy system is not operating. Components and subsystems may be used as parts of both systems where the component or subsystem is a recognized, acceptable product in the conventional building industry.

backward-inclined fan. Backward-inclined fans travel about twice the speed of the forward-curved fan. The normal selection range of the backward inclined fan is approximately 40-85 percent of wide open cubic feet per minute. Maximum static efficiency of about 80 percent generally occurs close to the edge of its normal operating range. Generally, the larger the fan the more efficient for a given selection.

bacteria. Single-celled microorganisms that lack chlorophyll. Some bacteria are capable of causing human, animal, or plant diseases; others are essential in pollution control because they break down organic matter in the air and in the water.

bacterial spoilage. Occurs when bacterial contaminants take over the fermentation process in competition with the yeast.

baffle. Any deflector device used to change the direction of flow or the velocity of water, sewage, or products of combustion such as fly ash or coarse particulate matter.

baffle plate. Holds the hot gases and heated air in a wood stove longer causing them to travel a longer path before exiting; brings and holds them in close contact with the stove plates. This results in more heat from the same amount of fuel.

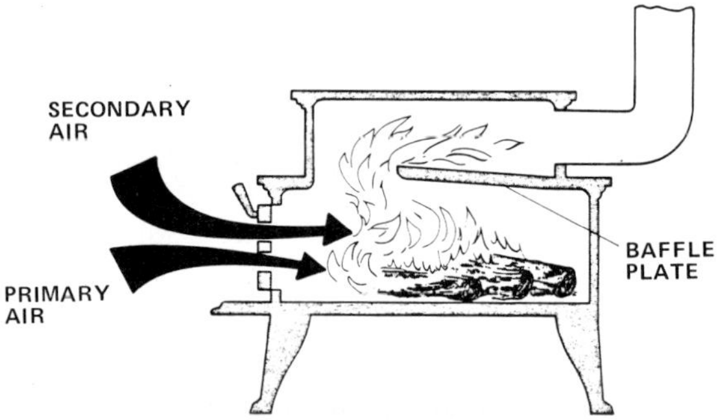

Courtesy N. Soderstrom, Heating Your Home with Wood, *Popular Science Skill Book.* Illustration of baffle plate in airtight wood stove. The baffle plate aids in the ignition of the remaining volatiles.

bagasse. The cellulosic fibrous material remaining after the extraction of the juice from sugarcane. It is used as a fuel and as a mix in making lightweight refractories; the dried and pulverized or shredded sugarcane fibers sometimes are added to a drilling fluid to plug crevices in and prevent loss of circulation liquid from a borehole.

bag filter. An apparatus for removing dust from dust-laden air, employing cylinders of closely woven material which permit passage of air but retain solid particles.

baghouse. Chamber in which exit gases from roasting, smelting, or calcining are filtered through membranes (bags) which arrest solids; a fabric filter used to separate particulates from an airstream; an air-pollution abatement device used to trap particulates by filtering gas streams through large fabric bags, usually made of glass fibers.

balance of system. Parts of a photovoltaic system other than the array: switches, controls, meters, power conditioning equipment, supporting structure for the array, and storage components, if any. The cost of land is sometimes included when comparing total system costs with the cost of other energy sources.

balance point. An outdoor temperature, usually 20° to 45°F (−7° to 7° C), at which a heat pump's output exactly equals the heating needs of the house. Below the balance point, supplementary heat is needed for indoor comfort.

baling. A means of reducing the volume of solid waste by compaction.

balling hydrometer or brix hydrometer. A triple-scale wine hydrometer designed to record the specific gravity of a solution containing sugar.

ballistic separator. A machine that separates inorganic from organic matter in a composting process.

band gap energy. The amount of energy (in electron volts) required to free an outer shell electron from its orbit about the nucleus to a free state, and thus to promote it from the valence level to the conduction level.

baroclinic atmosphere. An atmosphere in which the density depends on both the temperature and the pressure. In a baroclinic atmosphere the geostrophic wind generally has vertical shear, and this shear is related to the horizontal temperature gradient by the thermal wind equation.

barometer. Instrument used for measuring atmospheric pressure.

barrel. A liquid measure of oil, usually crude oil, equal to 42 American gallons (159.2 liters) or about 306 pounds (139 kilograms). One barrel equals 5.6 cubic feet (ft^3) or 0.159 cubic meters (m^3). For crude oil, one barrel is about 0.136 metric tons (MT), 0.134 long tons, and 0.150 short tons. The energy values in millions (10^6) of British thermal units (Btu) per barrel of petroleum products are: crude petroleum, 5.6; residual fuel oil, 6.29; distillate fuel oil, 5.83; gasoline, 5.25; jet fuel (kerosene type), 5.67; jet fuel (naphtha type), 5.36; kerosene, 5.67; petroleum coke, 6.02; and asphalt, 6.64.

barrier. See **cell barrier.**

barrier energy. The energy given up by an electron in penetrating the cell barrier; a measure of the electrostatic potential of the barrier.

bar screen. In wastewater treatment, a screen that removes large floating and suspended solids.

Barstow pilot plant. See **Solar One.**

basal metabolism. The amount of heat liberated by a person at rest in a comfortable environment (about 40 British thermal units per hour [Btu/hr]).

baseboard radiator. A heat-disseminating unit located at the lower perimeter of a room.

baseload. The minimum amount of electric power which a utility must supply in a 24-hour period. Utilities typically operate their most efficient generators (usually their newest and largest) to meet baseload demand. (See diagram p. 48.)

baseload power plant. In an electric utility system, baseload power plants supply the majority of total electric energy demand. They operate continuously except when shut down for planned maintenance or equipment failure. They are designed

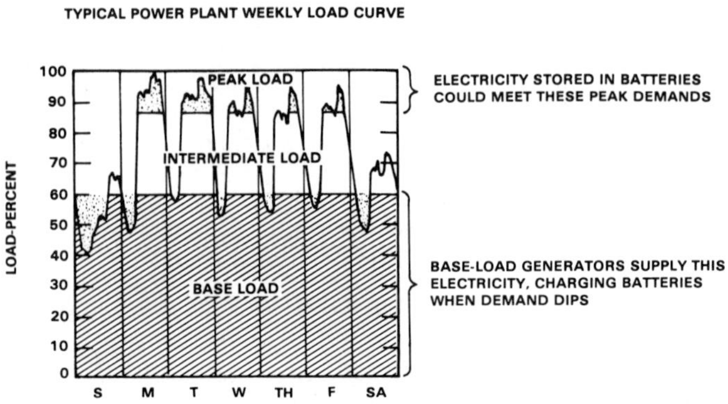

Courtesy DOE

Illustration of baseload versus intermediate and peak load.

to provide the combination of capital cost and efficiency which enables power to be generated at minimum cost.

basic hydrolysis. Decomposition or alteration of a chemical substance by water in the presence of alkali.

batch distillation. In alcohol fuel production, this is the easiest method of distillation. The entire charge (batch) is placed in the holding tank of the still and is heated to boiling. The distillate is taken, or withdrawn, only from the condenser. The traditional "moonshiner" still is of this design, but a properly constructed still with a packed column is far superior to this, and is the best method for small alcohol fuel applications. See **packed distillation column.**

batch fermentation. Fermentation conducted from start to finish in a single vessel.

batch process. A production process in which the material is introduced as discrete charges, each of which is processed to completion separately.

bathyclinograph. In oceanography, an instrument for measuring vertical currents in the deep sea.

bathymeter. An instrument that measures temperature, pressure, and sound velocity to depths up to 7 miles (11.3 kilometers). The devices are completely transistorized and use frequency modulation for telemetering.

battery. A device which converts energy from chemical compounds (called reactants) into electricity. A battery consists of one or more units (called cells) which are enclosed in an insulated casing. Each cell contains a positive and negative electrode separated by an electrolyte.

Battery Energy Storage Test Facility (BEST). An experimental laboratory in Hillsborough Township, NJ, for the testing of electric storage batteries; jointly funded by the Electric Power Research Institute and the Department of Energy.

Baudelot cooler. An arrangement of pipes one above another through which refrigerant flows and is vaporized as it absorbs heat from the water being cooled by trickling over the tubes.

Beadwall ™. A form of movable insulation developed by David Harrison of Zomeworks Corp. The system employs tiny polystyrene beads blown into the space between the two sheets of glass (or plastic) in a double-glazed window or skylight.

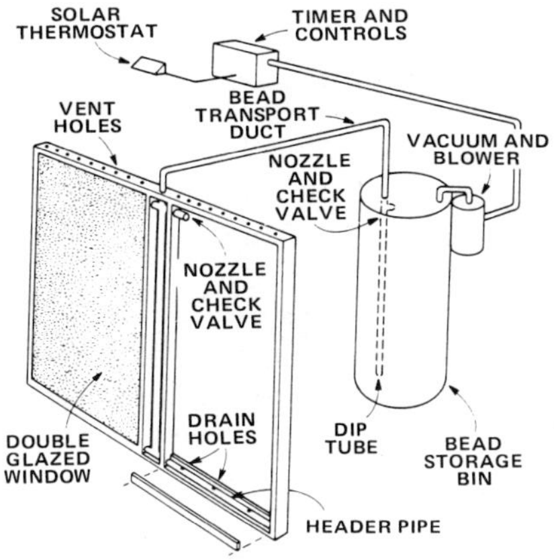

Courtesy A. J. Hand, Home Energy How-To, *Popular Science Book/Harper and Row.*
The beadwall system—patented by Zomeworks—fills or removes styrene beads between two glass panels for insulation of window areas.

beam alignment. The adjustment of individual mirror facets of a heliostat to place their images in the desired relationship to one another.

beam characterization subsystem. A system for the rapid and automatic measurement and characterization of flux delivered by any single heliostat.

beam radiation. Solar radiation which is not scattered by dust or water droplets. It is capable of being focused and casts shadows.

bearing angle. See **azimuth (solar).**

bearing wall. A wall which supports a vertical load in addition to its own weight.

Beaufort scale. A numerical scale, graded from 0 to 12, devised by Admiral Beaufort in the nineteenth century to indicate wind speeds. This scale has been adopted internationally.

bedplate. A base-plate for supporting a system component or structure.

bedrock. Any solid rock exposed at the surface of the earth or overlain by unconsolidated material.

beer. The product of fermentation by microorganisms; the fermented mash, which contains about 7 to 12 percent alcohol; usually refers to the alcohol solution remaining after yeast fermentation of sugars.

beer still. The stripping section of a distillation column for concentrating ethanol, or the first column of a two (or more) column system, in which the first separation from the mash takes place.

beer well. The surge tank used for storing beer prior to alcohol distillation.

bell-and-spigot joint. The commonly used joint in cast-iron pipe. Each piece is made with an enlarged diameter or bell at one end into which the plain or spigot end of another piece is inserted when laying. The joint is then made tight by cement, oakum, lead, or rubber caulked into the bell around the spigot.

bellows. An expandable metal device containing a fluid which will volatilize at some desired temperature, expand the device, and open or close an opening or a switch, as in controls and steam traps.

benthic region. The bottom of a body of water. This region supports the benthos, a type of life that not only lives upon but contributes to the character of the bottom.

benthos. The plant and animal life whose habitat is the bottom of a sea, lake, or river.

benzene (C_6H_6). A colorless liquid hydrocarbon, made from coal tar and by catalytic reforming of naphthenes. It is used in the manufacture of phenol, styrene, nylon, detergents, aniline, phthalic anhydride, and other compounds. It is also used as a solvent, and as a component of high-octane gasoline.

benzine. An obsolete term for light petroleum distillates covering the gasoline and naphtha range.

berm. A mound of earth either abutting a house wall to help stabilize temperature inside house, or positioned to deflect wind from house.

Courtesy DOE

North side of the solar-powered Shenandoah Community Center featuring earth graded up next to wall as a berm.

beta-amylase. Enzyme which converts dextrins into glucose.

Betz law. The theoretical maximum amount of wind energy that can be extracted by a machine was calculated to be 59.3 percent by A. Betz. In practice, the efficiency limit is approximately 40 percent.

bimetal. Two metals of different coefficients of expansion welded together so that the piece will bend in one direction when heated, in the other when cooled, and thus can be used in opening and closing electrical circuits, as in thermostats.

binary cycle. Combination of two turbine cycles utilizing two different working fluids in electrical generation plants. The waste heat from the first turbine cycle provides the heat energy for the second turbine cycle. The purpose of a binary cycle is to obtain higher efficiencies from the energy source.

binary-fluid systems. A system in which hot fluid is passed through a heat exchanger to transfer heat to a low-boiling-point fluid which is then used as the working fluid.

bin method. A method of predicting monthly or annual heating and/or cooling loads using an instantaneous load calculation at many different outdoor dry-bulb temperatures (q.v.), and multiplying the result by the number of hours of occurrence of each temperature.

bioassay. The employment of living organisms to determine the biological effect of some substance, factor, or condition.

bioconversion. A general term describing the conversion of one form of energy into another by plants or microorganisms. Synthesis of organic compounds from carbon dioxide by plants in bioconversion of solar energy into stored chemical energy. Similarly, digestion of solid wastes or sewage by microorganisms to form methane, to bioconversion of one form of stored chemical energy into another, more useful form.

biodegradable. Used in sewage disposal and water pollution to describe those substances that can be quickly broken down by the bacteria used for this purpose at sewage disposal plants.

biofouling. The formation of film and/or slime on a substrate from attachment of micro-and macroaquatic organisms; can occur under freshwater and saltwater conditions.The growth of small marine organisms which can decrease the heat-transfer efficiency of heat exchangers used in ocean thermal energy conversion systems.

biogas. The fuel gas product of anaerobic digestion, predominately methane (CH_4) and carbon dioxide (CO_2) (also called Gobar gas, marsh gas).

biogas process. An anaerobic digestion process for converting solid municipal waste and sewage into pipeline quality fuel gas and an odor-free, stable solid. (See diagram p. 52.)

biological control. A method of controlling pests by means of introduced or naturally occurring predatory organisms, sterilization, or the use of inhibiting hormones, etc., rather than by mechanical or chemical means.

biological energy conversion. Using a living plant or animal to convert one form of energy to another; for example, trees use solar radiation as an energy source to produce cellulose which can then be used as a fuel.

biological oxidation. The process by which bacteria and other microorganisms feed on complex organic materials and decompose them. Self-purification of waterways and activated sludge, and trickling filter wastewater treatment processes depend on this principle. The process is also called biochemical oxidation.

biological oxygen demand (BOD). An indirect measure of the concentration of

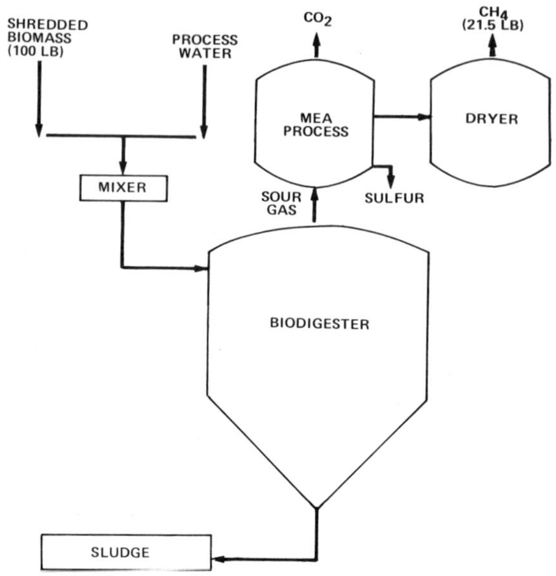

Courtesy EPRI

Flow diagram of biodigester functions which produce biogas.

biologically degradable material present in organic wastes. It is the amount of oxygen used by aerobic bacteria when allowed to degrade the organic matter in the presence of air at a constant temperature [68°F (20°C)] for a specific time (5 days). It is expressed in milligrams of oxygen used per liter of liquid waste (mg/liter) or in parts per million (ppm).

biomass. Any material derived from growing organisms such as agricultural products and residues, trees, wood and bark residues, animal manures, and algae. This word has come into common use for discussions of productivity of photosynthesis and energy use of cellulose materials irrespective of species. Organic matter, such as trees, crops, manure, and aquatic plants, that is available on a renewable basis.

Defined in the Energy Security Act (P.L. 96-294), 1980, as "any organic matter which is available on a renewable basis, including agricultural crops and agricultural wastes and residues, wood and wood wastes and residues, animal wastes, municipal wastes, and aquatic plants."

Biomass Catalytic Gasification. A conceptual process for converting biomass (wood) to gas by a catalytic process. The feed preparation process mixes the catalysts and wood feed material for injection into the reactor. The reactor is used to react the biomass with steam to catalytically convert the gaseous components to the desired products. The product gas from the reactor is purified by a conventional clean-up system to produce the desired pure final product. The clean-up step would include acid gas removal (primarily CO_2) and perhaps tar removal. Both can be accomplished with conventional equipment. The catalyst recovery step is intended to remove the solid particulate catalysts from the residue, and to recover the alkali metal compound gasification catalysts for recycling. The former would be accomplished mechanically while the latter may require aqueous leaching or flotation techniques.

biomass energy project. Defined in the Energy Security Act (P.L. 96-294), 1980, as "any facility (or portion of a facility) located in the United States which is primarily for (a) the production of biomass fuel (and byproducts); or (b) the combustion of biomass for the purpose of generating industrial process heat, mechanical power, or electricity (including cogeneration)."

biomass fuels. The production of synthetic fuels, such as ethanol, methane and hydrogen, from terrestrial and sea plants and from animal wastes. Defined in the Energy Security Act (P.L. 96-294), 1980, as "any gaseous, liquid, or solid fuel produced by conversion of biomass."

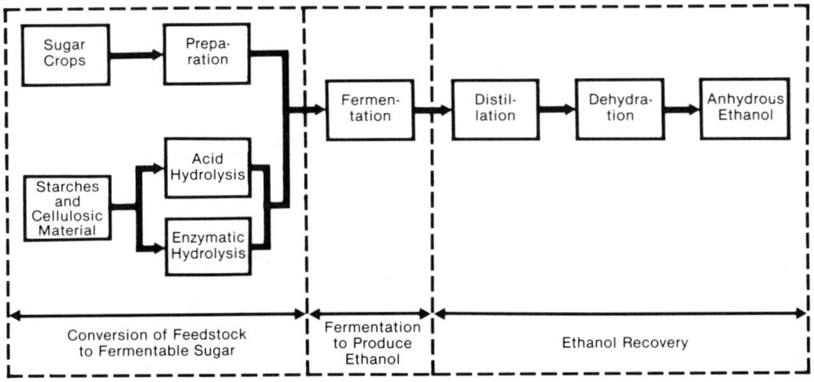

Courtesy U.S. National Alcohol Fuels Commission

Block diagram of steps to convert biomass to anhydrous ethanol for use as power alcohol or gasohol fuel.

biomethanation. The formation of methane by microorganisms from hydrogen, carbon monoxide, and carbon dioxide.

biomonitoring. The use of living organisms to test the suitability of effluent for discharge into receiving waters and to test the quality of such waters downstream from a discharge.

biophotolysis. The action of light on a biological system which results in the dissociation of a substrate, usually water, to produce hydrogen.

biosphere. Zone at, and adjacent to, the earth's surface where all life exists; all living organisms of the earth.

biostabilizer. A machine used to convert solid waste into compost by grinding and aeration.

biosynthesis. Production/transformation of substances from other compounds assisted by living organisms.

biota. All the species of plants and animals occurring within a certain area.

bitumen. Refers to solid and semisolid hydrocarbons.

blackbody. Term describing an ideal substance which would absorb all the radiation falling on it and reflect nothing. An alternative definition is a body which emits the maximum possible radiation, i.e., its emissivity is 1.0.

black pipe. Steel pipe that has not been galvanized.

blank flange. A flange in which the bolt holes have not been drilled.

blastcoil. Heat-transfer surface, most frequently of an extended surface arrangement, over which air is blown to be heated or cooled, depending on the temperature of the fluid within the pipelike surface.

blending. The process of mixing two or more oils having different properties to obtain a final blend having the desired characteristics. This can be accomplished by off-line batch processes or by in-line operations as part of continuous-flow operations.

blending stock. Any of the stock used to make commercial gasoline, including natural gasoline, straight-run gasoline, cracked gasoline, polymer gasoline, alkylate, aromatics, and ethanol mixed with gasoline to produce gasohol.

blind flange. A flange used to seal off the end of a pipe.

blocking. The interception of part of the reflected sunlight from one heliostat by the backside of a second heliostat. The process whereby a barrier to stratified flow prevents the dense fluid layers nearest the surface from crossing the barrier because of insufficient kinetic energy.

bloom. A proliferation of living algae and/or other aquatic plants on the surface of lakes or ponds. Blooms are frequently stimulated by phosphate enrichment.

blow. In connection with air distribution, the distance an air stream travels from an air outlet to the point where the air velocity is reduced to 50 feet per minute (ft/min).

body. That part of the valve which attaches to the pipe line or equipment either with screwed ends, flanged ends, or solder joint ends, and encloses the working parts of the valves.

bog. Wet, spongy land usually poorly drained, highly acid and rich in plant residue.

boiler. A closed vessel in which water is converted to pressurized steam.

boiler burner unit. Boiler designed especially for gas or oil and sold integrally with the burner.

boiler efficiency. Ratio of heat absorbed by the water in the boiler to the total heat supplied to the boiler.

boiler feedwater. The preheated water which is converted to steam in a boiler.

boiler fuel. Natural gas used as a fuel for the generation of steam.

boiler heating surface. That part of the interior surface of a boiler subjected to heat on the one side, and transmitting the heat to water (or steam or other fluid) within the boiler. Direct surface is that subjected to direct radiant rays of the burning fuel, other heating surface is termed indirect.

boiler horsepower. Represents the conversion of 34.5 pounds of water per hour (15.7 kg/hr) to steam at a pressure of 14.7 pounds per square inch (101 kilopascals) and a temperature of 212°F (100°C).

boiler pressure. The pressure of the steam or water in a boiler. It is generally expressed in pounds per square inch gauge (psig).

boiler rating. The heating capacity of a steam boiler expressed in British thermal units per hour. Sometimes expressed in horsepower or pounds of steam per hour.

boiling point. The temperature at which the transition from the liquid to the gaseous phase occurs in a pure substance at fixed pressure.

bond. The cohesion or adhesion that develops between particles; also the link between two atoms due to an electron pair resonating or rotating between them. If each atom contributes an electron, the bond is atomic and nonpolar; if held unequally, it is polar. If one atom contributes both electrons, the bond is molecular.

bond (BATF). A type of insurance which gives the government security against possible loss of distilled spirits tax revenue; not required for alcohol fuel plants producing less than 10,000 proof gallons per year.

bonnet. Part of a valve used to guide and support the valve stem.

booster. An accessory to increase capacity or output. In refrigeration, a compressor discharging into the suction line of another unit. In heating, the term is frequently applied to a hot water circulator.

booster mill. Second water-pumping windmill added in tandem to provide added capacity for transporting water some distance uphill or overland.

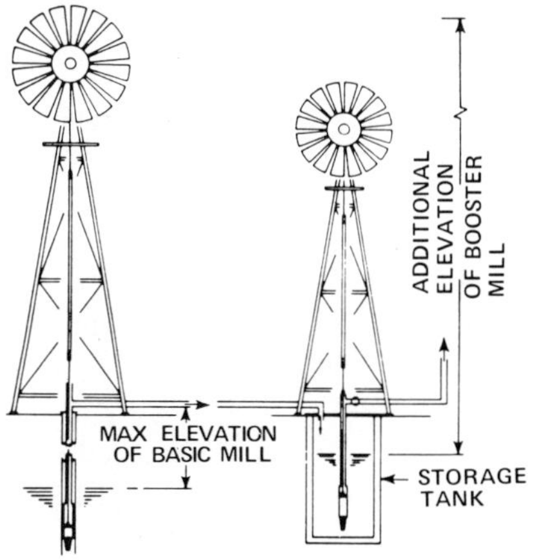

Courtesy DOE

Illustration of use of additional windmill to boost water additional height for storage or use.

boot. Sheet-metal transformation pieces used in warm-air heating, and connected to horizontal round leaders on one end and to vertical rectangular stacks on the other.

boron (B). A chemical element, atomic number 5, semimetallic in nature, used as a dopant to make *p*-silicon for photovoltaic cells.

bottled gas. The liquefied petroleum gases propane and butane, contained under moderate pressure (about 125 pounds per square inch [861.3 kilopascals] and 30 pounds per square inch [206.8 kilopascals] respectively) in cylinders.

bottoming cycle. A means to increase the thermal efficiency of a steam electric generating system by converting some waste heat from the condenser into electricity rather than discharging all of it to the environment. The lower temperature cycle in any energy conversion system where two (or more) separate cycles are used in cascade fashion (exhaust of one feeds input to another). See **topping cycle.**

boundary layer. The layer of the atmosphere where the direct effect of turbulent transfer of heat and momentum is observable. Above the boundary layer the free atmosphere is in approximate geostrophic or gradient wind balance.

box stove. A four-legged, rectangular, or oval-shaped stove with draft at one end. Generally not convertible to a fireplace.

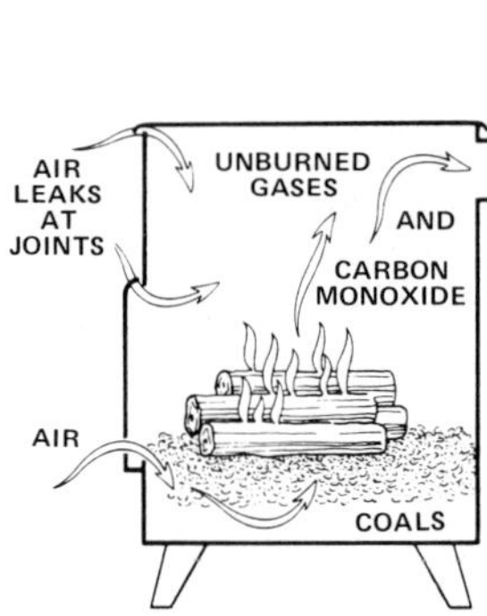

Courtesy A. J. Hand, Home Energy How-To, *Popular Science Book/Harper and Row.*
An ordinary box stove passes volatile exhaust directly out the smoke stack. These simple stoves are only 30% efficient.

brackish water. A mixture of fresh- and salt water.

brake horsepower (BHP). The power output delivered by the crankshaft of an engine (cfm = cubic feet per minute; TP = total fan power):

$$\text{BHP} = \frac{\text{cfm}}{6356} \times \frac{\text{TP}}{\text{Fan Total Efficiency}}$$

branch. The outlet or inlet of a fitting not in line with the run, and taking off at an angle with the run.

branch interval. In plumbing, a vertical length of waste pipe, usually one-story high, and to which the horizontal branches from that floor are connected.

branch tee. A tee having many side branches.

branch vent. A vent connecting one or more individual vents with a vent stack.

Brayton cycle. A thermodynamic cycle consisting of two constant-pressure processes interspersed with two constant-entropy processes. Known also as complete-expansion diesel cycle or Joule cycle. (See diagram p. 57.)

Brayton engine. Gas turbine engine employing a split shaft, with the compressor turbine physically separated from the power turbine connected to the transmission. With single-shaft design, the compressor and power turbines turn together. Air enters the compressor and is ducted through two heat exchangers heated by ex-

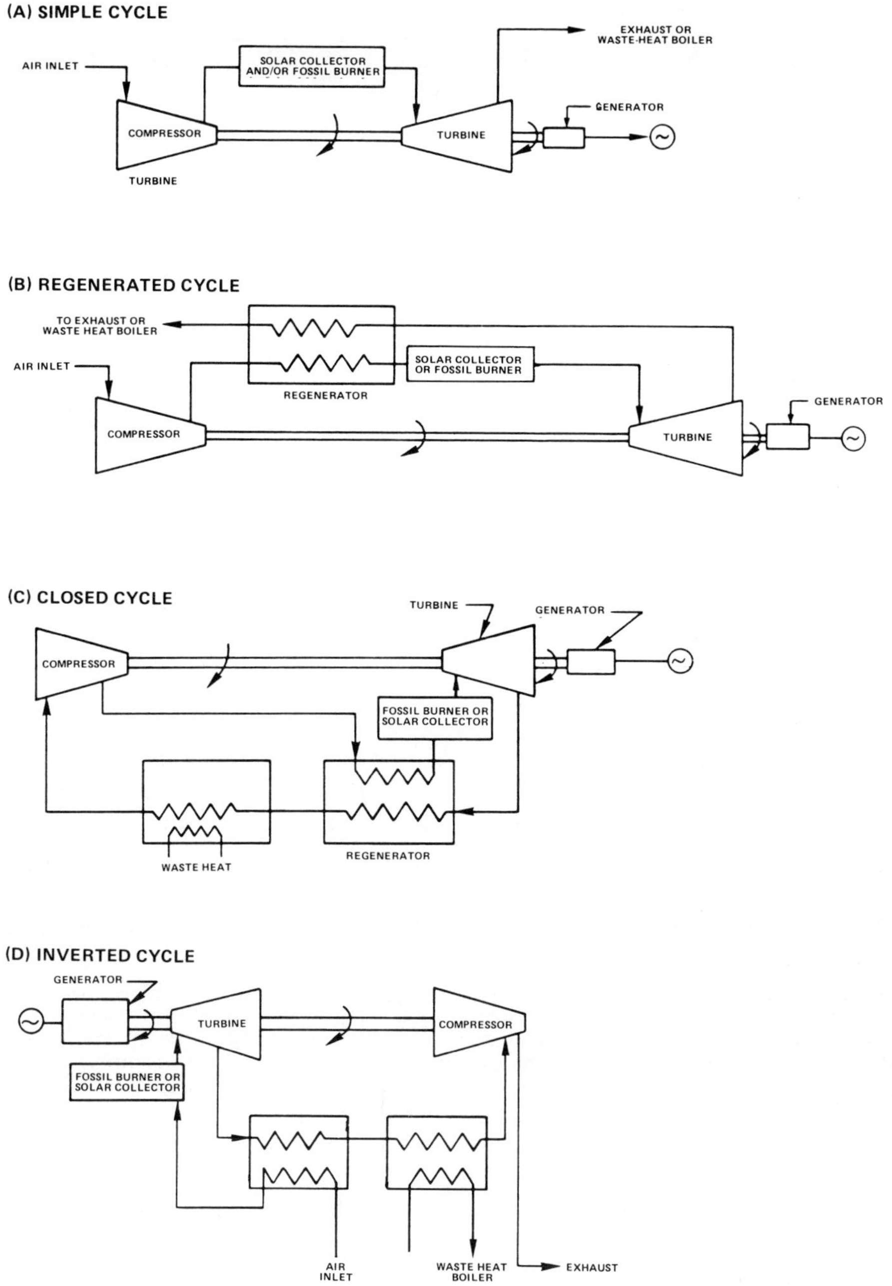

Prepared by OTA Using Manufacturer's Data

Block diagram for simple cycle, regenerated cycle, closed cycle, and inverted cycle Brayton systems compatible with solar applications.

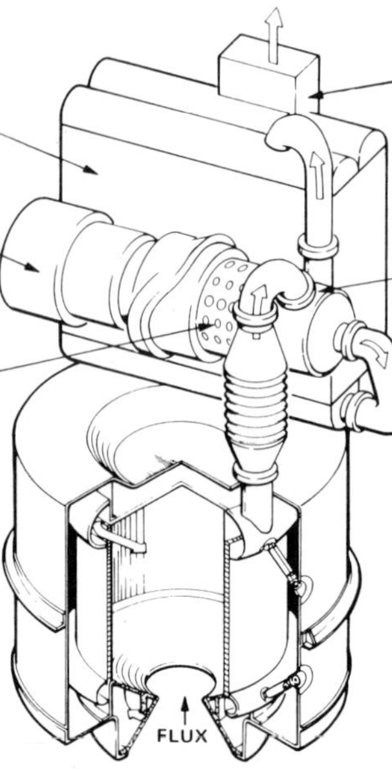

Courtesy DOE

Pictorial diagram of first generation Brayton engine and receiver.

haust gases. The heated, compressed air enters the combustor, where fuel is continuously injected. The mixture is ignited and expands. Expanding gas turns both the compressor and power turbines.

break-even costs. The system costs at which the price of a system's product is equal to the price of the equivalent energy product of another type of system.

breeze. In general, any light or moderate wind blowing from land to sea or from sea to land.

bright annealing. Annealing is a protective coating medium to prevent discoloration of a bright metal surface.

brine. A strong saline solution, such as calcium chloride, used in refrigeration. The natural salt concentration at the bottom layer of solar ponds.

British thermal unit (Btu). The amount of heat required to raise the temperature of 1 pound of water 1 degree Fahrenheit under stated conditions of pressure and temperature [equal to 252 calories (cal), 778 foot-pounds (ft.-lb), 1055 joules (J), and 0.293 watt-hours (Wh)]. It is the standard unit for measuring quantity of heat energy.

buffer storage. The use of some form of thermal energy storage for decoupling the transients associated with the energy source from the end use process of energy, typically less than one-half hour of storage.

building codes. Whether a solar system will comply with a building code is a matter that can be resolved only in the locale where the system will be installed. If your county or city has a building code, a call to your local building official is advised. If a state-wide building code is in effect, it will be enforced locally, and problems can be discussed with your local official. You should investigate building codes, as well as any applicable zoning regulations, before beginning construction. In some cases a building permit may be required before work can begin, even when retrofitting a solar system onto an existing building.

building energy ratio. The space-conditioning load of a building.

building envelope. The elements of a building (walls, roof, floors) that enclose conditioned spaces, and through which thermal energy may be transferred to or from the outside.

building heat-loss factor. Measure of the heating requirements of a building expressed in Btu per degree-day. This factor is multiplied by the number of degree-days (q.v.) in a given period to determine the energy required to heat the building during that period.

building orientation. The relationship of a building to south. A building's orientation is specified by the direction of its longest axis.

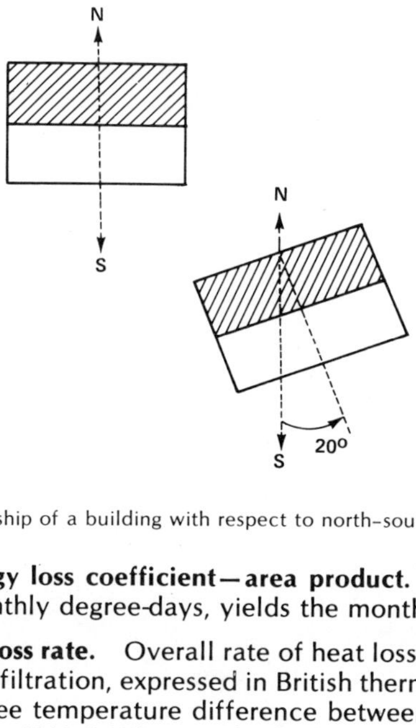

Courtesy HUD

Relationship of a building with respect to north–south orientation.

building overall energy loss coefficient—area product. The factor which, when multiplied by the monthly degree-days, yields the monthly space heating load.

building overall heat loss rate. Overall rate of heat loss from a building by means of transmission plus infiltration, expressed in British thermal units per hour or watts (Btu/h or W) per degree temperature difference between inside and outside; also known as the building "UA." When multiplied by 24 hours, the product is the building heat-loss factor.

building storm drain. A drain for carrying rain, surface water, condensate, or similar discharge to the building drain or sewer.

building subdrain. That part of the drainage system which cannot drain by gravity to the building sewer.

building trap. A trap in the building drain to prevent air circulation between the building drainage system and the building sewer.

bulk density. The mass (weight) of a material divided by the actual volume it displaces as a whole substance expressed in pounds per cubic foot (lb/ft^3); kilograms per cubic meter (kg/m^3); etc.

Bureau of Alcohol, Tobacco, and Firearms (BATF). Under the U.S. Department of Treasury. Responsible for the issuance of permits.

burner capacity. The maximum British thermal units per hour that can be released by a burner with a stable flame and satisfactory combustion.

bus. A major electrical interconnection or tie.

busbar. A heavy metal conductor for high-amperage electricity.

busbar price. The price of electricity at a generating plant; does not include price increment resulting from transmission and distribution costs.

bushing. Removable cylindrical linings for an opening, such as the end of a pipe. These linings limit the size of the opening, resist abrasion, and act as a watertight guide. Thus, when two different sizes of pipe are mated, one or more bushings might be used.

butt weld joint. A welded pipe joint made with the ends of the two pipes butting each other, the weld being around the periphery.

Buys Ballot's law. Statement of the principle of winds; the law that when the observer has his back to the wind, lower barometric pressure is to his left in the northern hemisphere, and to his right in the southern hemisphere. This is due to the earth's rotation.

bypass. An alternative path, in a duct or pipe, for a fluid to flow from one point to another, with the direction determined by the opening or closing of valves or dampers in the main line as well as in the bypass.

byproducts (residuals). Secondary products which have commercial value and are obtained from the processing of raw material.

cable system. Harvested tree or log retrieval system involving the transport along and/or by means of steel cables with the load being lifted partially or wholly off the ground.

cadmium (Cd). A chemical element, atomic number 48, used in making certain types of solar cells.

cadmium sulfide (CdS). A yellow-orange chemical compound produced from cadmium metal in large quantities; used as a pigment. As a semiconductor, CdS is always *n*-type.

calorie. Originally, the amount of heat energy required to raise the temperature of 1 gram of water 1 degree centigrade. Because this quantity varies with the temperature of the water, the calorie has been redefined in terms of other energy units. One calorie is equal to 4.2 joules (J). (When capitalized, Calorie means 1000 calories or 1 kcal. Therefore, 1 Calorie is equal to 4184 J.)

calorific intensity. The temperature of a fuel attained by its combustion.

calorific power. The quantity of heat liberated when a unit weight or a unit volume of a fuel is completely burned.

calorific value. The energy content of fuels; it represents the amount of heat released when the fuel is completely burnt. It is measured in Btu/ft^3 (or per lb) or J/cm^3 (or per kg).

calorifies. The science of heating.

calorimeter. Any apparatus for measuring the amount of heat generated in a body, or emitted by it.

candlepower. The international standard unit of luminous intensity, defined as 1/60th of the intensity of 1 square centimeter (cm^2) of a blackbody radiator at the temperature of solidification of platinum. It is used as a measure of illuminants.

canopy. A rigid housing or cover that encloses the hardware (e.g., roller) associated with the uppermost portion of a shade or curtain.

canopy cover. A numerical description of the extent to which branches and leaves have developed and cover the sunlit area.

cap. The top part of the housing of a check valve (equivalent to the bonnet of a gate or globe valve) which may be either screwed or bolted onto the main body.

capacitance rate. Mass flow rate multiplied by the specific heat of the fluid flowing through a component such as a heat exchanger.

capacitor. Electrical component working on the condensor principle. Two conducting plates are separated by an insulating layer. When alternating current is applied, the capacitor is adjusted so that its leading current balances the lag of the circuit giving a high power factor.

capacity. The maximum power output or load for which a machine, apparatus, station, or system is rated.

capacity credit. The amount of conventional generating capacity displaced by a solar power plant, expressed in megawatts (MW) or as a fraction of the nominal solar plant output.

capacity factor. The actual amount of electricity generated by a power plant during a time interval divided by the amount of electricity that would be generated by the plant during the same interval if it operated at maximum capacity. Fraction of year that the plant delivers energy at its equivalent rated power capacity, e.g., annual energy output, kilowatt hours (kWh) divided by rated end use power, kilowatts (kW), times 8760 hours in 1 year. See **plant factor**.

capacity of the wind. The total amount of detrital material of a given kind that can be sustained (per unit volume of air) by a wind of a given velocity. In the aggregate, wind transports more material than water, although water at the same speed of flow is capable of transporting much larger particles. During a dust storm, the wind may carry from 160 short tons up to 126,000 short tons per cubic mile of air.

capillary water. Water held above the water table in soil by capillary force.

capital cost. Cost of equipment, construction, land, and other items required to construct a facility; different from recurrent operating and maintenance costs.

capital intensive. Indicates that a relatively large percentage of the total costs of production, as for solar technologies, are associated with the initial costs rather than the operating costs. Also used in another sense to differentiate from technologies which are labor-intensive.

capital recovery factor. A factor that gives the uniform annual payment necessary to repay a debt in *n* years at discount rate *i*.

carbohydrate [$C_x(H_2O)_y$]. The most abundant class of all organic compounds is that of the carbohydrates, which constitute three-fourths of the dry weight of all vegetation. They consist of (1) monosaccharides (simple sugars): fructose (levulose) and its isomer, glucose (dextrose); (2) disaccharides: sucrose, maltose, cellobiose, and lactose; and (3) polysaccharides: starch and cellulose. The simple sugars are crystalline and water-soluble, with a sweet taste; starches are water-soluble, tasteless, and amorphous; and cellulose is insoluble in water and organic solvents, and is only partially crystalline.

carbon (C). A nonmetallic element existing in diamonds, graphite, coal, petroleum, asphalt, limestone, and other carbonates, and in all organic compounds. It is also obtained artificially in varying degrees of purity.

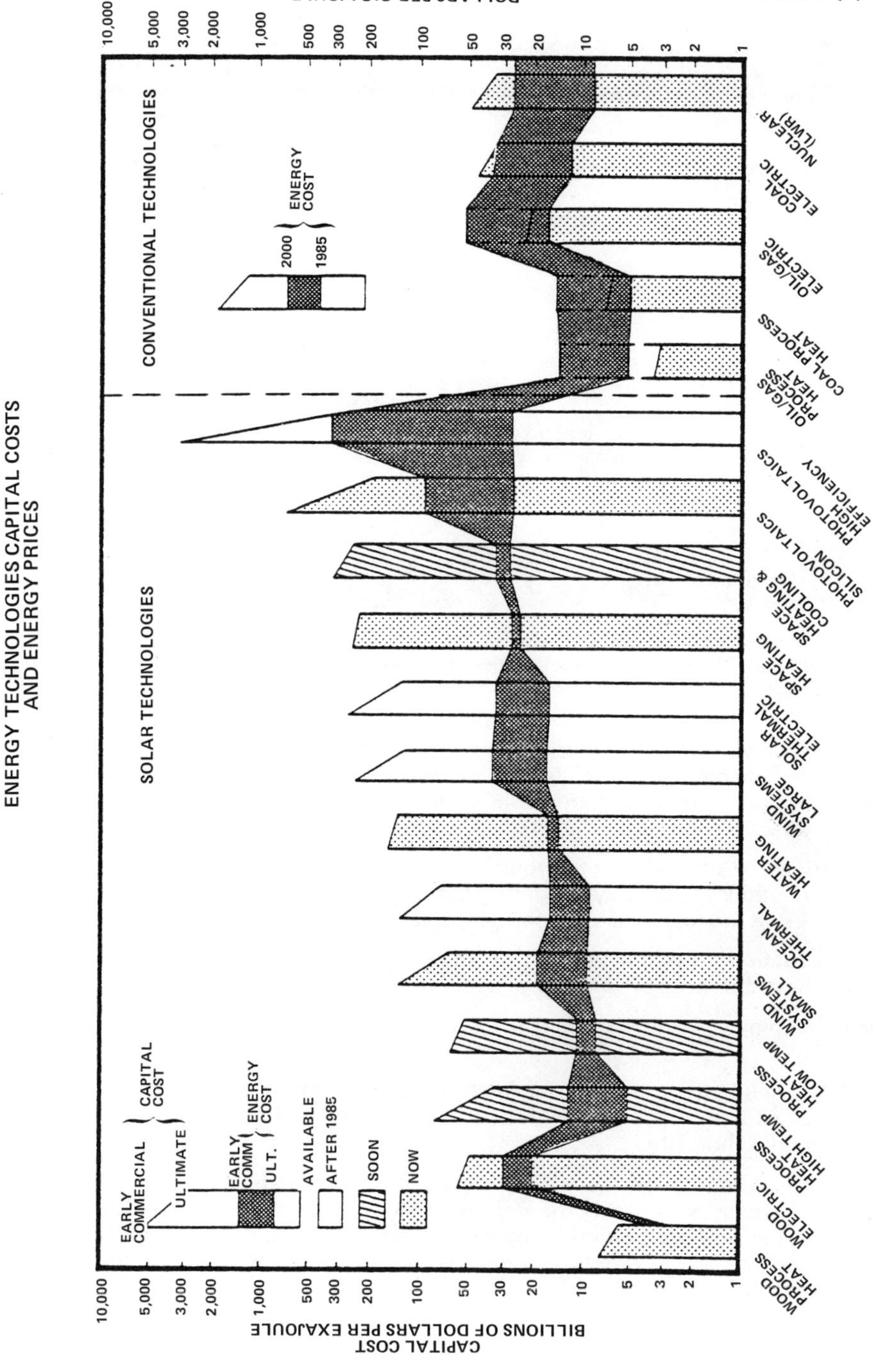

Courtesy Mitre Corporation

Illustration of capital costs and energy prices for solar and conventional energy technologies.

carbon dioxide (CO_2). A colorless, odorless, nonpoisonous gas that is a normal part of the ambient air. Carbon dioxide is a product of fossil fuel combustion, and some researchers have theorized that excess CO_2 raises atmospheric temperatures.

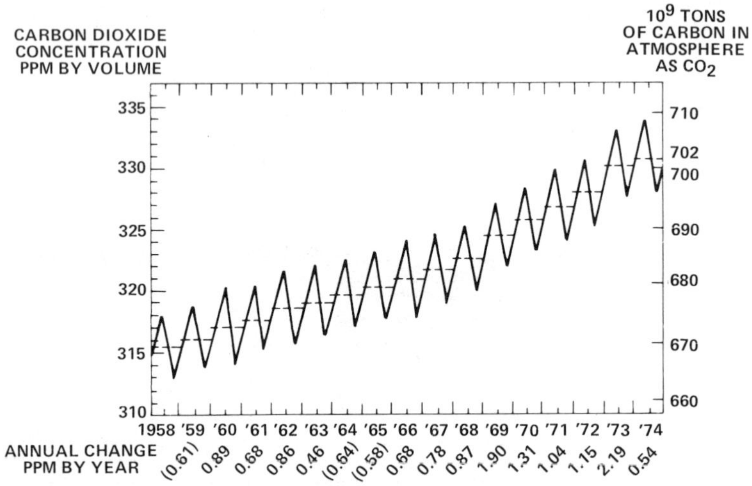

Courtesy DOE

Levels of atmospheric carbon dioxide concentration at Mauna Loa Hawaii Observatory show increases over a 16-year period.

carbon monoxide (CO). A colorless, odorless, highly toxic gas that is a normal byproduct of incomplete fossil fuel combustion. Carbon monoxide, one of the major air pollutants, can be harmful in small amounts if breathed over a certain period of time.

carbon oxides. Compounds of carbon and oxygen produced when the carbon of fossil fuels combines with oxygen during burning. The two most common such oxides are carbon monoxide, a very poisonous gas, and carbon dioxide.

carbon reduction. Fixation of CO_2 as an energy-rich carbohydrate, usually via photosynthesis but possible also by a synthetic process.

carbon steel pipe. Steel pipe which owes its properties chiefly to the carbon which it contains.

carbon zinc cell battery. A cell for the production of electric energy by galvanic oxidation of carbon. It is commonly used in household appliances and radios, and has a relatively short life.

Carnot cycle. An ideal heat engine cycle, conceived by Sadi Carnot, in which the sequence of operations forming the working cycle consists of isothermal expansion, adiabatic expansion, isothermal compression, and adiabatic compression back to its initial state. An ideal Carnot-cycle engine converts heat into work with the maximum theoretical efficiency.

Carnot efficiency. The maximum efficiency with which work can be produced from heat in ideal processes. Carnot efficiency is only dependent upon the maximum and minimum temperatures available. (See diagram p. 65.)

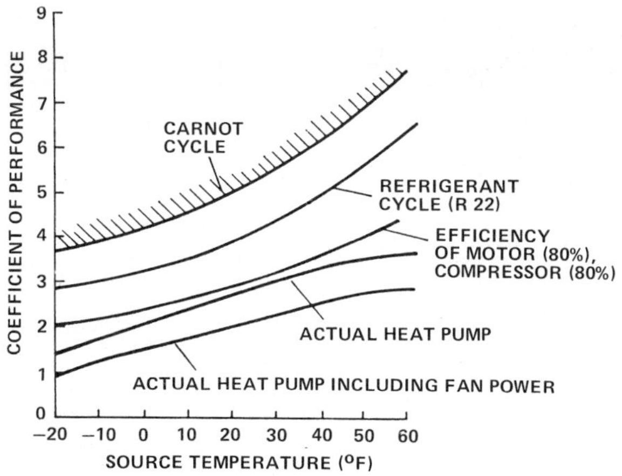

Courtesy F. Kreith and J. F. Kreider, Principles of Solar Engineering, *Hemisphere Publishing Corp.*
Illustration of various heat pump heating efficiencies versus ideal Carnot coefficient of performance.

carry through. The number of days without sunshine during which the heat storage system in a building heated by solar energy can provide adequate space and/or water heating.

cash flow analysis. Comparison of solar-energy costs and savings on a month-by-month basis. Positive (or favorable) cash flow begins when monthly fuel bill savings exceed monthly solar loan payments.

cassava. A starchy root crop used for tapioca; can be grown on marginal croplands along the southern coast of the United States. In some areas it is being considered as a feedstock for alcohol fuel production.

catabolism. Chemical and physical processes within living organisms involving the release of energy.

catalysis. Modification (especially an increase in the rate) of a chemical reaction induced by material unchanged chemically at the end of the reaction; any reaction brought about by a separate agent.

catalyst. A substance that changes the rate of a reaction without itself undergoing any net change; a substance that induces catalyst.

cathode. The electrode where electrons enter (current leaves) an operating system such as a battery, an electrolytic cell, or an x-ray or a vacuum tube. In a battery or electrolytic cell, it is the electrode where reduction occurs.

cathodic protection. A method of preventing oxidation (rusting) of exposed metal structures such as bridges by imposing between the structure and the ground a small electrical voltage that opposes the flow of electrons, and is greater than the voltage that is present during oxidation.

caulking. Latex or elastomeric materials used to reduce air leaks by packing cracks or air-space openings. Caulking is commonly used wherever two different materials or parts of the house meet, such as around windows, door frames, and siding joints. (See table p. 66.)

MATERIAL	RECOMMENDED USES	SHRINKAGE	ADHESION	REMARKS
SILICONE HOUSEHOLD	SEALS JOINTS BETWEEN BATH AND KITCHEN FIXTURES AND TILE; ADHESIVE FOR TILES AND METAL FIXTURES; SEALS METAL JOINTS AS IN PLUMBING AND GUTTERS	LITTLE OR NONE	GOOD TO EXCELLENT	READILY AVAILABLE. FLEXIBILITY OF CURED SILICONE ALLOWS STRETCH OF JOINTS UP TO THREE TIMES NORMAL WIDTH OR COMPRESSION TO ONE-HALF THE WIDTH. COST: HIGH
SILICONE CONSTRUCTION	SEALS MOST DISSIMILAR BUILDING MATERIALS (I.E., WOOD AND STONE; METAL FLASHING AND BRICK)	SAME AS ABOVE	SAME AS ABOVE	REMAINS FLEXIBLE FOR LIFE AFTER CURING. PERMITS JOINTS TO STRETCH OR COMPRESS. SILICONES WILL STICK TO PAINTED SURFACES, BUT PAINT WILL NOT ADHERE TO CURED SILICONE. COST: HIGH
BUTYL RUBBER	SEALS MOST DISSIMILAR MATERIALS (GLASS, METAL, PLASTIC, WOOD, CONCRETE). SEALS AROUND WINDOWS AND FLASHING, OR BONDS LOOSE SHINGLES	FROM 5 TO 30 PERCENT	GOOD	LESS RESILIENT THAN SILICONES. ALLOWS FOR JOINT MOVEMENT BUT DOES NOT BECOME BRITTLE WITH AGE. CAN BE PAINTED AFTER SKIN FORMS. APPLY WHEN TEMPERATURE IS ABOVE 40° F. COST: HIGH
LATEX	SEALS JOINTS AROUND TUB AND SHOWER; FILLS CRACKS IN TILE, PLASTER, GLASS, AND PLASTIC; FILLS NAIL-HOLES	FROM 5 TO 10 PERCENT	GOOD TO EXCELLENT	EASY TO USE. SEAMS CAN BE TRIMMED OR SMOOTHED WITH MOIST FINGER OR TOOL. GOOD WATER RESISTANCE WHEN DRY. CAN BE SANDED AND PAINTED. LESS ELASTIC THAN ABOVE MATERIALS. EASY TO CLEAN UP. COST: MODERATE
OIL-BASE CAULKS	SEALS EXTERIOR SEAMS AND JOINTS ON BUILDING MATERIALS	FROM 10 TO 20 PERCENT	GOOD	READILY AVAILABLE. LEAST EXPENSIVE OF THE FOUR TYPES. ROPE AND TUBE FORM. OILS DRY OUT AND CAUSE MATERIAL TO HARDEN AND FALL OUT. COST: LOW

Courtesy USDA

Comparison of various caulking materials for home insulation applications.

caustic soda (NaOH). Sodium hydroxide, a strongly alkaline, caustic substance used as the cleaning agent in some detergents.

cavity receiver. A solar energy receiver in the form of a cavity where the solar radiation enters through one or more openings (apertures) and is absorbed on the internal surfaces of the structure. (See diagram p. 67.)

ccf. One hundred cubic feet. A unit of measure used for small amounts of natural gas.

cell. (1) A battery unit consisting of two electrodes separately contacting an electrolyte so that there is a potential difference between them. (2) With respect to solid waste disposal, earthen compartments in which solid wastes are dumped.

cell barrier. A very thin region of static electric charge along the interface of the positive and negative layers in a photovoltaic cell. The barrier inhibits the movement of electrons from one layer to the other, so that higher energy electrons from one side diffuse preferentially through it in one direction, creating a current, and thus a voltage across the cell. Also called the depletion zone, or the cell junction.

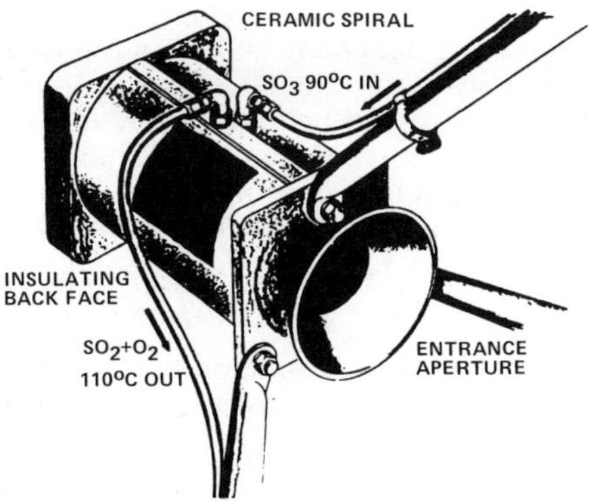

Courtesy DOE

Cavity converter–heat exchanger for solar thermal energy system.

cell junction. The area of immediate contact between two layers (positive and negative) of a photovoltaic cell. The junction lies at the center of the cell barrier or depletion zone.

cellobiose ($C_{12}H_{22}O_{11}$). Cellobiose is composed of two D-glucose molecules. It is a disaccharide, but it is not a fermentable sugar. When cellulose is hydrolyzed, one of the products is cellobiose.

cell recycle. The process of separating yeast from fully fermented beer and returning it to ferment a new mash; can be done with clear worts in either batch or continuous alcohol production operations.

cellular glass. Foamlike material made from glass, with many fine bubbles uniformly distributed throughout, having moderately high strength for light weight. It is used widely for thermal insulation.

cellulase. The enzyme complex, produced by fungi and bacteria, which is capable of decomposing cellulose into smaller fragments, primarily glucose. Fungi such as *Trichoderma viride, Aspergillus niger,* and others are being mutated to produce larger quantities of cellulase and will soon be of prime importance in the production of alcohol fuel.

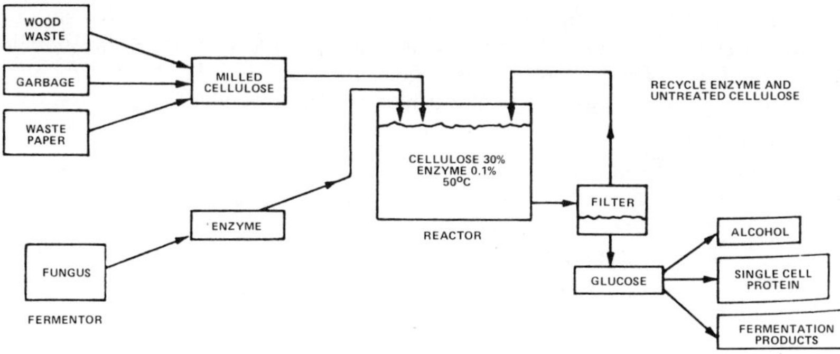

Reprinted with permission from Russell E. Anderson, Biological Paths to Self-Reliance, *Van Nostrand Reinhold*

Block diagram of fermentation of cellulose to produce glucose. [See **cellulose ($C_6H_{10}O_5$)**.]

cellulose ($C_6H_{10}O_5$). Cellulose is the fundamental constituent of all vegetable tissues and therefore is the most abundant organic material in the world. It is a natural polysaccharide with polymerized units of glucose from 1000 to 3500. These units can be hydrolyzed but, because of cellulose's physical structure, the process is difficult to accomplish. This structure is unusual in that it is not a single crystal, but consists of crystalline areas embedded in amorphous areas. Cellulose, due to its purpose of adding structural strength to plants, is insoluble in organic liquids.

cellulosic feedstocks. Materials, such as wood, crop stalks, and newsprint, containing sugar units linked by bonds that are not easily ruptured.

cellulosic fiber. A form of loose home insulation that is blown in for installation. It is more effective than mineral fiber insulations such as rock wool and glass fiber.

centigrade (or Celsius) scale. A temperature scale that takes the melting point of ice as 0°C and the boiling point of water as 100°C. To convert centigrade to fahrenheit, multiply the reading by 9, divide by 5, and add 32.

central fan system. An indirect system of heating in which the air is heated by steam or hot water at a central location, and carried to or from the rooms to be heated by a fan and a system of ducts. Also termed hot blast system.

central heating plant. Either (a) one heating plant in a building serving all or most of the rooms in the building, as distinguished from individual room heaters; or (b) one heating plant serving two or more buildings. The latter is also termed district heating.

central power. The generation of electricity in large power plants with distribution through a network of transmission lines (grid) for sale to a number of users. Opposite of distributed power.

central receiver. One of three principal types of concentrating solar thermal power systems in which a single, tower-mounted receiver is heated by the solar flux impinging on it by reflection from a field of independent tracking mirrors.

central receiver systems. The central receiver system focuses all of the incoming direct solar radiation on a single point rather than along the length of an absorber pipe. Consequently, significantly higher temperatures are attainable with this system than with line-focusing cylindrical distributed collectors. These higher temperatures, currently upwards of 1004°F (540°C), have resulted in greater overall conversion efficiencies of approximately 24 percent. The basic system configuration uses arrays of tracking mirrors called heliostats, which direct solar energy to a single large receiver mounted atop a tower. The working fluid contained in the receiver is heated by the focused solar energy and then piped to the turbogenerator unit (usually located at the base of the tower) where it powers, either directly or indirectly, a conventional turbine and thus generates electricity.

central receiver technology. The central receiver concept consists of a field of tracking mirrors (heliostats) which intercept and redirect sunlight to a centrally located receiver mounted on top of a tower. The redirected energy is used to heat or vaporize a fluid (water in present systems; gas, molten salt, or liquid sodium in future systems) which is circulated through the receiver. The fluid is then either passed through a heat exchanger to heat a working fluid, or as in some water systems, used directly as the working fluid. The working fluid when heated is converted into a high-temperature [650° to 2500°F (343° to 1371°C)] high-pressure gas which can be used as process heat for industrial applications or for driving turbines

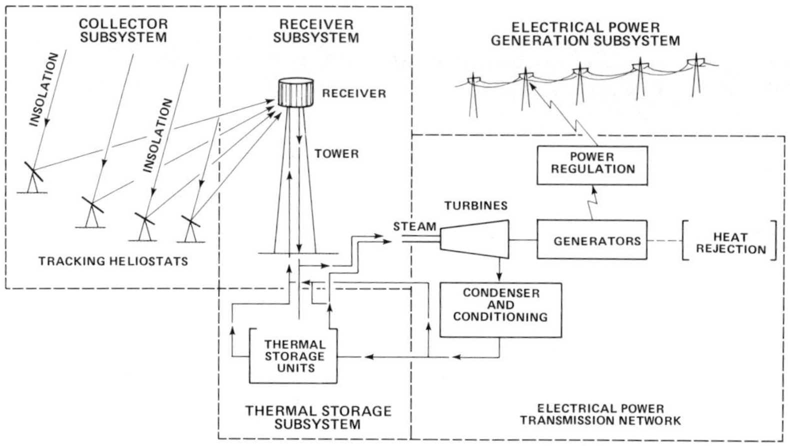

Courtesy DOE

Block diagram of central receiver solar thermal energy system.

which in turn can provide electrical power to a utility grid. The central receiver concept is well suited for large-scale, high-temperature systems, and in addition shows great promise for achieving cost effectiveness for smaller systems also. It is believed that central receiver technology has the potential to displace significant quantities of fossil energy by the year 2000.

Central Receiver Test Facility. The Central Receiver Test Facility was constructed at Albuquerque, NM and began full operation in October, 1978. The Central

Courtesy DOE

Seventy-two heliostat array and tower for central receiver at DOE's Solar Thermal Test Facility operated by Sandia Laboratories.

Receiver Test Facility is managed by Sandia National Laboratories. The facility consists of 222 heliostats which can deliver over 5 megawatts (MW) (thermal energy) to various targets mounted on top of a 200-foot (61-meter) central tower and can generate temperatures greater than 4000°F (2200°C). Each of the heliostats consists of 24 4-foot (1.2-meter) by 4-foot (1.2-meter) mirrors. The tower includes four separate test bays, air and water/steam cooling systems, and computer data acquisition/monitoring capabilities. The major function of the Central Receiver Test Facility is to offer flexibility in testing of a variety of new heliostat/receiver combinations, as well as to investigate advanced applications for central receiver technology and to acquaint utility and industrial technicians with solar facility operations.

central station power. Production of power—usually electrical—in large quantities at a generation plant as opposed to production at the point of consumption.

centrifugal collector. Any of several mechanical systems using centrifugal force to remove aerosols from a gas stream.

centrifugal pump. High-speed pump that drives water with a rotating impeller.

centrifuge. A rotating device for separating liquids of different specific gravities or for separating suspended colloidal particles according to particle-size fractions by centrifugal force.

cetane number. A measure of a fuel's ease of self-ignition; the higher the number the better the fuel for a diesel engine.

channelization. The straightening and deepening of streams to permit water to move faster, to reduce flooding, or to drain marshy acreage for farming. However, channelization reduces the organic waste assimilation capacity of the stream and may disturb fish breeding and destroy the stream's natural beauty.

Chapman engine. Chapman engines have developed a new positive displacement, vane-type orbital bottoming cycle engine. After flowing through the engine, hot coolant enters the heat exchanger. Hot exhausts and oil also flow through this heat exchanger. Heat from these sources, normally wasted in conventional automobiles, heat Freon flowing through the exchanger. Freon passes through the throttle valve, expanding in the orbital engine and turning the engine's main shaft. Used Freon condenses in a heat exchanger, giving up its heat to cooled coolant coming from the radiator.

char. A porous, solid residue resulting from the incomplete combustion of organic material. If produced from coal, it is called coke; if produced from wood or bone, it is called charcoal. It is closer to pure carbon than the coal, wood, or bone from which it is produced.

charcoal. A dark-colored or black porous form of carbon made from vegetable or animal substances (i.e, by charring wood in a kiln or by excluding air from retort) and used for fuel. Wood after all volatile substances have been heated and driven off. Consists mainly of carbon and gives intense efficient heat.

charge. As a noun, the quantity of something introduced into a vessel. In refrigeration, the quantity of refrigerant in a system. Putting heat into storage through radiant absorption or convective heat transfer (blown in).

charged particle. An ion; an elementary particle that carries a positive or negative electric charge.

check valve. Mechanical valve that prevents water spilling back into a wind-driven well. A one-way valve for the water going through a solar-controlled hot-water system (or pool). A check valve is used to prevent thermosiphoning.

chemical compatibility. The ability of materials and components in contact with each other to resist mutual chemical degradation, such as that caused by electrolytic action or plasticizer migration.

chemical heat pump. A heat pump is a device which extracts thermal energy from a low-temperature source and rejects it to a higher temperature demand. The seemingly impossible task of moving heat "uphill," i.e., from a low temperature to a higher temperature, may be accomplished only by expending energy to "drive" the process. Heat pumps based on the mechanical compression and expansion of a fluid such as ammonia or Freon are in common use in home refrigerators, air conditioners, and heat-pump heating–cooling systems. In a chemical heat pump, the mechanical process is replaced with a thermochemical one, involving two separate reactions in which a solid or liquid is decomposed to form a vapor and another solid or liquid. The vapor (or gas) must be common to both reactions, as it is the movement of this vapor which accomplishes the heat pumping action.

chemical oxygen demand (COD). A measure of the amount of oxygen required to oxidize organic and oxidizable inorganic compounds in water. The chemical oxygen demand test, like the biological oxygen demand test, is used to determine the degree of pollution in an effluent.

chilling effect. The cooling of the earth's temperature because of the increase of atmospheric particulates which prevent penetration of the sun's energy.

chimney. An upright tube for conducting smoke and gases from the house. Can be heavy masonry or a pipe consisting of two or more layers of steel which are insulated.

chimney effect. The tendency of heated air or gas to rise in a duct or other vertical passage, as in a chimney, a small enclosure, or building, due to its lower density compared to the surrounding air or gas.

chimney lid. Device used on a chimney, which operates by remote control to open or close and prevent heat loss through some opening such as a fireplace.

chinook. A warm, dry wind that descends the eastern side of the Rocky Mountains, and generally blows from the southwest.

Chipboard®. A window insulation product made from many small wooden chips bonded together.

chlorination. The application of chlorine to drinking water, sewage, or industrial waste for disinfection or oxidation of undesirable compounds.

chlorine-contact chamber. In a waste-treatment plant, a chamber in which effluent is disinfected by chlorine before it is discharged to the receiving waters.

chloroplast. A small portion of a plant cell which contains the light-absorbing pigment chlorophyll, and converts light energy to chemical energy.

chlorosis. Yellowing or whitening of normally green plant parts. It can be caused by disease organisms, lack of oxygen or nutrients in the soil, or by various air pollutants.

chord. The distance from the leading to the trailing edge of an airfoil.

chubs. A type of packaging for latent-heat storage materials which involves sealing storage material in tubular, sausagelike containers.

circuit. A conducting part, or a system of conducting parts, through which an electric current flows or is intended to flow.

circuit breaker. An overload protective device to interrupt the flow of electric current when it becomes excessive or exceeds a predetermined value.

clapper. A common term which is used to describe the disk of a swing-type check valve.

clarification. In wastewater treatment, the removal of turbidity and suspended solids by settling, often aided by centrifugal action and chemically induced coagulation.

clarifier. In wastewater treatment, a settling tank which mechanically removes settleable solids from wastes.

class. In refrigeration, a system is classified according to a safety code as Class A if the refrigerant charge is 1000 lb (454 kg) or more, Class B 100 to 1000 (45 to 454 kg) lb, Class C 20 to 100 lb (9 to 45 kg), Class D 6 to 20 lb (3 to 9 kg) and Class E 6 lb (3 kg) or less.

clearance. In an engine or compressor, the space in the cylinder beyond the piston when the latter is at its extreme position in that end.

clearness number. Ratio of clear-day direct radiation intensity at a location to the intensity calculated for the standard atmosphere for the same location and date.

clear octane. The octane number of a gasoline before the addition of antiknock additives.

clerestory. A window located high in a wall near the eaves, used for light, heat gain, and ventilation.

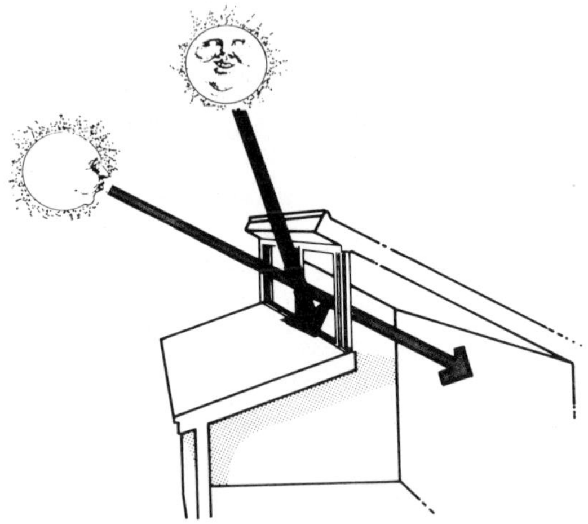

Reprinted with permission from E. Mazria, The Passive Solar Energy Book, *Rodale Press*
A vertical window, above the line of sight, provides natural lighting.

climate. Prevailing or average weather conditions of a geographic region as shown by temperature and meteorological changes over a period of years.

clo. A unit of clothing insulation defined as the insulation necessary to keep a sitting person comfortable in a normally ventilated room at 70°F (21°C) and 60 percent relative humidity. In physical terms it is equal to 42.7 Btu per sq ft per hour.

close coupled. An energy system in which the fuel production equipment is in close proximity to the fuel use equipment.

closed cycle. Where the fluid is used over and over without introduction of new fluid, as in a hot-water heating or mechanical refrigeration system.

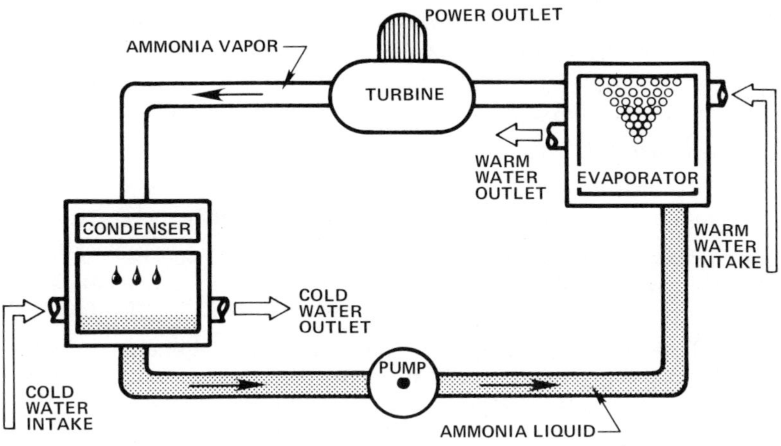

Courtesy DOE

A schematic diagram of a closed cycle ocean thermal energy conversion system.

closed-cycle ocean thermal energy. Energy conversion where the OTEC fluid is recirculated in a closed-loop system. See **open-cycle ocean thermal energy.**

closed-loop. System in which heat-transfer liquid from collectors circulates through a heat exchanger immersed in heat-storage liquid, passing its heat to heat-storage liquid while remaining isolated from it. A system in which no part is vented to the atmosphere.

closed-loop solar heating system. Active system in which the sun's heat passes to household water indirectly. Solar energy is collected by a fluid such as antifreeze solution or a special oil that is repeatedly recirculated through a closed pipe loop. Heat from this fluid is transferred to the household water through the wall(s) of a heat exchanger at the storage tank.

closed system. A process or device which maintains a constant amount of material within itself, neither accepting matter from nor rejecting matter to its surroundings.

closed water piping system. A heating system which uses an air tank as a means of pressurizing the system, and circulating water as a heat medium.

close nipple. A nipple which is twice the length of a standard pipe thread.

cloud cover. That portion of the sky cover which is attributed to clouds, usually measured in tenths of sky covered.

Cloud gel®. A window sheet insulation product of Suntak Research Associates which is transparent when cool and opaque when hot.

cluster. A group of any number of wind turbines sited within a sufficiently close proximity that they may be considered, for most purposes, as a single power-producing unit; also called a Wind-Energy Conversion System farm.

coach screw. Like a carriage bolt, it is usually a wood screw with a round head and a square shank above the screw threads. The square shank is useful for a firm fastening as it penetrates the wood.

coagulation. The clumping of particles in order to settle out impurities; often induced by chemicals such as lime or alum.

coastal zone. Coastal waters and adjacent lands that exert a measurable influence on the uses of the sea and its ecology.

coefficient of expansion. The factor which expresses the change per unit length of any material for each degree of temperature.

coefficient of haze. A measurement of visibility interference in the atmosphere.

coefficient of heat transmission. The rate of heat loss in Btu's per hour through a square foot of a wall or other building surface, when the difference between indoor and outdoor air temperatures is 1°F (0.6°C). Often called a U-value (q.v.).

coefficient of performance (COP). The ratio of the useful energy output from a device to the incoming energy. For heat pumps the COP can be as large as 4 or 5; one unit of electricity can transfer 3 or 4 units of outside energy indoors while still converting the unit of electricity to usable thermal energy.

coefficient of thermal expansion. The fractional change in the length of a material per degree of temperature change.

coefficient permeability. The rate of flow of water under laminar flow conditions through a unit cross-section of a porous medium under a unit hydraulic gradient and a standard temperature, usually 68°F (20°C).

cofiring. The use of a mixture of two fuels within the same combustion chamber, e.g., wood and coal.

cogeneration. The generation of electricity with direct use of the residual heat for industrial process heat or for space heating. (See diagram p. 75.)

cohesion. That property by which molecules of the same kind, or of the same body, are held together in opposition to forces tending to separate them.

cold night sky. The low effective temperature of the sky on a clear night. Most of the heat radiated from a body outdoors is given off to the cold night sky. This process is used by radiative natural cooling systems having roof ponds. Sky access for such systems is not crucial, with over 80 percent of radiant heat loss occurring to the sky 30° above the horizon.

cold 70. That condition during the heating season, in spaces where 70°F (21°C) is the desired air temperature, when the air temperature at the thermostat is at 70°F (21°C) but the temperature at the floor or below the thermostat is lower than 70°F (21°C) This stratification results in relatively long "off" periods of the heating

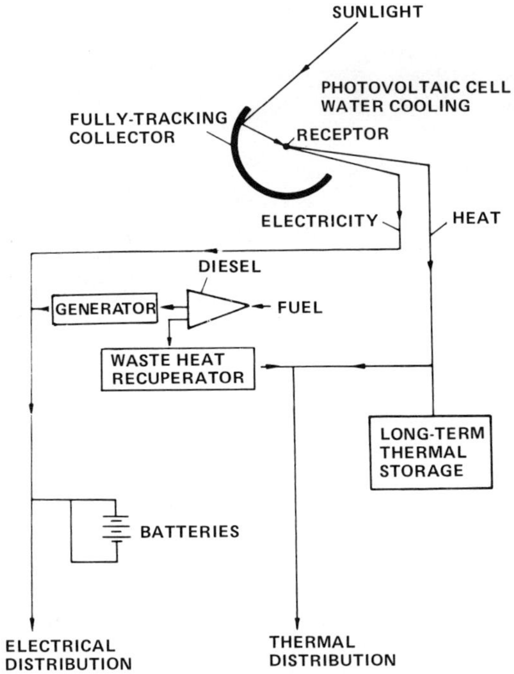

Courtesy DOE

Block diagram of photovoltaic cogeneration system.

system and a feeling of discomfort even though 70° (21 °C) is maintained at the control.

coliform index. An index of the purity of water based on a count of its coliform bacteria.

coliform organism. Any of a number of organisms common to the intestinal tract of humans and animals whose presence in wastewater is an indicator of pollution and of potentially dangerous bacterial contamination.

collateral value. The resale value of alcohol equipment and/or plants; specifically the ability of the equipment to derive from sale the amount necessary to pay off the debt borrowed against the unit.

collection. The process of trapping solar radiation and converting it to heat.

collector. The component of a solar system which collects solar radiation. It may consist of an insulated box (metal, wood, plastic, or fiber glass) with or without glazing containing an absorber plate and flow passages which carry the transfer medium (either liquid or gaseous) to be heated. The collector may also be integrated with storage, for example, south glass with water drums or water tubes. (See photograph p. 76.)

collector aperture. The glazed opening in a collector which admits solar radiation.

collector/concentrator subsystem. The collector/concentrator subsystem has as

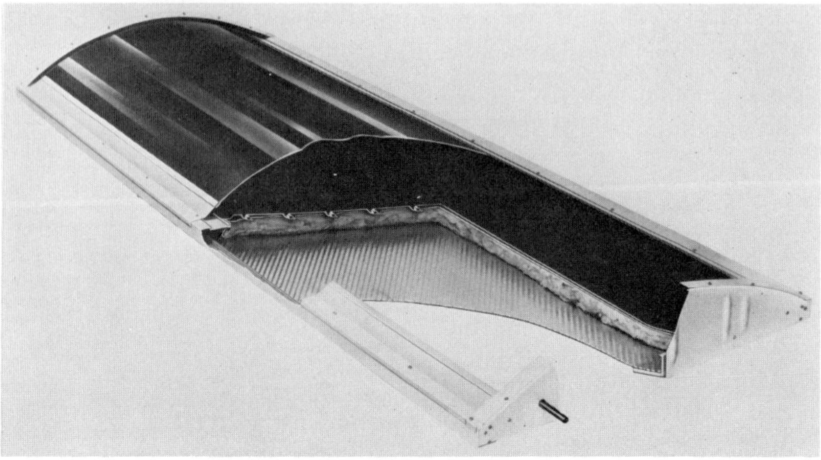

Courtesy DOE

Arched acrylic-covered solar collector with aluminum extrusions locked around copper tubes.

its basic function the interception, redirection, and concentration of direct solar radiation to a receiver/heat-transfer subsystem. This subsystem consists of a field of heliostats and a tracking control system to maintain continuous focus on the central receiver. Unlike distributed collectors, the heliostats of the central receiver system must track the sun in two dimensions.

collector efficiency. The ratio of incoming solar radiation captured by the collector to the radiant energy incident on the collector. For example, if the system captures one-half of the incoming radiation, the system is said to be 50 percent efficient. Efficiency is the capability of a collector to capture heat under various climatic conditions. There is no way a collector can be 100 percent efficient, that is, capture all the heat that falls on the collector; 55 percent is good under desirable weather conditions.

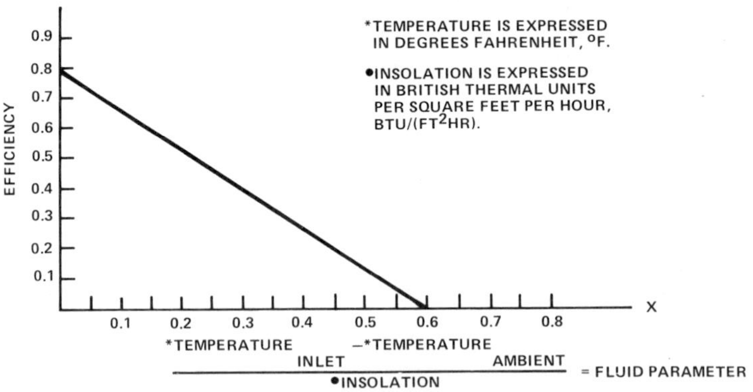

Courtesy DOE

Diagram of solar collector efficiency.

collector efficiency (instantaneous). The ratio of the amount of energy removed by the heat transfer fluid per unit of aperture over a time period of 5 minutes or less to the total incident solar radiation on the collector for the same time period under steady state conditions (test method described in ASHRAE 93-77).

collector fluid. A collector will usually use either a liquid or a gas as the transfer fluid. The most common liquids are water or a water–ethylene glycol solution. The most common gas is air.

collector—heat exchanger correction factor. An index ranging in value from 0 to 1 indicating the penalty in useful energy collection resulting from heat exchange between the collector and the storage tank in liquid solar heating systems.

collector heat removal efficiency factor. The ratio of the actual useful energy gain of a flat-plate solar collector to the energy gain if the entire collector plate were at the temperature of the inlet fluid.

collector mounting. A collector can be mounted to remain stationary, be adjustable as to tilt angle (measured from the horizontal) to follow the change in solar declination, or be designed to track the sun. Tracking is done by employing either an equatorial mount or an azimuth mounting, for the purpose of increasing the absorption of the daily solar irradiation.

collector overall energy loss coefficient. A parameter characterizing the energy losses of the collector to the surroundings.

collector plates. In a solar-collector system, the primary function of the solar-collector plates is to absorb as much of the radiation reaching it as possible, and to lose as little heat as possible when transferring the retained heat to the transport fluid.

collector subsystem. The assembly used for absorbing solar radiation, converting it into useful thermal energy, and transferring the thermal energy to a heat transfer fluid. An array of individually controlled heliostats, including the wiring and controls, that redirects the available solar insolation onto a receiver.

collector tilt. The angle at which a solar heat collector is tilted up from the horizontal plane to face the sun for better performance.

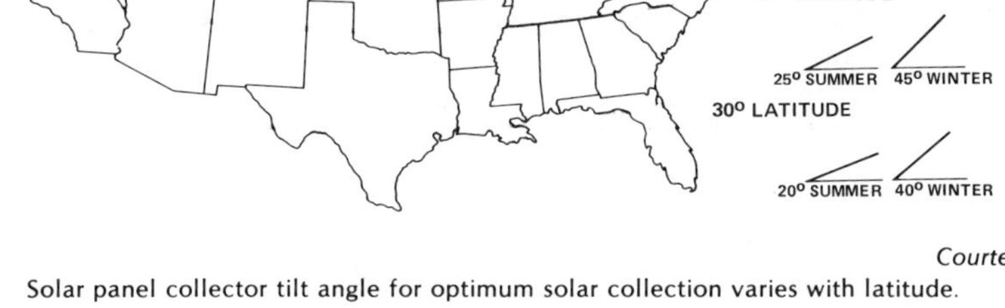

Courtesy DOE

Solar panel collector tilt angle for optimum solar collection varies with latitude.

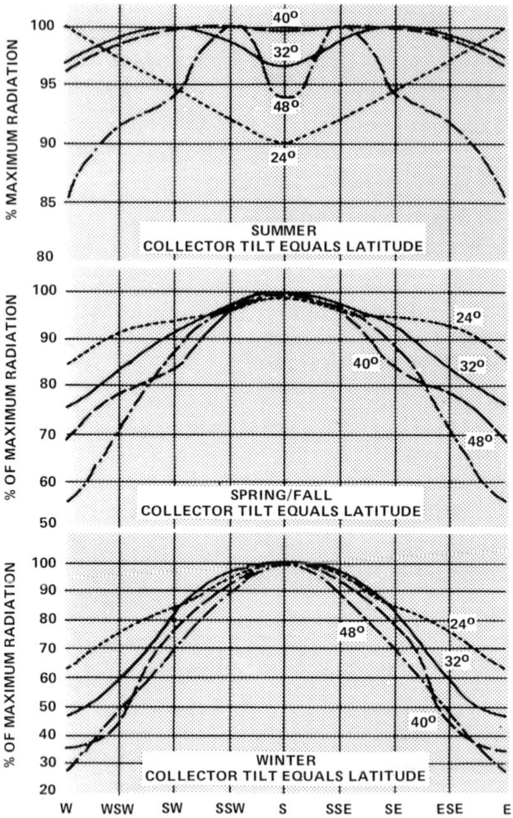

Reprinted with permission from William C. Dickinson and Paul Cheremisinoff, Solar Energy Technology Handbook, *Marcel Dekker*

Charts show effects of collector orientation. (See **collector tilt.**)

collimated light. Parallel rays of light; the direct or beam component of solar radiation.

colloid. A substance composed of extremely small particles, ranging from 0.2 to 0.005 micron which, when mixed with a liquid, does not settle, but remains permanently suspended. The colloidal suspension thus formed has properties that are quite different from the simple, solid–liquid mixture or a solution.

column. Vertical, cylindrical vessel containing a series of perforated plates or packed with materials through which vapors may pass, used to increase the degree of separation of liquid mixtures by distillation or extraction.

combination utility. A utility which supplies more than one service (gas, electricity, water, etc.).

combined collector. A photovoltaic device or module that provides useful heat energy in addition to electricity. See **photovoltaic-thermal system**.

combined-cycle plant. A plant which achieves higher efficiency by employing two thermodynamic cycles in tandem. For example, heat rejected from a gas-fired turbine is used to generate steam to operate a steam turbine.

combined sewer. A sewerage system that carries both sanitary sewage and stormwater runoff. During dry weather, combined sewers carry all wastewater to the treatment plant. During a storm, only part of the flow is intercepted because of plant overloading; the remainder goes untreated to the receiving stream.

combustion. Burning. Technically, a rapid oxidation accompanied by the release of energy in the form of heat and light. It is one of the three basic contributing factors causing air pollution; the others are attrition and vaporization.

comfort energy. Used to describe any form of energy whose end-use is the heating and cooling of buildings and homes.

comfort air conditioning. The conditioning of air for the comfort or well-being of human beings, as distinguished from that for the furtherance of a manufacturing or other industrial process.

comfort zone. That area on the psychrometric chart covered by combinations of temperatures and humidities at which, in subjective tests, more than 50 percent of the subjects were comfortable.

commercial solar buildings. Solar heating and cooling systems for commercial buildings are similar in design and components to those for residential use, though the sizing criteria are different. As with any solar application, each system must be individually designed. You should seek the services of an architect or engineer to determine the feasibility and cost-effectiveness of solar energy for your particular project.

comminution. Mechanical shredding or pulverizing of waste, a process that converts it into a homogeneous and more manageable material. Used in solid waste management and in the primary stage of wastewater treatment.

compaction. Reducing the bulk of solid waste by rolling and tamping.

companion flange. A pipe flange to connect with another flange or with a flanged valve or fitting. It is attached to the pipe by threads, welding, or other method and differs from a flange which is an integral part of a pipe or fitting.

complete combustion. There are two terms in common use for ideal combustion: *perfect combustion,* where all combustibles in the fuel are burned with no excess air, so that the products contain only CO_2, H_2O (plus SO_2 if there is sulfur in the fuel), N_2; complete combustion, where all combustibles are burned, but with excess air, so that the flue products contain free O_2 in addition to CO_2, H_2O, and N_2, but with no CO. The former is a theoretical unattainable ideal, the latter is easily maintained only with fairly high proportions of excess air. The most successful fuel burning equipment from an operating standpoint is that which attains complete combustion with a minimum of excess air, CO, and smoke with the minimum of attention.

completely denatured alcohol. Ethyl alcohol which is at least 160 proof blended, pursuant to formulas prescribed by BATF, with sufficient quantities of various denaturants to make it unfit for and not readily recoverable for beverage use; this may then be distributed through retail outlets without permits. (Compare to **specially denatured alcohol.**)

components. An individually distinguishable product that forms part of a more complex product (i.e., subsystem or system).

compost. Relatively stable decomposed organic material.

composting. A controlled process of degrading organic matter by microorganisms. (1) Mechanical—a method in which the compost is continuously and mechanically mixed and aerated; (2) ventilated cell—compost is mixed and aerated by being dropped through a vertical series of ventilated cells; (3) windrow—an open-air method in which compostable material is placed in windrows, piles, or ventilated bins or pits and occasionally turned or mixed. The process may be anaerobic or aerobic.

compound. A chemical term denoting a combination of two or more distinct elements.

compound compression. Compression by stages in two or more cylinders.

compound engine. A steam engine in which high-pressure steam expands and does work in a succession of cylinders or turbine chambers.

compound parabolid collector (CPC). A form of concentrating collector which does not track the sun.

compressed air storage. A method of storing reserve energy by compressing air in large underground caverns. A small water reservoir is necessary to maintain constant air pressure, and the air system is used only in conjunction with gas turbine generators. Off-peak electricity is used to run a large air compressor which compresses air to about 40 atmospheres about 1500 feet (457 meters) underground. This air is later released to run the gas turbine.

compression. The action on a material which decreases its volume as the pressure to which it is subjected increases.

compression ignition engine. An internal combustion engine in which ignition of the liquid fuel injected into the cylinder is performed by the heat of compression of the air charge.

compression joint. A multipiece joint with cup-shaped threaded nuts which, when tightened, compress tapered sleeves so that they form a tight joint on the periphery of the tubing they connect.

compression ratio. The ratio of absolute pressure before and after compression.

compression refrigerating system. A refrigerating system in which the cooling effect results from expansion of a refrigerant after mechanical compression.

compressor. A mechanical device for raising the pressure of a gas; also a machine which compresses air.

Compton effect. The elastic scattering of a photon by an electron. In each such process the electron gains energy and recoils, and the photon loses energy. This is one of three ways photons lose energy upon interacting with matter.

concentrating collectors. A device which uses reflective surfaces to concentrate the sun's rays onto a smaller area, where they are absorbed and converted to heat energy. Concentration ratios over 10,000 have been achieved. (See diagram p. 81.)

concentration. The percentage of a specified constituent in a mixture to the total quantity of the mixture.

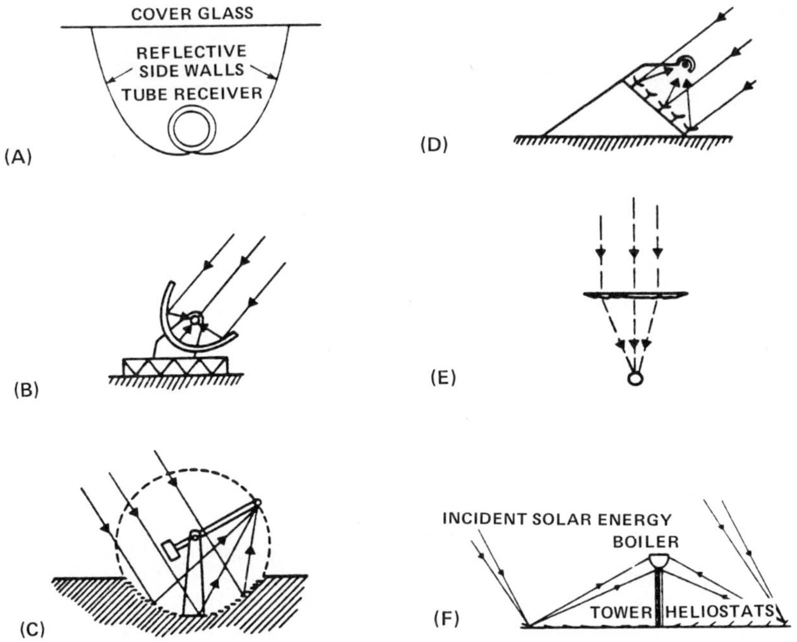

Courtesy DOE

Principal concentrating collectors include (A) compound parabolic collector (CPC), (B) parabolic trough, (C) fixed reflector moving receiver, (D) fixed receiver moving reflector, (E) Fresnel lens, (F) central receiver.

concentration ratio. The ratio of the received energy on a small area from a curved mirror or multiple flat surfaces with perfect reflectivity to that arriving from the sun. Often measured in "suns." Commonly used to refer to the ratio of aperture to receiver areas.

concentrator. A device or structure that concentrates a windstream. Solar collector that focuses or funnels solar radiation onto a relatively smaller absorber at the focal point or line. A photovoltaic array which includes an optical component such as a lens or focusing mirror to direct incident sunlight onto a solar cell of smaller area.

concentrator cell assembly. The smallest complete assembly of solar cells—which may be environmentally protected—designed to generate direct current (DC) power under concentrated terrestrial sunlight.

condensate. The liquid resulting when water in vapor form contacts cool surface.

condensation point. The temperature at which the vapor changes into its liquid state (i.e., steam into water).

condenser. A heat-transfer device that reduces a fluid substance from its vapor phase to its liquid phase by reducing its temperature as it contacts cooling surfaces in its path. (See diagram p. 82.)

condensers, air-conditioning. Condensers are used to liquefy a gas. There are two commonly used types of condensers for air-conditioning work. The shell and tube

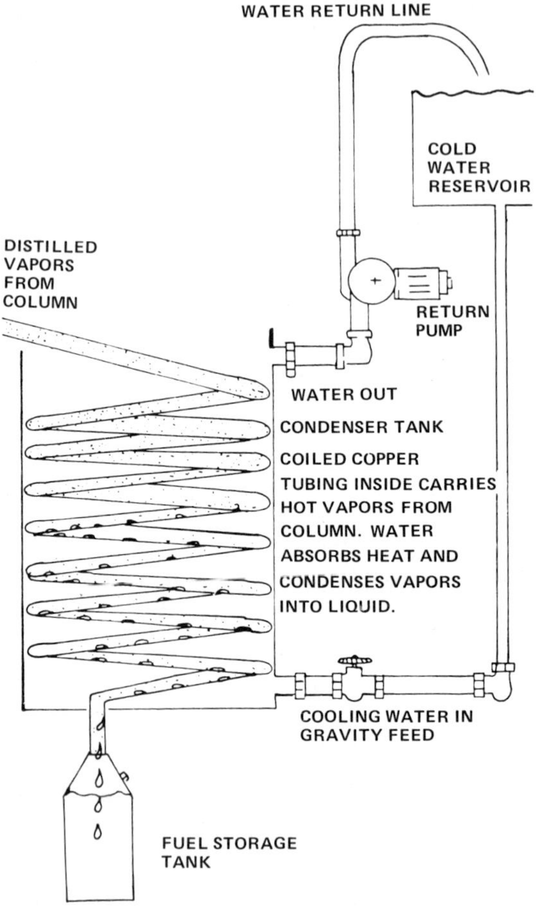

Courtesy Larry W. Carley, How to Make Your Own Alcohol Fuel, *TAB Books Inc.*
Illustration of operation of condenser used in production of ethanol for fuel.

condensers consist of a large number of tubes inside a steel shell or cylinder. Cooling water circulates inside the tubes while vapor flows into the shell, circulates around outside the tubes, is condensed and drawn off the bottom as a liquid. The evaporative condenser consists of a coil over which water is sprayed while air is blown across the tubes of the coil. The refrigeration vapor is cooled, or condensed, inside the tubes due to the extraction of heat caused by the evaporation of the water on the outside.

condensing heat rejection effect. That portion of the heat rejected at the refrigerating condenser used for condensing the refrigerant vapor to a saturated liquid at the entering refrigerant pressure.

condensing unit. An assembly attached to one base, and including refrigerating compressor, motor, condenser, receiver, and necessary accessories.

condition line. The infinite number of combinations of wet- and dry-bulb temperatures which will satisfy the requirements of an air supply for a given room condition form what is known as the condition line on the psychrometric chart (q.v.).

conductance. The quantity of heat (Btu) which will flow through 1 square foot (0.093 m^2) of material in 1 hour, when there is a 1°F (0.6°C) temperature difference between both surfaces. Conductance values are given for a specific thickness of material, not per inch of thickness. For homogeneous materials, such as concrete, dividing the conductivity (*K*) of the material by its thickness (*X*) gives the conductance (*C*).

conduction. The transfer of heat through matter by the transfer of kinetic energy from particle to particle rather than by a flow of heated material; the transmission of energy directly from molecule to molecule. It is the way in which electricity travels through a wire or heat moves from a warm body to a cool one when the two bodies are placed in contact. Transmission of energy through a medium which does not involve movement of the medium itself. The flow of heat between a hotter material and a colder material that are in direct physical contact.

conduction band; conduction level. Energy level at which electrons are not bound to (orbiting) a specific atomic nucleus, but are free to wander among the atoms.

conductivity (K). The ease with which heat will flow through a material as determined by the material's physical characteristics. The quantity of heat (Btu) that will flow through 1 square foot (0.093 m^2) of material, 1 inch thick (25.4 mm), in 1 hour, when there is a temperature difference of 1°F (0.6°C) between its surfaces.

connected load. The sum of the capacities or ratings of the gas or electric power-consuming apparatus connected to a supplying system, or any part of the system under consideration.

connecting socket (for pipe). See **pipe, connecting socket.**

conservation of energy. The total energy of an isolated system remains constant irrespective of whatever internal changes may take place. Energy disappearing in one form reappears in another.

conservation strategy. The actions taken by the federal government to reduce demand for energy, particularly petroleum, and including the setting of minimum mileage standards for new automobiles, and the provision of incentives and standards to increase residential insulation and energy-use efficiency. (See diagram p. 84.)

constant dollars. Dollar estimates from which the effects of changes in the general price level have been removed, reported in terms of a base year value.

consumer conservation products. The Department of Energy has initiated programs to improve product energy efficiency, and in so doing, incur substantial savings in both energy and money. Such products include additional insulation for water heaters; dual-purpose appliances in which waste heat from one will operate the other; more efficient electric and gas heaters and motor-compressor units for refrigerators, freezers, and room air conditioners; long-lasting fluorescent light bulbs; and new types of windows that, in winter, will allow the sun's rays to pass through and keep the heat from bouncing back outside and, in summer, will automatically change to reflect heat and light.

consumption charge. That portion of a utility charge based on energy actually consumed, as distinguished from the demand charge.

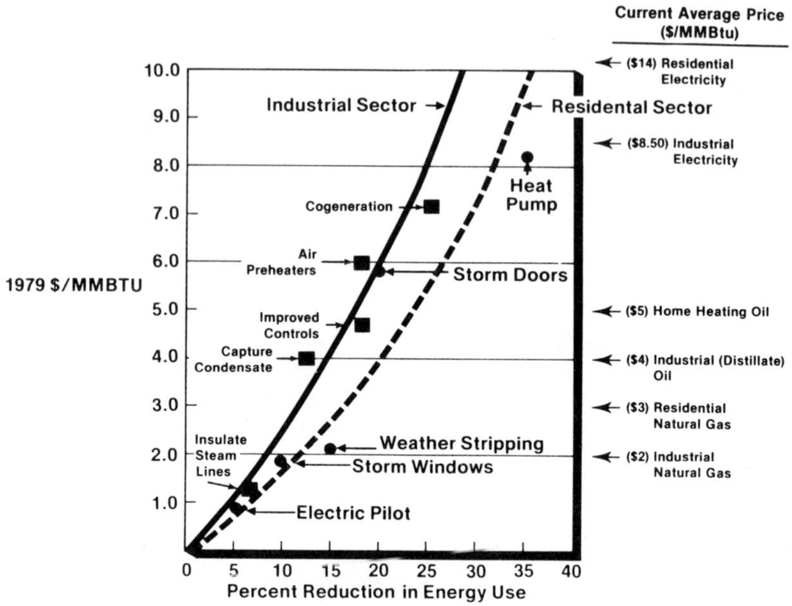

Courtesy Mitre Corporation

Graph of industrial and residental conservation measures utilized as part of national conservation strategy.

contaminants. Materials which, when added to the potable water supply system, cause the water to be unfit for human or animal consumption.

continental shelf. The submerged shelf of land sloping gradually from the exposed edge of a continent. It is usually defined as those areas where the water is less than 600 feet (200 meters) deep.

continuous distillation. For alcohol fuel production, continuous distillation is carried out in a fractionating column. The mash is introduced into the column at its midpoint in a one-column still, and at the top of the stripper column in a two-column still. The feed is at a constant rate and of a constant composition. The distillate is taken off at a predetermined proof and, if the column is held in equilibrium, it will remain at that proof throughout the run. The spent mash (bottoms) is also taken off at a constant rate. For small distilleries, no intermediate products will be taken out of the still, but in large commercial plants, other products may be withdrawn at various points in the rectifier section of the column.

continuous fermentation. A steady state of fermentation system that operates without interruption; each stage of fermentation occurs in a separate section of the fermenter, and flow rates are set to correspond with required residence times.

continuous power. Power available from a Wind-Energy Conversion System (WECS) facility on a continuous basis under the most adverse wind conditions contemplated (usually significant only if storage is an integral part of the WECS facility).

contrails. Long narrow clouds caused by the disturbance of the atmosphere during passage of high-flying jets. Proliferation of contrails may cause changes in the weather. In the midwest excessive contrails could cause cloudy conditions hampering solar system performance.

control strategy. The algorithm of the wind turbine microprocessor that takes wind data and wind-turbine status or performance data to determine what, if any, changes in machine operation should be initiated (e.g., startup, shutdown, yaw maneuver, pitch control).

control subsystem. Assembly of devices and their electrical, pneumatic, or hydraulic auxiliaries used to regulate the processes of collecting, transporting, storing, and utilizing energy in response to the thermal, safety, and health requirements of the building occupants.

convection. The transfer of heat by means of the upward motion of the particles of a liquid or a gas which is heated from beneath; transmission of heat by moving masses of air. Heat transfer by warmed room air moving by gravity to other parts of the house. One method by which a stove heats your home.

convector. In heat transfer, a surface desinged to transfer its heat to a surrounding fluid largely or wholly by convection. Such a surface may or may not be enclosed or concealed.

conversion efficiency. A figure of merit for energy conversion devices such as a steam turbine-generator. It consists of the amount of energy produced expressed as a percentage of the energy consumed.

conversion efficiency (cell). The ratio of the electric energy produced by a solar cell (under full sun conditions) to the energy from sunlight incident upon the cell.

conversion fuel factor. A number of units of one system in corresponding values in another system. The following energy equivalents are among those commonly used:

Coal:
Anthracite = 26.0 million Btu/ton.
Bituminous = 24.8 million Btu/ton.
Subbituminous = 19.0 million Btu/ton.
Lignite = 13.4 million Btu/ton.

Petroleum:
Crude petroleum = 5.60 million Btu/bbl (42 gal).
Residual Fuel Oil = 6.29 million Btu/bbl.
Distillate Fuel Oil = 5.83 million Btu/bbl.
Gasoline (including aviation) = 5.25 million Btu/bbl.
Jet Fuel (kerosene type) = 5.67 million Btu/bbl.
Jet Fuel (naphtha type) = 5.36 million Btu/bbl.
Kerosene = 5.67 million Btu/bbl.
Asphalt and Road Oil = 6.64 million Btu/bbl.

Natural Gas:
Dry = 1031 Btu/ft^3 at STP.
Wet = 1103 Btu/ft^3 at STP.
Liquids (avg.) = 4.1 million Btu/bbl.

conversion process. A process by which energy is converted from one form to another, such as radiant energy to heat or electric energy.

conversion system. A device or process that converts a raw energy form into another more useful form of energy. For example, conversion of wood into methanol or sunlight into electricity.

conversion unit. A unit used to convert heating equipment from one fuel to another.

converter. An apparatus for transforming the quality or quantity of electrical energy. A class of devices for performing DC/AC power conversion or "inversion."

convertor. Usually, a heat exchanger for transferring heat from steam to water.

cooker. A tank or vessel designed to cook a liquid or extract or digest solids in suspension; the cooker usually contains a source of heat, and is fitted with an agitator; its purpose is to aid in breaking down starches into fermentable sugars.

cooling capacity. The quantity of heat that a room air conditioner is capable of removing from a room in 1 hour's time.

cooling coil. A coil of pipe or tubing used as a heat exchanger to cool material inside or outside the coil.

cooling degree day. A measure of the need for air conditioning (cooling based on temperature and humidity).

cooling load. Amount of heat and humidity that must be removed from a building in order to maintain occupant comfort.

cooling pond. An artificial pond used to receive and dissipate waste heat, usually from a steam-electric power plant. Approximately 1 acre (0.41 hectare) of pond surface is needed per megawatt of electric output for a modern steam-electric power plant. A large body of water that loses heat from its surface, largely by evaporation but also by convection and radiation.

cooling season. Portion of year (usually June to September in the northern hemisphere) when outdoor heat makes indoor cooling desirable to maintain comfort.

cooling system (solar). Assembly of subsystems and components necessary to convert solar energy into thermal energy and use this energy in combination with auxiliary energy, whenever required, for space cooling.

cooling tower. A device to remove excess heat from water used in industrial operations, as in electric power generation. A device for cooling water by evaporation.

cool north sky. The area of north sky with relatively low temperature on clear days. Heat can be dissipated during the day from a surface shaded from the sun and facing north. This is made possible by the cool spot in the sky that occurs at the point opposite from and at a right angle to the sun. Averaged over the day, the coolest spot is due north at an angle of elevation from the horizon equal to 90° minus the altitude of the sun at noon. This can be an effective method for natural cooling using shaded roof pond systems.

coppice system. Silvicultural system in which the crops of trees originate mainly from the shoots of stumps left after harvest.

coproducts. The resulting substances and materials that accompany the production of ethanol by the fermentation process.

cord of wood. A stack of wood 4 feet (1.2 m) by 4 feet (1.2 m) by 8 feet (2.4 m) and containing a gross space of 128 cubic feet (3.7 cubic meters). Burned, it produces approximately 5 million calories of energy.

Coriolis force. An apparent force caused by the earth's rotation which deflects a moving body on the surface of the earth to the right in the northern hemisphere and to the left in the southern hemisphere. Named for the French engineer and mathematician, G. G. Coriolis. Because the first complete explanation of this apparent force was by the American meteorologist, W. Ferrel, it is also called Ferrel's law.

corrosion. The gradual destruction of a material by a chemical process such as oxidation or solution. The deterioration of a substance or its properties because of a chemical or electrochemical reaction with its environment.

cosine loss. The reduction of projected heliostat area visible to the sun caused by the tilt of the heliostat; proportional to the cosine of the angle of inclination of the normal of the heliostat surface to the sun's rays.

cosmic rays. Radiation of many sorts that stem directly or indirectly from sources outside the earth's atmosphere. Cosmic radiation is part of the natural background radiation, and some of its constituents have extremely high energies.

coulomb. The Système International (SI) practical meter-kilogram-second unit of electrical charge equal to the quantity of electricity transferred by a current of 1 ampere in 1 second.

counterflow. In a heat exchanger, where the fluid absorbing heat and the fluid losing heat are so directed that lower and higher temperatures of the one are adjacent to the lower and higher temperatures of the other, respectively. Ordinarily, the one fluid is flowing in the opposite direction from the other, hence the term.

counterflow heat exchanger. Device in which two fluids flow in opposite directions, with one fluid transferring heat to the other.

coupling. A connecting device, such as a threaded sleeve used to connect two pipes; also an arrangement for transferring electrical energy from one circuit to another.

cover material. Soil that is used to cover compacted solid waste in a sanitary landfill.

cover plate. Transparent material covering the absorber aperture of a flat-plate solar collector. Purpose is to reduce the heat loss from the absorber to the surroundings and to protect the absorber. (See diagram p. 88.)

cps. Cycles per second.

cracking still. The combined equipment—furnace, reaction chamber, fractionator—for the thermal conversion of heavier charging stock to gasoline.

crankshaft. The engine shaft that converts the reciprocating motion and force of pistons and connecting rods to rotary motion and torque.

creep. Time-dependent deformation resulting from sustained loads; can be influenced by local environmental factors such as heat, cyclic stress, and solar radiation.

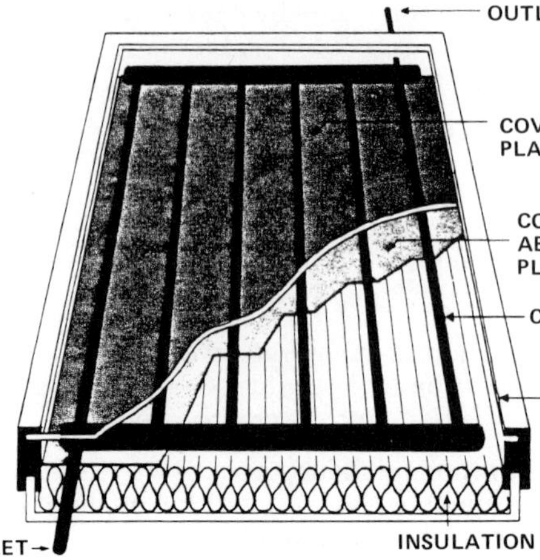

Courtesy National Copper Association

Cover plate protects solar collector from the elements and provides heat retention through the "greenhouse effect."

creosote. The black tarlike substance which is deposited on the inside of stoves, smoke pipe, and chimneys.

crevice corrosion. An intense form of localized corrosion which frequently occurs within crevices, under deposits or gaskets. It is usually associated with small volumes of stagnant solution, e.g., undergasket corrosion of stainless steel in seawater service.

critical point. The condition where the liquid and gas of a substance have the same properties. Above the critical point, the phenomena of boiling does not exist.

critical temperature. That temperature above which a substance can exist only in the gaseous state, no matter what pressure is exerted.

critical velocity. Reynold's critical velocity is that velocity at which the flow changes from laminar to turbulent, and where friction ceases to be proportional to the first power of the velocity and becomes proportional to a higher power—practically the square. Belanger's critical velocity is that condition in open channels for which the velocity head equals one-half the mean depth. Kennedy's critical velocity is that velocity in open channels which will neither deposit nor pick up silt.

Crosbyton hemispherical bowl. The Crosbyton Project is the lead effort for the research and development of hemispherical bowls. Hemispherical bowls focus sunlight onto a linear receiver which precesses to follow the movement of the sun across the sky. The objectives of this project are to develop the fixed mirror, distributed focus collector concept and to demonstrate the feasibility of the concept. Construction was completed and operation was begun of a 65-foot (20-meter) diameter prototype bowl in 1981. (See photograph p. 89.)

cross connection. Any physical connection or arrangement between two otherwise separate piping systems—one of which contains potable water and the other either water of unknown or questionable safety, or steam, gas, chemicals, or other substances—whereby there may be a flow from one system to the other. The direction of flow depends on the pressure difference between the two systems.

Courtesy DOE

Photograph of operational hemispherical bowl solar thermal collector in Crosbyton, Texas.

cross-over. A small fitting with a double offset or shaped like the letter U with the ends turned out. It is only made in small sizes and used to allow pipes to pass each other when the pipes are in the same plane.

cross valve. A valve fitted on a transverse pipe so as to open communication between two parallel pipes.

crosswind. A wind blowing in a direction not parallel to a course.

crystal. A solid body which possesses a natural symmetry is called a crystal. Crystals have sharp edges and flat surfaces, and are given unique shapes and special properties by the orderly arrangement of their atoms, molecules, or ions into repeatable patterns called lattices.

cube factor. The ratio of the cube root of the mean cube of the wind speed to the mean wind speed, the cube root of the energy pattern factor.

cubic. A measure of volume which is calculated by multiplying width times length times height.

cubic foot. The most common unit of measurement of gas volume. It is the amount of gas required to fill a volume of 1 cubic foot (ft^3) under stated conditions of temperature, pressure, and water vapor. One cubic foot of natural gas equals 1000 British thermal units (Btu) under standard conditions of atmosphere and temperature.

cubic meter. A measure of volume in the metric system. One cubic meter (m^3) equals 8.6 barrels (U.S. liquids); 35.3 cubic feet (ft^3); 1.3 cubic yards (yd^3); 264.1 gallons (U.S.) (gal); and 999.9 liters.

current meter. An instrument, such as a galvanometer, for measuring the strength of an electric current; also an instrument for measuring the velocity of moving water or fluid.

curtain. A flexible sheet (e.g., of cloth) or a set of several such sheets in series, used to cover a large portion of a window or cover a frame or space beside a window. The device may be movable; if it is, the motions are mainly horizontal. The purpose of the device may be to provide visual privacy, to diffuse daylight that enters the room, to reduce heat-loss, or to make the window region more attractive. Some curtains are translucent; some are opaque. Ordinarily, curtains are drawn aside at the start of the day.

cut-in speed. The lowest wind speed below which no usable power is produced by a wind turbine. Care should be taken to define whether the cut-in speed is expressed at hub height or at some reference level, e.g., 33 feet (10 meters).

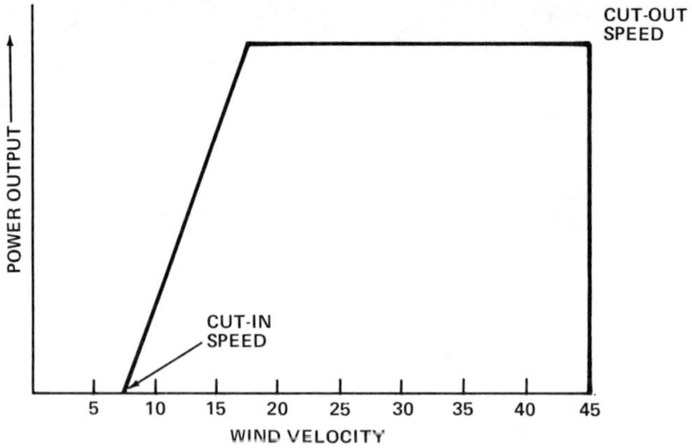

Courtesy DOE

Diagram illustrating initial wind velocity which produces power. This point is called the cut-in wind speed.

cut-out speed. The highest wind speed above which a wind turbine produces no power (avoid the term "furling speed" unless actual furling of sails is involved). Care should be taken to define whether the cut-out speed is expressed at hub height or at some reference level; e.g., 33 feet (10 meters).

cutting. A short length cut from a young, living stem, branch, or root used for propagation, i.e., producing a whole new plant in soil or other media.

cycle. An operation of a number of events which, when completed, starts the same series of events over in the same order. Thus a mechanical refrigeration cycle consists of compression, condensing, expansion, and evaporation, repeated over and over.

cyclic compound. Organic compound whose structure is characterized by a closed ring; its three main groups are alicyclic, aromatic, and heterocyclic.

cyclone collector. A device used to collect large-size particulates from polluted air by centrifugal force.

cycloolefin. An alicyclic hydrocarbon with two or more double bonds.

cycloparaffin. An alicyclic hydrocarbon having three or more of its carbon atoms in each molecule united in a ring structure. Each of these ring carbon atoms is bound to two hydrogen atoms or alkyl groups.

Courtesy Pinson Energy Corp./DOE-Rocky Flats

Photograph of a Pinson 1-kW small scale cycloturbine wind-energy conversion system.

cycloturbine. The cycloturbine small wind-energy conversion system is a vertical axis wind-generation system.

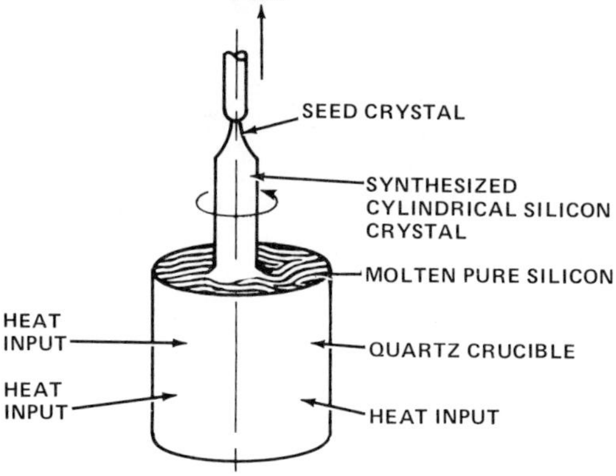

Courtesy SERI

A crystal growth process developed by Czochralski is used to produce silicon cut into wafers for photovoltaic cells. (See **Czochralski wafers.**)

cylinder. The bottom element in a water-pumping windmill. The cylinder contains a check valve and plunger and valve. The pump rod connects to the plunger and valve, and moves vertically in sync with the wind.

Czochralski wafers. Most solar cells are currently made by slicing thin wafers cut from so-called Czochralski silicon ingots. This produces reliable, high-efficiency solar cells of the type used by space satellites. The current problem with this approach is that the process is too expensive for large-scale photovoltaic residential system use. Also, method of growing a perfect crystal of large size by slowly lifting a seed crystal from a molten bath of the material under careful conditions of cooling. (See diagram p. 91.)

D

daily control. Controlling solar heat over a day's period: not too much heat in late afternoon but heat stored for use at night; for example, a masonry wall.

daily storage option. An alternative residential photovoltaic system with minimum battery storage required for daily operation (15 kWh capacity).

damper. A device used to vary the volume of air passing through an air outlet, inlet, or duct.

Darcy's law. In fluid dynamics, the velocity of flow of liquid through a porous medium, because of a difference in pressure, is proportional to the pressure gradient in the direction of flow.

Darrieus machine. A vertical-axis machine that has long, thin blades in the shape of loops connected at the top and bottom of the axle; often called "eggbeater" windmill because of its appearance. (See photograph p. 94.)

data base. A set of numbers, variables, and information that is used to provide the operational criteria for decisionmaking.

day temperature. The mean temperature assumed for the design day.

dead air space. A confined space of air. A dead air space tends to reduce both conduction and convection of heat. This fact is used in virtually all insulating materials and systems, such as double glazing, Beadwall®, fiberglass batts, rigid foam panels, fur and hair, and loose-fill insulations like pumice, vermiculite, rock wool, and goose down.

deadband (thermostat). Difference between heating and cooling thermostat settings in conditioned space.

decentralized systems. In solar systems, refers to establishment of autonomous units, such as households or neighborhoods, to provide their own electricity or heat.

deciduous trees. Trees which shed their leaves each winter at the end of the growing season. (See diagram p. 94.)

declination. The angular position of the sun at solar noon with respect to the plane of the equator, north positive, i.e., the angular position of the sun north or south of the equator; a function of the time of year. (See diagram p. 95.)

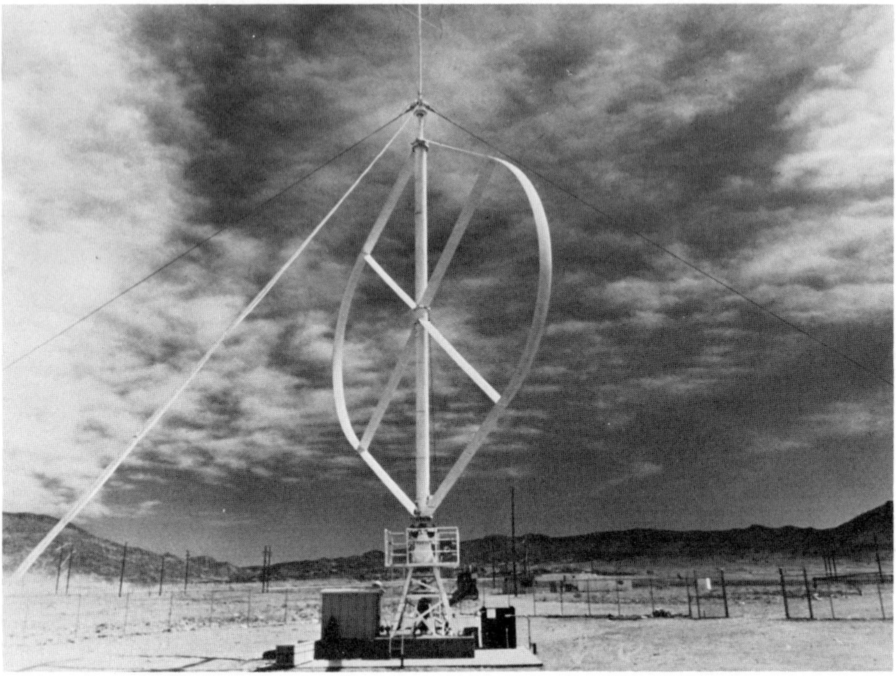

Courtesy DOE

Sandia Laboratories operates this 56-ft (17-m) diameter vertical axis Darrieus wind-energy conversion system.

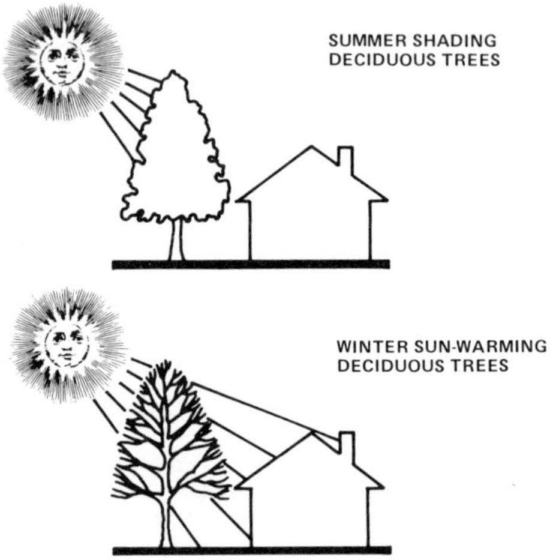

Courtesy HUD/SERI

During winter, deciduous trees and shrubs shed their leaves and let sun help to warm the house; during summer, the trees provide passive solar shade.

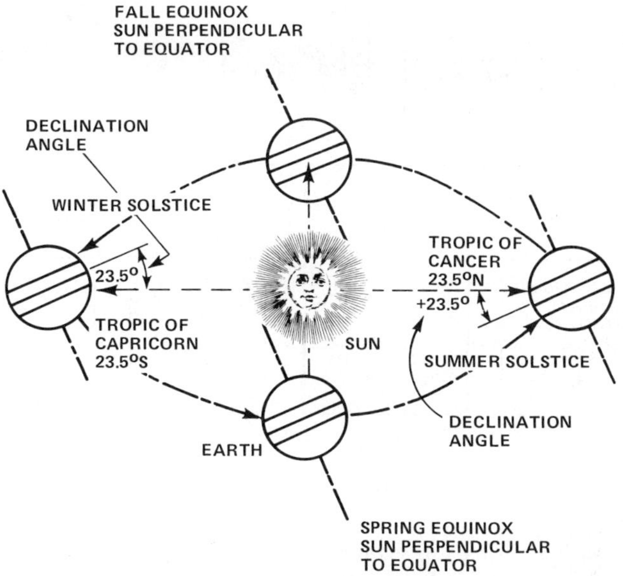

Courtesy National Copper Association

Illustration of solar declination in each season.

declining block rates. A utility rate structure in which the charge for energy decreases as the amount of energy consumed increases.

decomposition. Reduction of the net energy level and change in chemical composition of organic matter because of the actions of aerobic or anaerobic microorganisms.

dedicated firm. A business firm engaged exclusively in solar or other single-product production.

dedicated storage. An energy storage system charged solely from a single-energy source.

deep discharge. Discharging a battery to 20 percent or less of its full charge.

deep well irrigation project. This project, located near Coolidge, AZ, is the first large-size application of solar energy to pump irrigation water from a deep [380 foot (116-meter)] well. It is being conducted jointly by DOE and the state of Arizona. The system uses trough-type collectors with an improved organic Rankine-cycle heat engine to provide 150 kilowatts (kW) of electric power to the utility grid. The utility in turn supplies power for the irrigation pump. The design objective of the system is to provide output for 16 hours a day.

definite proportions law. A chemical compound always contains the same elements in the same proportions by weight.

degree-day. A unit that represents a 1°F deviation from some fixed reference point (usually 65°F) in the mean daily outdoor temperature. If the average outdoor temperature is 40°F for one day, then 25 (65° minus 40°) degree-days (DD) result. Used to determine the demand of a heating season for different locales. (See p. 96.)

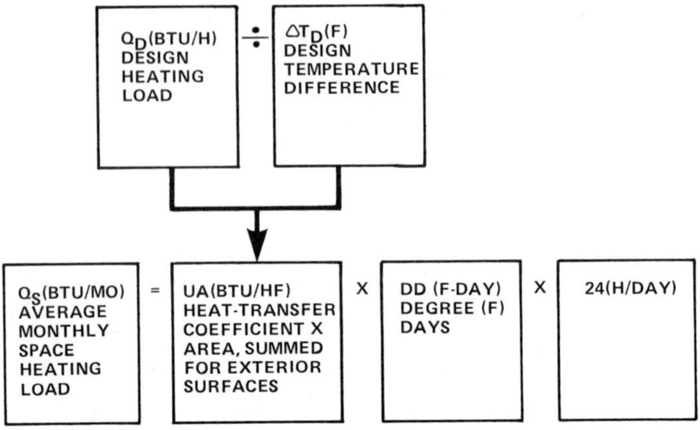

Courtesy F. Kreith and J. F. Kreider, Principles of Solar Engineering, *Hemisphere Publishing Corp.*
The number of degree-days can be determined by using this diagram.

degree-day heating. An expression of a climatic heating requirement expressed by the difference in degrees F below the average outdoor temperature for each day and an established indoor temperature base of 65°F. (The assumption behind selecting this base is that average construction will provide interior comfort when the exterior temperature is 65°F.) The total number of degree-days over the heating season indicates the relative severity of the winter in that area. If the minimum and maximum ambient temperatures are 30°F and 50°F, the number of heating degree days for that day is 65 − 40 = 25.

degree-hour. The product of 1 hour and (usually) the number of degrees F the hourly mean temperature is above a base point, usually 85°F. Thus, in an hour, when the temperature is 92°F, there are 7 degree-hours on an 85°F base. The degree-hour is used in measuring, roughly, the cooling load in summer for cases where process heat, heat from human beings, and humidity are relatively unimportant as compared with dry-bulb temperature of the outside air.

degrees balling. A calibration of a hydrometer in percentage of sugar. One degree equals 1 percent dissolved sugar solids. For example, 14° balling indicates that the alcohol mash has a 14 percent sugar concentration.

degrees Kelvin. A thermometric scale in which the unit of measurement equals 1°C (centigrade) but the initial point is different. That is, the Kelvin scale is so constructed that absolute zero corresponds to −273.16°C. Hence 0°C (the freezing point of water at sea level) is equal to +273°K.

degrees Rankine. Absolute temperature on the Fahrenheit scale, or degrees F plus 459.6.

dehumidify. To reduce, by any process, the quantity of water vapor contained in a solid or gas.

dehydration. The process of removing 95 percent or more of the water from any substance by exposure to high temperature or by chemical means.

dehydrator. A device or material which will remove water from a substance. In refrigeration, a device for removing moisture from refrigerants; also called a drier.

dehydrogenation. Process in which hydrogen is removed by chemical means.

deionization. Removal of ions from solution by chemical means.

demand. The rate at which electric energy is delivered to or by a system expressed in kilowatts, kilovolt-amperes, or another suitable unit, at a given instant or averaged over any designated period of time. The amount of energy required to satisfy the needs of a stated sector of the economy. See **end-use demand.**

demand charge. That part of a utility service charged on the basis of the possible demand as distinguished from the actual energy consumed.

demand day. That 24-hour period specified by a supplier-user contract for purposes of determining the purchaser's daily quantity of energy used. This term is primarily used in pipeline-distribution company agreements. It is similar to, and usually coincides with, the distribution company "sendout day."

demand diversity. The overall variation in the time at which individual demands occur.

demand factor. The ratio of the maximum demand of a system, or part of a system, to the total connected load of a system or part of the system under consideration.

demand interval. The period of time during which the flow is averaged in determining demand.

demand load. Domestic water heating needs to be supplied by solar or conventional energy.

demand meter. System using a mechanical mechanism to register the highest level of power consumed between meter readings. This maximum-kilowatt figure determines the electric rate the consumer pays for the billing period.

denature. The process of adding a substance to ethyl alcohol to make it unfit for human consumption; the denaturing agent may be gasoline or other substances specified by the Bureau of Alcohol, Tobacco, and Firearms.

dendrite. A slender threadlike spike of pure crystalline material, such as silicon.

dendritic-web growth. A ribbon-pulling process used to make solar cell ribbons. The filament crystal widens as it grows, and narrow filaments or dendrites at its edges grow into hot silicon. As the seed pellet is pulled up from the silicon, the dendrites continue to grow. A smooth web of a single-crystal silicon, six thousandths of an inch thick, forms between the dendrites. (See diagram p. 98.)

density. The weight of a unit volume of a substance which is expressed in pounds per cubic feet.

Department of Agriculture. Responsible for several portions of the solar program including some wind machines and some biomass activities. Congress mandated a major USDA solar demonstration program in 1977.

Department of Energy. In October 1977, DOE was created to consolidate the multitude of energy-oriented government progams and agencies. The Federal Energy Administration and the Energy Research and Development Administration are now part of DOE. The department will carry out the National Energy Plan policy through a unified organization that will coordinate and manage energy conserva-

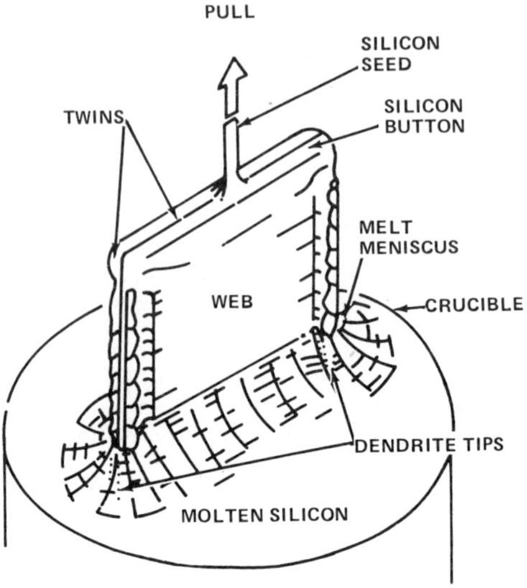

Courtesy SERI

Illustration of dendritic web ribbon growth to produce silicon for photovoltaic collectors.

tion, supply development, information collection and analysis, regulation, research, development, and demonstration. An Energy Information Administration within the department will organize and analyze information so that it can be used by governments, industry, and the public. In September, 1981, President Reagan recommended abolishing the DOE.

Department of Housing and Urban Development. The federal organization responsible for most of the residential solar heating and cooling demonstration programs.

dependable capacity. The load-carrying ability of the wind-energy conversion system facility under adverse conditions for the time interval and period specified when related to the characteristics of the load to be supplied.

depletion allowance. A tax allowance extended to the owner of exhaustible resources based on an estimate of the permanent reduction in value caused by the removal of the resource.

depletion zone. Same as cell barrier. The term derives from the fact that this microscopically thin region is depleted of charge carriers (free electrons and holes).

deposit. Anything laid down. Formerly applied only to matter left by the agency of water, but now includes mineral matter in any form that is precipitated by chemical or other agencies, as the ores in veins. Mineral deposit or ore deposit is used to designate a natural occurrence of a useful mineral or ore in sufficient extent and degree of concentration to invite exploitation.

depreciation. Allocation of initial cost of a facility or system to the time period during which it is used.

depth. The level of liquid in a container.

depth difference. One level subtracted from another level; difference between two levels or depths; analogous to temperature difference.

depth of stratum. The vertical distance from the surface of the earth to a stratum.

derating. The production of energy at a level less than the nominal capacity of a particular boiler.

desalinization. Salt removal from sea or brackish water. Any process for making potable water from seawater or other salt waters. Reuse of vapors through compressive distillation of multiple-effect evaporation is practiced in order to limit heat consumption. Distillation is the oldest method. Electrodialysis is another. Other methods are: freezing by direct contact of refrigerant with seawater, foam separation, liquid-liquid extraction, various nonelectric membrane processes, and ion exchange.

desertification. The often irreversible increase in desert size due to removal of trees and other vegetation, and the resulting decreased ability to retain soil, nutrients, and water.

desiccant. A substance having an affinity for water; used for drying purposes.

desiccant (adsorption) system. A cooling system in which solar energy regenerates a desiccant used as the cooling medium.

desiccant cooling. Cooling provided by evaporatively cooling air that has been dried with a desiccant (a liquid or solid with a strong affinity for water). The desiccant is then "regenerated" for reuse by heating it to a high temperature to drive off the water.

desiccant-type spacers. Small objects which hold two sheets of glass, or other materials, apart. Desiccant-type spacers include a water-absorbing material such as calcium chloride to keep condensation from forming on the glass plates. Condensation in a collector panel would keep out some of the sun's energy.

designated solar surface. A south-facing wall within 45° of true south, or a south-tilted surface within 60° of true south, or a horizontal surface all of which receive adequate direct solar radiation to provide for the intended use. The designated surface need not be a wall or roof but can include the hypothetical plane under the collector as it rests on mounting brackets.

design conditions. Indoor and outdoor temperatures, humidities, and wind speeds used to predict maximum expected heating and cooling loads for a building.

design day. A 24-hour period of the greatest theoretical energy demand, used as a basis for designing purchase contracts, and/or production facilities, and/or delivery capacity.

design day availability. The amount of each type of energy arranged to be available on the design day and the maximum combination of such supplies.

design heating load. The total heat loss from a building under the most severe winter conditions likely to occur.

design horsepower. The specified horsepower for a chain-drive multiplied by a service factor. It is the value used to select the chain size for the drive.

design life. Period of time during which a heating or hot-water system is expected to function without requiring major maintenance or replacement.

design load. The load generally taken as the worst combination of forces and loads which a structure is calculated to sustain. The term is similarly applied to projects such as air conditioning.

design outside. The lowest outdoor temperature expected during a heating season.

design point. The time of year for which system performance is specified.

design pressure. The maximum operating pressure permitted by various codes, as determined by the design procedures applicable to the material and location involved.

design temperature. A designated temperature close to the most severe winter or summer temperature extremes of a climate, used in estimating a house's heating or cooling requirements.

design temperature difference. The maximum probable difference between the indoor and the ambient temperatures.

design tip speed ratio. Tip speed ratio for which the power coefficient is a maximum.

design voltage. The nominal voltage for which a line or piece of equipment is designed; a reference level of voltage for identification and not necessarily the level at which it operates.

desorption. The process of removing a sorbed substance by the reverse of adsorption or absorption.

desulfurization. Process in which sulfur and sulfur compounds are removed from gases of liquid hydrocarbon mixtures by chemical or catalytic means.

detergent. Synthetic washing agent that, like soap, lowers the surface tension of water, emulsifies oils, and holds dirt in suspension.

detonation. The destructive reaction that occurs when a fuel explodes prematurely or erratically inside an engine. Detonation can be caused by too much compression, overadvanced timing, or too low an octane in the fuel.

detritus. Dead organic tissues/organisms in an ecosystem.

developed length. Length of a section of pipe along the center line through the pipe fittings.

dewatering. Removal of the free water from a solid substance.

dewpoint. The temperature, at constant pressure and constant water vapor content, to which air must be cooled in order for saturation or condensation to occur. Also called saturation point.

dewpoint temperature. The temperature at which a vapor begins to condense and deposit as liquid.

dextrin. A polymer of D-glucose which is intermediate in complexity between starch and maltose and formed by partial hydrolysis of starches.

dextrose. The same as glucose.

dialysis. The separation of substances in solution by means of semipermeable membranes (such as parchment, cellophane, or living cells) through which the smaller molecules and ions diffuse readily while the larger molecules and colloidal particles diffuse very slowly or not at all. Such separations are important in nature (i.e., in living organisms and soil).

diastase. Originally, the term diastase was used to describe the hydrolysis of starch to sugars, but today it is used to mean the enzymes associated with barley malt.

diatomaceous earth. A fine siliceous material resembling chalk used in wastewater treatment plants to filter sewage effluent to remove solid. May also be used as inactive ingredients in pesticide formulations applied as dust or powder.

dielectric. An insulator; a material which offers relatively high resistance to the passage of an electric current but through which magnetic or electrostatic lines of force may pass. Most insulating materials (i.e., air, porcelain, mica, and glass) are dielectrics. A perfect vacuum would constitute a perfect dielectric.

diesel engine. A type of internal-combustion engine that burns heavy oil and needs no spark to ignite the fuel. It works in the following way: Air is introduced into the cylinders and compressed until it reaches a temperature above the ignition point of the fuel; fuel oil is sprayed into the air; it is ignited by the high temperature and burned; the pressure of the hot expanding gases pushes the piston downward; the gases produced by combustion are evacuated through the exhaust valves.

diesel oil. Oil fraction remaining after petroleum and kerosene have been distilled from crude oil.

difference of potential. The difference in electrical pressure existing between any two points in an electrical system or between any point of such a system and the earth. Measured by a voltmeter.

differential thermostat. Automatic device that responds to temperature differences (between collectors and heat storage, for example) in regulating operation of an active solar system.

diffused air. A type of sewage aeration. Air is pumped into the sewage through a perforated pipe.

diffused sky radiation. Indirect sunlight.

diffuse insolation. Sunlight scattered by atmospheric particles and gases; it arrives on the earth's surface from all directions and cannot be focused.

diffuse sunlight. Sunlight that reaches the earth after being reflected off atmospheric particles. On a cloudy day, diffuse light may account for all the sunlight received at the surface. Diffuse sunlight comes along no set path; it generally comes from the entire skyvault, the most coming from the area of the sky near the sun.

diffuser. A device or structure that diffuses a windstream.

diffuse radiation. Solar radiation, scattered as it passes through atmospheric molecules, water vapor, dust, and other particles, so that it appears to come from entire sky, as on a hazy or overcast day.

diffusion. The movement of individual molecules through narrow spaces by the molecular velocity and bouncing associated with individual gaseous or liquid molecules.

diffusion length. The mean distance a charge carrier (free electron or hole) moves before recombining with another hole or electron. Distances are short, typically several micrometers to a few hundred micrometers. Cell efficiency improves with increasing minority carrier diffusion length.

digester. A device using anaerobic bacteria to decompose organic wastes to produce methane.

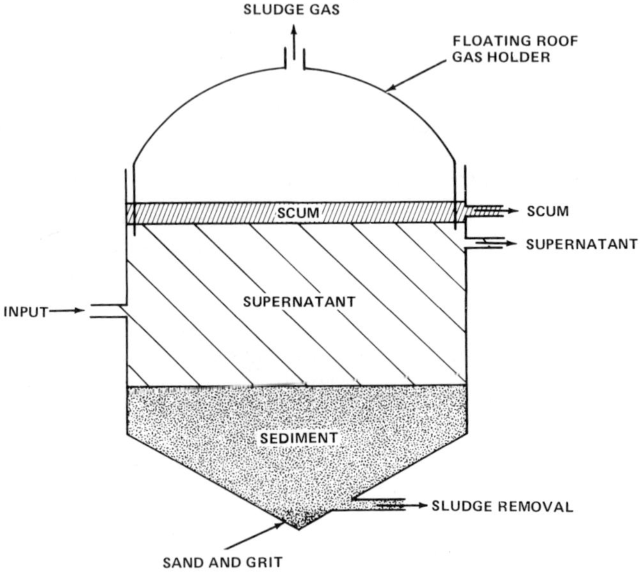

Courtesy DOE/SERI

A conventional sewage sludge digestor showing the various fractions of digesting biomass.

digestion. The biochemical decomposition of organic matter. Digestion of sewage sludge takes place in tanks where the sludge decomposes, resulting in partial gasification, liquefaction, and mineralization of pollutants.

dilution ratio. The ratio of the volume of water of a stream to the volume of incoming waste. The capacity of a stream to assimilate waste is partially dependent upon the solution ratio.

diode. An electric device which changes alternating current to direct current. See **rectifier.**

diode function. The method by which a solar heating system allows heat to go to storage when it is available, but prevents heat loss from storage at night; for example, a controller in an active system provides a diode function.

diol. Chemical compound consisting of two hydroxyl groups.

direct beam radiation. Solar radiation that arrives in a straight line from the sun and is not scattered by the atmosphere. It casts shadows and can be focused.

direct conversion. Transformation of sunlight to electricity without an intervening thermodynamic cycle.

direct current (DC). An electric current, such as that produced by a battery, in which the electrical potential does not change its sign, and the voltage is often invariant with time. In a direct current, therefore, energy is carried by a continuous, unidirectional flow of electrons through a conductor.

direct energy conversion. The generation of electricity from an energy source which does not involve transference of energy to a working fluid. The conversion methods have no moving parts and usually produce direct current. Some methods include thermoelectric conversion, thermionic conversion, and magnetohydrodynamic conversion.

direct expansion. Generic term used to describe refrigeration systems in which the cooling effect is obtained directly from the refrigerant (e.g., refrigerant is evaporated directly in a cooling coil in the airstream).

direct-fired. A heating unit in which the combustion products are mixed with the air or liquid being heated.

direct-fired heater. A fuel-burning device in which the heat from the fuel is transferred through metal to air which is then introduced to the space to be heated. A warm-air furnace is direct fired; a boiler is indirect.

direct-gain system. A passive system where the sun heats the floor or walls of the house directly through south-facing windows.

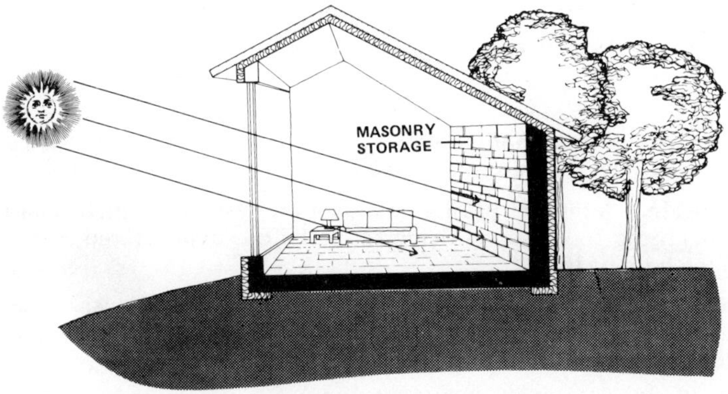

Courtesy DOE/HUD

Direct solar radiation on masonry wall stores energy.

direct insolation. The solar energy incident on a surface that comes from within the solid angle subtended by the solar disk, i.e., that sunshine capable of casting a sharply defined shadow. The solar power density from the sun falling on a surface of given orientation (W/m^2). Often, the direct insolation is measured with equipment whose acceptance angle is as large as 6 degrees, rather than the 1/2 degree of the solar disk.

direct normal insolation. The direct insolation on a surface perpendicular to the sun's rays.

direct radiation. Light that has traveled a straight path from the sun, as opposed to diffuse sky radiation. Direct radiation casts shadows on a clear day. Also called beam radiation.

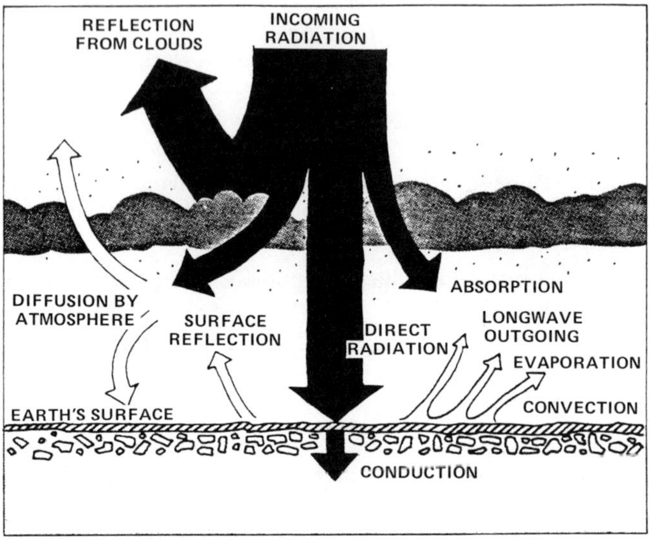

Courtesy National Copper Association

Distribution of incoming solar radiation shows relationship of direct radiation to overall radiation.

direct solar conversion (photovoltaics). The process of directly converting sunlight to electricity through solar cells.

direct solar gain. A type of passive solar-heating system in which solar radiation passes through the south-facing living space before being stored in the thermal mass for long-term heating.

direct sunlight. Sunlight that comes straight from the sun. Sky-space angles and most solar-planning guidelines are based on direct sunlight. Direct sunlight has higher intensity than diffused sunlight.

direct system. See **open-loop system.**

direct use. Using the sun's radiation directly as in heating and generating electricity.

disability glare. Spurious light from any solar-energy source which impairs a viewer's ability to discern a given object.

disaccharide. Maltose, sucrose, lactose, and cellobiose are disaccharides. They are the combination of two monosaccharide units. With the exception of cellobiose, these are fermentable sugars.

disc. The disc-shaped device which is attached to the bottom of the stem and which is brought into contact with, or lifted off of the seating surfaces to close or open a globe valve.

disc fan. An axial-flow fan with blades formed of flat sheets or plates so bent that when rotated the blades impart to the air a motion along the axis of the fan shaft.

discharge. Removing heat from storage by radiation or convective heat transfer (blown out). See **charge; emission; supercharge.**

discharge/charge rate. The time rate for transferring energy to or from storage rated power.

discomfort index. See **temperature-humidity index.**

discounted cash flow. The present worth of a future payment.

discount factor—energy real growth rate factor. The product of the discount factor (DF) and energy real growth rate factor (ERGR), summed from the present to the year in question equals the discounted energy real growth rate factor.

discount rate. A factor used to discount future recurring costs to present value cost for decisionmaking. It is the interest rate that the capital could be earning if invested rather than used to purchase the solar system, or it is the annual interest charge on a loan made to finance the system.

disinfection. Effective killing by chemical or physical processes of all organisms capable of causing infectious disease. Chlorination is the disinfection method commonly employed in biomass treatment processes.

dispersed systems. Term used to describe applications in which solar systems are installed at many separate locations.

dispersion. The distribution of finely divided particles in a medium.

dissipator. Any device used to dissipate or reject heat in natural cooling systems. Dissipators typically work by radiation, evaporation, or conduction. They range from operable windows used for night ventilation to more complex roof ponds.

dissociation. Heat is given off in the ordinary combustion reactions; that is, the action is exothermic. The chemical action can, however, be reversed; in this case, heat is necessary to complete the reaction (this action is termed endothermic). For example, when carbon burns to carbon dioxide, heat is given up when the molecule of CO_2 is broken up into carbon and oxygen again, heat is absorbed.

When the temperature of combustion has reached the equilibrium temperature, the breaking up of molecules is taking place at such a rate that the heat absorbed is equal to that given up by the combination. Consequently, the equilibrium temperature is a limiting temperature beyond which a higher temperature cannot be obtained. The breaking up of the molecules at the equilibrium temperature is known as dissociation.

dissolution. The taking up of a substance by a liquid with the formation of a homogeneous solution.

dissolved oxygen. The oxygen dissolved in water or sewage. Adequately dissolved oxygen is necessary for the life of fish and other aquatic organisms and for the prevention of offensive odors. Low dissolved oxygen concentrations generally are due to discharge of excessive organic solids having high BOD, the result of inadequate waste treatment.

dissolved solids. The total amount of dissolved material, organic and inorganic, contained in water or wastes. Excessive dissolved solids make water unpalatable for drinking and unsuitable for industrial uses.

distillate. That portion of a liquid which is removed as a vapor and condensed during a distillation process.

distillation. The process of separating a mixture of two or more components. The mixture is heated and either one or both of the components vaporize and are condensed into different containers. Distillation works best when one of the components has a higher degree of volatility than the other. Ethanol (EtOH) has a boiling point of 172.9°F (78.3°C), while that of water (H_2O) is 212°F (100°C). This makes it easy to distill EtOH from H_2O. See also **continuous distillation; simple distillation; solar distillation.**

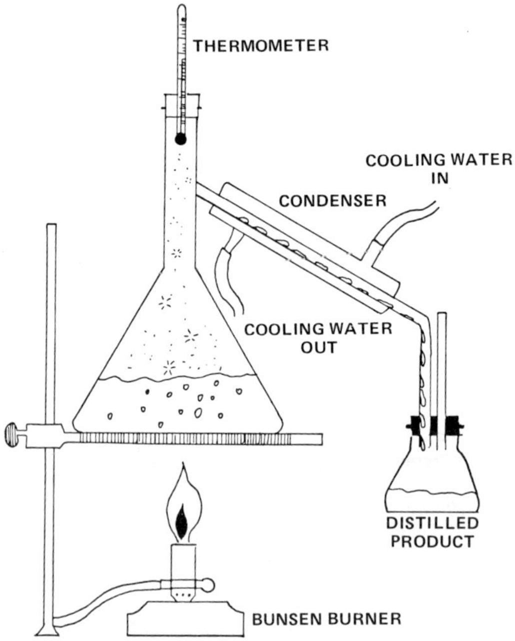

Courtesy Larry W. Carley, How to Make Your Own Alcohol Fuels, *TAB Books Inc.*

Process diagram for distillation.

distilled spirits plant. A plant, including fuel alcohol plants, authorized by the Bureau of Alcohol, Tobacco, and Firearms to produce, store, or process ethyl alcohol in any of its forms.

distiller's dark grains. See **distiller's dried grains with solubles (DDGS).**

distiller's dried grains. The water-insoluble, dried distiller's grains (DDG) coproduct of the grain fermentation process which may be used as a high-protein (28 percent) animal feed. See **distiller's grain.**

distiller's dried grains with solubles (DDGS). A grain mixture obtained by mixing distiller's dried grains and distiller's dried solubles.

distiller's dried solubles (DDS). A mixture of water-soluble oils and hydrocarbons obtained by condensing the thin stillage fraction of the solids (DDS) obtained from fermentation and distillation processes.

distiller's feeds. Coproducts resulting from the fermentation of cereal grains by the yeast *Saccharomyces cerevisiae;* the nonfermentable portion of grain mash.

distiller's grain. The nonfermentable portion of a grain mash comprised of protein, unconverted carbohydrates and sugars, and mineral material.

distributed collectors. One basic configuration of the distributed collector system is a series of modular cylindrical (parabolic) or segmented mirror focusing collectors with an interconnected absorber pipe network to carry the solar-heating working fluid to a heat exchanger unit. There the working fluid generates steam in order to power the turbogenerator. Since distributed collectors focus direct solar radiation on the absorber pipe, they must incorporate at least a one-dimensional tracking mechanism for purposes of tracking the sun. Distributed collector systems of the parabolic or segmented mirror configuration are generally designed to operate with working fluid temperatures in the vicinity of 572°F (300°C). Another distributed collector system is the paraboloidal dish. This system consists of a series of circular dish-shaped collectors in the form of a parabola, with the absorber located at the focus. The absorber units are interconnected with insulated piping. Paraboloidal dish systems are capable of generating temperatures of approximately 1472°F (800°C). These systems generally have overall conversion efficiencies in the range of approximately 20 to 25 percent.

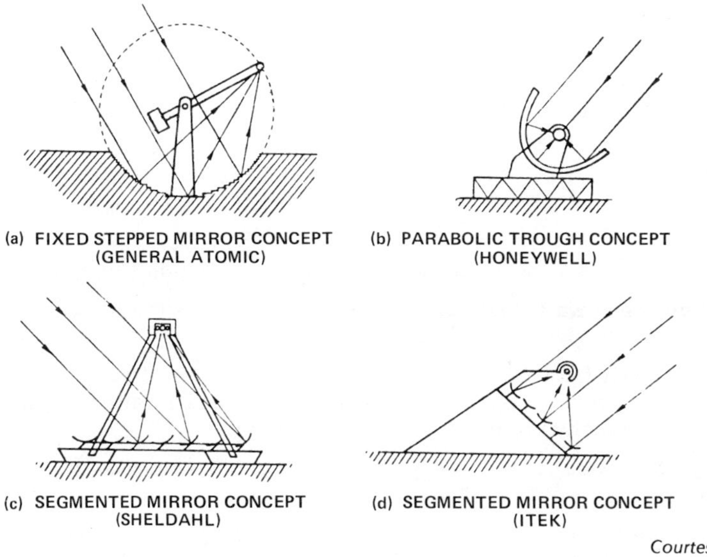

Courtesy SERI

Illustration of distributed solar collector concepts.

distributed power. Generic term for any power supply located near the point where the power is used. Opposite of central power. See **remote site; stand-alone.**

distributed system. See **decentralized systems.**

distribution. Getting heat to the living space. In a passive house, distribution to north-facing rooms is often difficult.

district heating. The heating of a multiplicity of buildings from one central boiler plant, from which steam or hot water is piped to the buildings. Can be a private function as in some housing projects, universities, hospitals, etc., or a public utility operation.

diurnal (or day-night). Diurnal means a daily event as opposed to nocturnal happenings. When used in solar-energy work, it refers to the effect of moisture in the atmosphere upon the collection of incident radiation by the collector. Because of diurnal cycling, periods of high humidity generally occur in the early morning hours near sunrise. The diffuse component of solar energy is therefore a reduced factor in the early morning hours of operation. In the late afternoon the humidity is usually lower on the diurnal cycle and collector efficiency stays higher than it does in early morning.

diurnal temperature range. The variation in outdoor temperature between day and night.

diurnal variation. The daily variation in the earth's magnetic field; in tides, having a period or cycle of approximately 1 lunar day (24.84 solar hours).

diversified firm. Business firm engaged in multiple production/R&D activities.

diversified load. A mix of different types of power-consuming devices, in residential use, various appliances, motors, etc., as opposed to space heating or water heating loads.

diversity. That characteristic of variety of electric loads whereby individual maximum demands usually occur at different times. Diversity among customer's loads results in diversity among the loads of distribution transformers, feeders, and substations, as well as between entire systems.

diversity factor. The ratio of the sum of the individual maximum loads during a period to the simultaneous maximum loads of all the same units during the same period.

doldrums. A part of the ocean near the equator abounding in light shifting winds.

dollars per peak watt. Dollars per peak watt (peak kilowatt) at peak insolation potential output; given for cell, panel, or system.

dollars per rated watt. Dollars per watt of delivered (rated) peak power, depending upon specific location's average insolation, array and system efficiencies, and system capacity factor.

domestic hot water. The hot water used for conventional purposes such as bathing and washing. (See diagram p. 109.)

Domestic Policy Review. A cabinet-level, multiagency review of the national solar-energy program, initiated by President Carter on May 3, 1978.

domestic water heating system. A solar heating system which supplies a portion of the energy needed to heat water used for domestic purposes, such as bathing and washing. (See diagram p. 110.)

domestic water pressure. The pressure of the potable water within the building from sources not related to the solar domestic hot-water system.

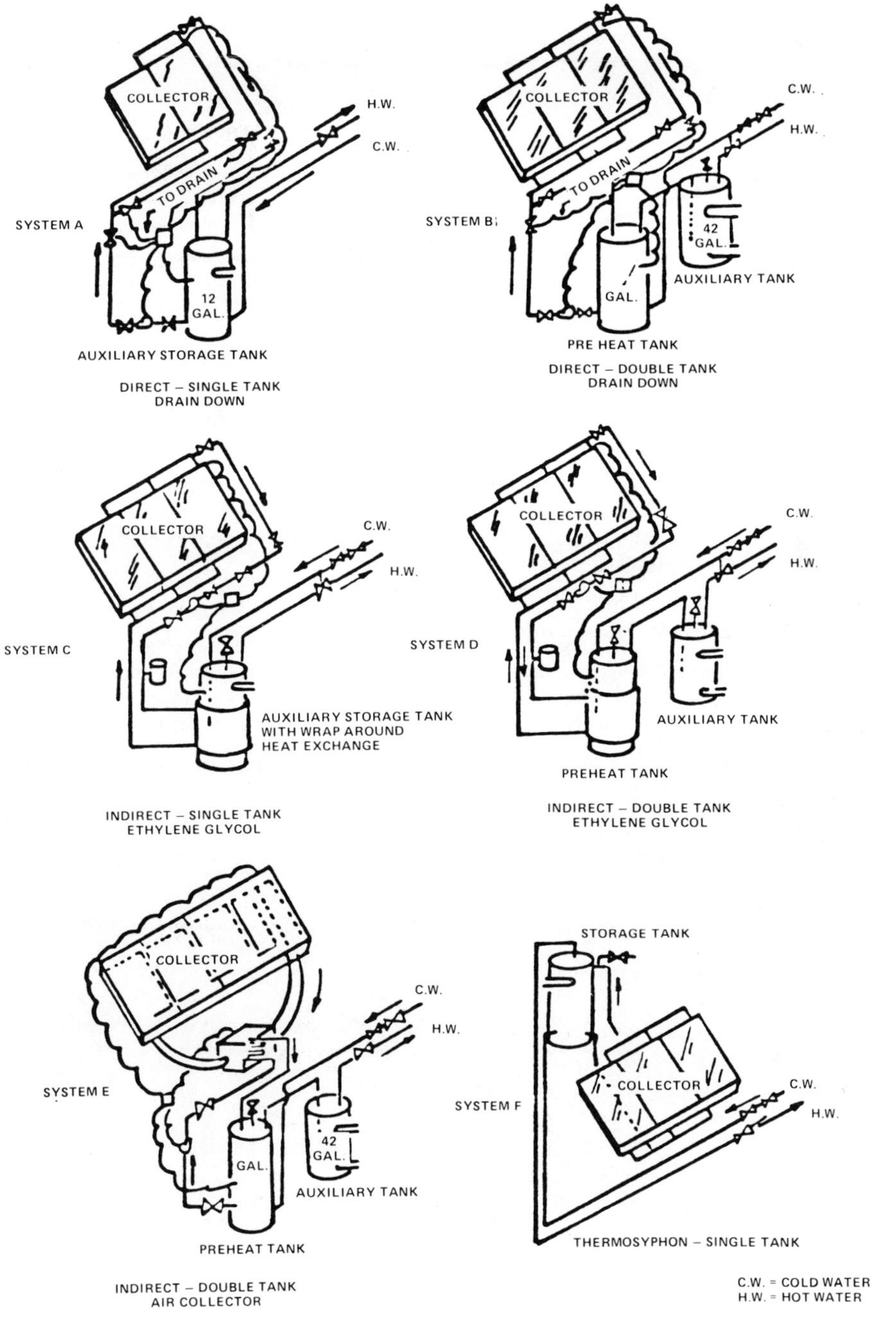

Courtesy DOE

Diagrams of six domestic hot water heating systems.

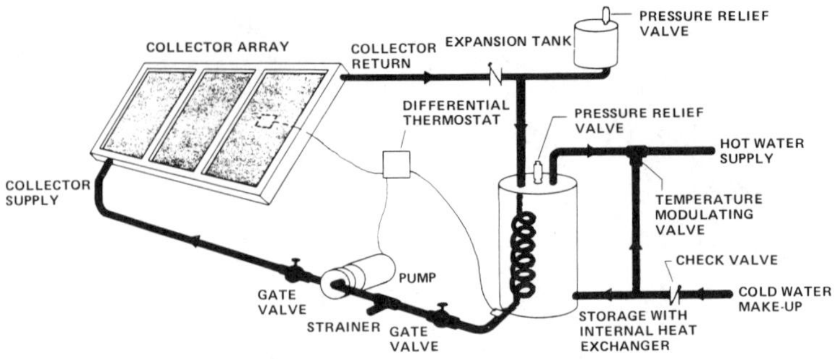

Courtesy National Copper Association

System diagram of pumped solar hot-water heating system with heat exchanger.

donor. A dopant, such as phosphorus, which supplies an additional electron to an otherwise balanced crystal structure.

dopant. A chemical element added in small amounts to an otherwise pure crystal to modify its electrical properties. An *n*-dopant introduces more electrons than are required for the perfect structure of the crystal. A *p*-dopant creates electron vacancies in the crystal structure.

doping. Controlled addition of impurities to a semiconductor to alter its electrical properties; a necessary step in making solar cells.

double-bundle condenser. Condenser (usually in a refrigeration machine) that contains two separate tube bundles allowing the option of rejecting heat either to the cooling tower or to another building system requiring heat input.

double extra-strong pipe. A schedule of steel or wrought-iron pipe weights in common use, generally Schedule 120.

double-glazing. Using two layers of glass on windows or solar collectors to reduce heat loss. The still air in the space between the windows acts as a good insulator.

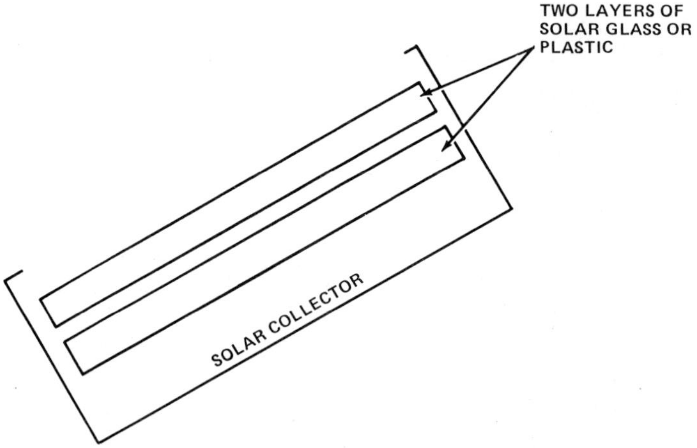

Two plates of glass or plastic are sealed above solar collector absorber surface.

double offset. Two offsets in succession in the same pipeline.

double pipe. In a heat exchanger, one in which two concentric pipes, one within the other, are used with one fluid flowing through the inner pipe, the other through the annular space between.

double sweep tee. A tee made with easy (long radius) curves between body and branch.

double-wall heat exchanger. One that provides two distinct walls between the heat-transfer fluid and the household water. Such a separation is necessary when the heat-transfer fluid contains nondrinkable substances.

double wall separation. Heat exchangers using nonpotable heat transfer fluids are separated from the potable water system by use of two walls between the fluids.

downdraft. A flow of air down the chimney or flue because of adverse draft conditions; a downward current of air or other gas as in a mine shaft, kiln, or carburetor.

downhill. Direction of heat flow (from a higher temperature to a lower one); direction of volume flow (from a higher depth to a lower one).

downpoint. The temperature at which useful heat energy can no longer be taken from storage. In an air-type solar system, useful heat can be taken from the rock stoarage bin at temperatures as low as 75°F (24°C). In most liquid-type solar systems, the downpoint is considerably higher—at least 90°F (32°C) and preferably about 110°F (43°C).

downstream. Any point in the direction of flow of a fluid or gas from the reference point.

downwind. Horizontal axis wind-energy conversion system configuration in which the rotor is downwind of the tower. (Wind passes first the tower, then the rotor plane.)

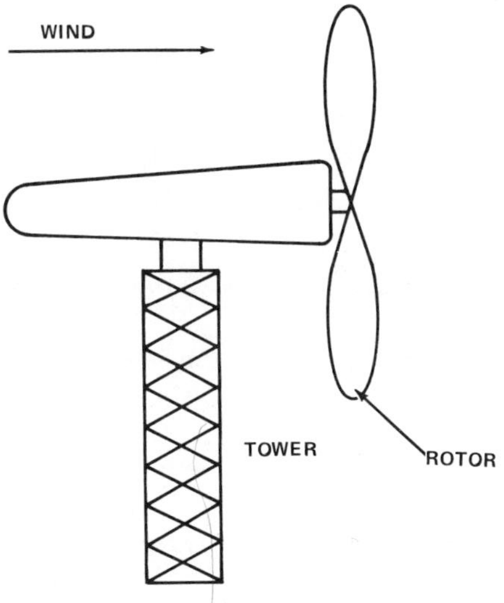

Courtesy DOE

Illustration of wind blowing downwind through tower to blade of wind-energy conversion system.

draft. A difference of pressure which tends to cause a flow of air and/or flue gases through the boiler, flue connector, breeching, flue, or chimney.

draft controller. Any device used to change the amount of combustion air admitted to the firebox. Sometimes called a draft regulator.

draft hood. A device built into an appliance, or made a part of the flue or vent connector from an appliance, which is designed to (a) assure the ready escape of the products of combustion from the combustion chamber in the event of no draft, back draft, or stoppage beyond the draft hood; (b) prevent a back draft from entering the combustion chamber of the appliance; and (c) neutralize the effect of stack action of the chimney or gas vent upon the operation of the appliance.

draft regulator. A device which functions to maintain a desired draft in the appliance by automatically reducing the draft to the desired value.

draft stabilizer. A device to maintain a constant draft in a fuel burning device.

drag. A force which "slows down" the motion of wind turbine blades, or actually causes motion and power to be produced by drag-type wind machines. Aerodynamic force on a surface which is parallel to the airflow that is producing that force.

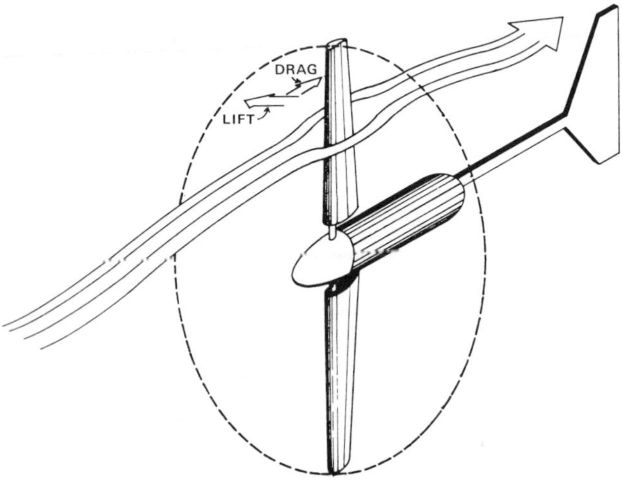

Courtesy DOE

Illustration of drag on wind-energy conversion system.

drag-type devices. Devices, such as the Savonius-type rotor, that are actuated by aerodynamic drag in a windstream. Device which extracts energy from the wind by using primarily drag forces.

drain. A pipe carrying wastewater in a building drainage system.

drainage system. The piping system within the building or premises to the point of disposal to a public sewer or disposal system.

drainback. A closed-loop water-heating system in which the heat transfer fluid drains into a holding tank when the pump is turned off, thereby protecting the collectors from freezing.

draindown. An open-loop water-heating system in which water from the collectors is dumped into a sump or drain when the pump is turned off, thereby preventing freeze damage.

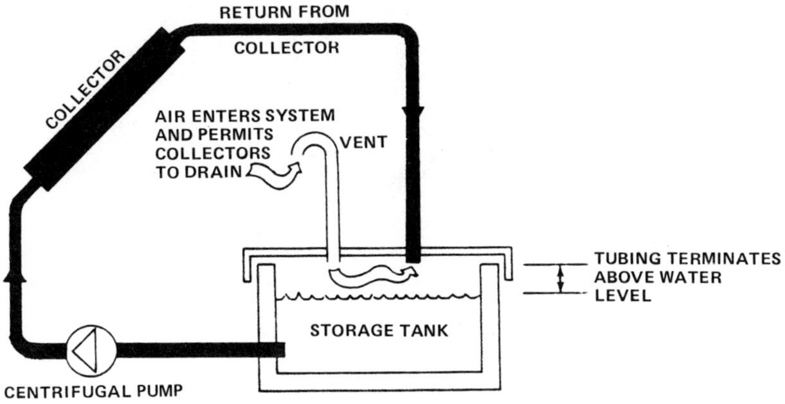

Courtesy National Copper Association

Diagram of drain down solar collector system with air vent.

drape. A curtain, especially a long window curtain.

drawback. A refund of part of the tax given when tax-paid alcohol is used to produce approved products unfit for beverage purposes.

dredging. A method for deepening streams, swamps, or coastal waters by scraping and removing solids from the bottom. The resulting mud is usually deposited in marshes in a process called filling. Dredging and filling can disturb natural ecological cycles. For example, dredging can destroy oyster beds and other aquatic life; filling can destroy the feeding and breeding grounds for many fish species.

drip. A container, or segment of piping, placed at a low point in a system to collect condensate, dust, or foreign material, enabling their removal.

drive train. A wind-energy conversion subsystem which transmits torque between the rotor main shaft and the generator or other load.

drop pipe. A galvanized pipe with smooth interior walls of 2-in. (50.8-mm) diameter which provides the tube in which the pump rod moves. The drop pipe fits onto the lowest element of the water pumping windmill, which is the cylinder.

drumwall. A passive system which uses water-filled drums for heat storage and a hinged, insulating wall to prevent nighttime heat loss. (See diagram p. 114.)

dry atmosphere. An atmosphere in which condensation of water vapor has not or is not taking place and therefore where the thermodynamics of transition between vapor, liquid, and ice phases of water may be ignored.

dry-bulb temperature. The temperature registered by the dry-bulb thermometer of a psychrometer. It is identical with the temperature of the air.

dry cell. A primary cell which does away with the liquid electrolyte so that it may be used in any position.

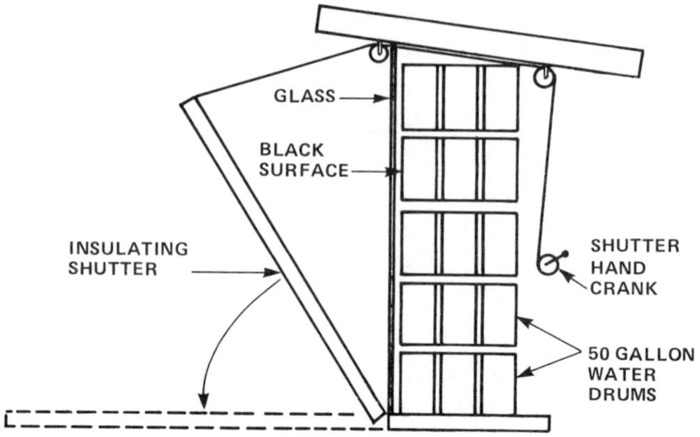

Courtesy SERI

Water-filled drums inside glass wall store heat.

dry gas. A natural gas consisting principally of methane and ethane and devoid of the heavier hydrocarbons; a gas which does not contain fractions that may easily condense under normal atmospheric conditions; also applied to gas that has been produced and from which liquid components have been removed.

dry limestone process. A method of controlling air pollution caused by sulfur oxides. The polluted gases are exposed to limestone which combines with oxides of sulfur to form manageable residues.

dry milling. A process of separating various components of grains, such as germ, bran, and starch without using water.

dry return. A condensate line in a steam-heating system carrying both water and air, usually located above the boiler waterline.

dry steam. Steam containing no moisture—either saturated or superheated steam; an energy source obtained when hot water boils in an underground reservoir, steam rises, some of it condensing on surrounding rock and dry steam approaches the surface to be tapped and used in a turbine.

dry steam energy system. A geothermal energy source; a vapor-dominated hydrothermal convective system generally having a temperature greater than 300°F (149°C); a natural steam field producing superheated steam. The major geothermal power plants in the United States are located at a natural steam field in Northern California known as "the Geysers."

dry ton. Biomass dried to a constant weight of 2000 pounds in a ventilated oven at a temperature above the boiling point of water, generally 212 ± 3.6°F (103 ± 2°C).

DTL. Direct-to-load photovoltaic power generation.

dual price market policy. Possibility that separate price levels will exist for general photovoltaic array markets and for federal government procurement market over time.

dual vent. In plumbing, a vent serving two fixtures and connected at the junction of the fixture drains from the two fixtures.

duct. A passageway made of sheet metal or other suitable material, not necessarily leak-tight, used for conveying air or other gas at low pressures.

duct board. An insulation material consisting of a 1- or 2-inch (25.4- to 50.8-mm) plate consisting of fiberglass faced on one side with aluminum foil and scrim cloth and faced on the other side with white paper.

duct fan. An axial flow fan mounted in, or intended for mounting in, a section of duct. See **tubeaxial fan; vaneaxial fan.**

dump. A land site where solid waste is disposed of in a manner that does not protect the environment.

Dura-Shade®. A window-insulating product of the Duracote Corporation which is a thin vinyl sheet with aluminum foil on one side.

dust. Fine-grain particulate matter that is capable of being suspended in air.

dustfall jar. An open-mouthed container used to collect large particles that fall out of the air. The particles are measured and analyzed.

duty cycle. The duration and periodicity of operation of a device.

dynamic head. The pressure equivalent of the velocity of a fluid. Customary units of "head" are usually taken in feet or inches of H_2O.

dynamo. A machine for converting mechanical energy into electrical energy by magnetoelectric induction. A dynamo may also be used as a motor.

Dyna-ship®. A clipper-type, wind-powered ship, designed by Wilhelm Prolss, which has, in effect, a continuous airfoil from the top of the mast to the bottom. The angle of the sails is changed by turning the whole mast hydraulically.

dyne. Unit of force; the force that will give a mass of 1 gram (g) an acceleration of 1 centimeter per second squared (cm/sec^2).

dystrophic lakes. Lakes between eutrophic and swamp stages of aging. Such lakes are shallow and have high humus content, high organic matter content, low nutrient availability, and high BOD.

E

earth berm. A mound of earth either abutting a house wall to help stabilize the temperature inside the house, or positioned to deflect wind from the house.

Courtesy DOE/SERI

Illustration of contemporary home with walls covered with earth.

earth current. A light electric current which appears to traverse the earth's surface but which in reality exists in a wire grounded at two ends, due to small potential differences between the two points at which the wire is gounded.

earth-sheltered buildings. An earth-sheltered building is one which is banked on one or more sides with earth, or built partially or entirely underground. Below the surface, earth temperatures are moderate and relatively constant the year-round. The effect of earth sheltering on the building is to reduce air infiltration, provide protection from winds, and reduce the temperature difference between the inside and outside surfaces of the walls. Keeping the interior of the house more temperate results in sharply reduced energy costs for heating and cooling. The construction of such a building, however, is a more demanding task than is the case for an aboveground structure, and usually requires the services of a professional, such as an architect or engineer. The most important factors to consider in planning for an earth-sheltered building include the following: the building site, the construction materials, the waterproofing system, the local building codes, and resale market.

easement. A form of private agreement with the potential to protect solar access. Easements are interests in property, which can be bought and sold like property itself. A common example is the utility easement.

ebb current. The movement of the tidal current away from shore or down a tidal stream.

eccentric. A device for converting continuous circular motion into reciprocating rectilinear motion, consisting of a disk mounted out of center on a driving shaft, and surrounded by a collar or strap connected with a rod. Rotation of the driving shaft gives the rod a back-and-forth motion.

ecological food chain. The stages (species) through which food energy is consumed, from sun to green plants, to animal 1, to animal 2, . . . , and finally to low-energy products, e.g., water and carbon dioxide.

ecological impact. The total effect of an environmental change, either natural or man-made, on the ecology of the area.

ecological niche. A set of conditions (temperature, physical location, point in food chain, etc.) which suits the growth requirements for a particular organism and gives it a unique survivability.

ecology. The interrelationships of living things to one another and to their environment or the study of such interrelationships.

economic optimal solar system cost. The cost of the solar system based on the collector area that gives the greatest annual savings due to the solar system.

economies of scale. A production process is characterized by economies of scale when costs per unit output fall as plant size is increased.

economizer. A heat exchanger for recovering heat from flue gases and using it to heat feedwater or combustion air.

economizer cycle. A method of operating a ventilation system to reduce refrigeration load. Whenever outdoor air conditions are more favorable (lower heat content) than return air conditions, outdoor air quantity is increased.

ecosystem. The interacting system of a biological community and its nonliving environment.

ECUBE. A building energy analysis computer model developed by the American Gas Association.

eddy. A circular movement of water or air. Eddies may be formed where currents pass obstructions or between two adjacent currents flowing counter to each other.

eddy loss. Energy lost by eddies as distinct from that lost by friction.

edge-defined film-fed growth. Method of growing thin "ribbons" of single crystal material by pulling through a die. (See diagram p. 118.)

edge retaining system. The metal channel which holds in place the edges of the various layers of a solar collector panel.

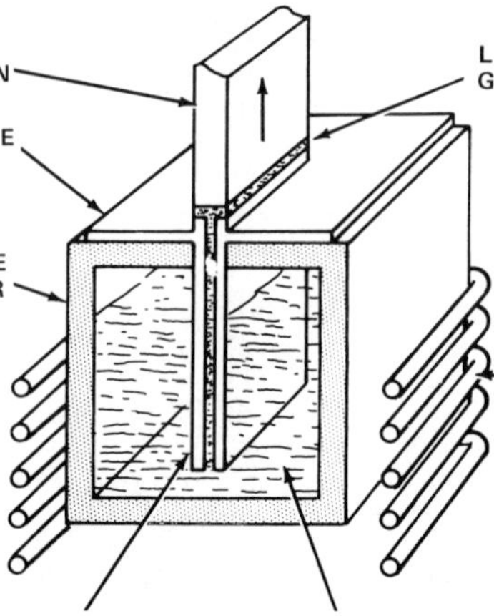

Courtesy SERI

Process diagram of edge-defined film-fed silicon ribbon growth technique.

effective capacity. The maximum load which a system is capable of carrying under existing service conditions.

effective carrying capacity. The power capacity that can be reliably furnished from storage.

effective horsepower. The amount of useful energy that can be delivered by an engine.

effectiveness. The ratio of actual heat transfer in a heat exchanger to the maximum possible heat transfer.

effective opening. The smallest cross-sectional area in the water-supply discharge in terms of the diameter of the circle of a round opening or, if the opening is not round, the diameter of a circle having that cross-sectional area.

effective temperature. An arbitrary index which combines into a single value the effect of temperature, humidity, and air movement on the sensation of warmth or cold felt by the human body. The numerical value is that of the temperature of still, saturated air which would induce an identical sensation.

efficacy of fixtures. Ratio of usable light to energy input for a lighting fixture or system (expressed in terms of lumens per watt).

efficiency. That percentage of the total energy content of a power plant's fuel which is converted into electricity, the remaining energy being lost to the environment as heat. The efficiency of an energy conversion is the ratio of the useful work or energy output to the work or energy input.

$$E = \frac{\text{Work or Energy Out}}{\text{Work or Energy In}} \times 100 \text{ percent}$$

Ratio of the heat delivered by a system to the solar energy supplied to it. Efficiency can be measured with respect to the system as a whole, or the focus can be on a particular component's performance. For example, when one-half of the incoming sunlight is passed along as heat, a collector is said to be 50 percent efficient; if one-half of that heat is then lost from the pipes and storage tank, system efficiency would be 25 percent.

efficiency of a solar cell. The ratio of the electrical power output of a solar cell to the solar power that it intercepts. For example, a solar cell 3 inches (76.2 mm) in diameter intercepts about 2.39 watts (W) of solar power under full sun conditions. If the electrical output of this cell is 0.34 watt (W), the efficiency will be

$$\frac{0.34}{2.39} = 0.14, \text{ or } 14 \text{ percent}$$

efflorescence. In geology, the formation of crystals by the evaporation of water from solutions brought to the surface by capillarity. See **capillary water**.

effluent. A discharge of pollutants into the environment, partially or completely treated or in its natural state. Generally used in regard to discharges into waters.

ejector. A device in which a high-velocity jet acts to entrain mechanically a second fluid to withdraw it from some region of like pressure and to deliver the mixture, with low turbulence, to a region of higher pressure.

Ekman spiral. An idealized mathematical description of the wind distribution in the planetary boundary layer, within which the earth's surface has an appreciable effect on the air motion. The model is simplified by assuming that within this layer the eddy viscosity and density are constant, the motion is horizontal and steady, the isobars are straight and parallel, and the geostrophic wind is constant with height.

elasticity. The property or quality of being elastic, as when an elastic body returns to its original form or condition after a displacing force is removed. A measure of the percentage impact of a change in one economic variable on another.

elbow (ell). A fitting that makes an angle between adjacent pipes. The angle is 90° unless another angle is specified. See **street elbow**.

electrical conductivity. The transfer of electricity within a substance from points of higher potential to points of lower potential. Conductivity is usually measured by its inverse—resistivity. The unit of conductivity is the mho.

electrical energy. The energy of moving electrons.

electrical heat. When a current flows in a circuit which contains resistance, heat is produced and the resistance and conductors of the circuit are raised in temperature.

electrical system. A system in which all the conductors and apparatus are electrically connected to a common source of electromotive force.

electric charge. A property of matter resulting from an imbalance between the number of protons and the number of electrons in a given piece of matter. The electron has a negative charge; the proton, a positive charge. Like charges repel each other; unlike charges attract.

electric circuit. Path followed by electrons from a power source (generator or battery) through an external line, including using devices, and returning through another line to the source.

electric current. A flow of electrons; electricity.

electric energy. The energy associated with electric charges and their movements. Measured in watt-hours (Wh) or kilowatt-hours (kWh). One watt-hour equals 860 calories (cal). Available heat in electricity; 1 kWh equals 3412.97 British thermal units (Btu).

electric furnace. A heater where forced air is blown past electrically heated coils. The air is heated by convection from the coils' surface.

electric generator. A source of electricity, especially one that transforms mechanical or heat energy into electric energy.

electricity. The general name for all phenomena that arise out of the flow or accumulation of electrons in matter. The unit of quantity is the coulomb (C); that of pressure, the volt (V); and that of flow, the ampere (A).

Electric Power Research Institute (EPRI). Located in Palo Alto, CA; the principal research arm of U.S. electric utilities. It has a sizable solar program.

electric power transmission. The transmission of energy through conductors is a flow of electrons under pressure. The flow is measured in amperes (A), and the pressure or electrical potential is measured in volts (V). The amount of energy transmitted is proportional to the product of the flow and the pressure.

electric utility company fuel costs. The total cost of fuel consumed divided by its total Btu content, multiplied by 1 million (10^6).

electrochemical. Chemical action employing a current of electricity to cause or to sustain the action.

electrochemical cell (battery). A battery consisting of a combination of two or more cells, each cell containing two conducting electrodes, one positive and the other negative, and made of unlike materials (metals or their salts). These electrodes are immersed in chemical solutions called electrolytes, whose function is mainly to carry positive ions from the negative pole to the positive. To neutralize the charge that results, a current of electrons will then flow through external conducting wires from one pole to the other. This process can be made to give up energy if sent through a proper energy-converting device such as an electric motor.

electrode. Reactive materials, such as metals and metal oxides, attached to grids that conduct electricity.

electrodialysis. A process employing an electrical current and an arrangement of permeable membranes for separating soluble minerals from water.

electrolysis. The use of an electric current to produce hydrogen and oxygen from water.

electrolyte. A substance that conducts electricity by the transfer of ions. These include aqueous solutions of salts, acids, or bases, and some molten salts.

electrolytic cell. An assembly consisting of a vessel, electrodes, and an electrolyte in which electrolysis can be carried out.

electromagnet. A core of magnetic metal (as soft iron) that is surrounded wholly or in part by a coil or wire that is magnetized when a electric current is passed through the wire, and that retains its power of attraction only while the current is flowing.

electromagnetic radiation. The wide range of radiant energy in the universe that makes up the total electromagnetic spectrum.

electromagnetic spectrum. The entire range of electrical energy extending from the extremely long rays of radio and electricity at one end to the extremely short x-rays at the other.

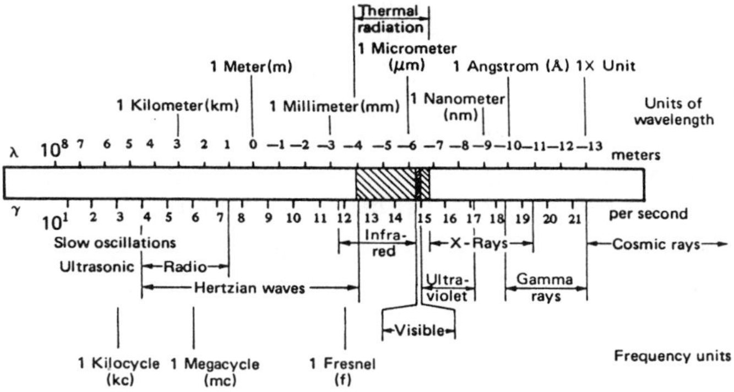

Courtesy F. Kreith and J. F. Kreider, Principles of Solar Engineering, *Hemisphere Publishing Corp.*
Electromagnetic wave spectrum diagram.

electromagnetism. Every electric current generates a magnetic field which is in a plane perpendicular to the current. The strength of the field is proportional to the current and, in the case of a long, straight wire is inversely proportional to the distance from the wire. This principle is important in magnetic prospecting insofar as it forms the basis for certain types of geomagnetic instruments.

electromotive force. A measure of the intensity of electrical energy needed to produce a current in a circuit. The practical unit is the volt (V).

electron. An elementary particle with a rest mass of 9.1×10^{-28} grams (g), bearing a negative electric charge. Electrons orbit the atomic nucleus; their transfer or rearrangement between atoms underlies all chemical reactions.

electronic cleaner. An air cleaner in which matter in the air stream is electrically charged, then attracted to surfaces oppositely charged.

electron volt (eV). The amount of kinetic energy gained by an electron when it is accelerated through an electric potential of 1 volt (V). It is equivalent to 1.603×10^{-12} erg.

electroplating. The process of putting a metallic coating or plating on a base material, which is usually metal or plastic, by means of electrodeposition. That is, the metallic plating is deposited on the base by passing an electric current through a plating bath in which are dipped the objects to be plated.

electrostatic precipitator. An air-pollution control device that removes particulate matter by imparting an electrical charge to particles in a gas stream for mechanical collection on an electrode.

element. A substance consisting entirely of atoms of the same atomic number.

Emergency Blanket®. A window-insulating product of King-Seeley Thermos Co.

which is a low-cost, compactly folded Mylar film on which aluminum has been vacuum deposited.

emission. The ability to radiate heat in the form of long-wave radiation.

emission factor. The average amount of a pollutant emitted from each type of polluting source in relation to a specific amount of material processed. For example, an emission factor for a blast furnace (used to make iron) would be a number of pounds of particulates per ton of raw materials.

emission inventory. A list of air pollutants emitted into a community's atmosphere, in amounts (usually tons) per day, by type of source. The emission inventory is basic to the establishment of emission standards.

emission standard. The maximum amount of a pollutant legally permitted to be discharged from a single source, either mobile or stationary.

emissivity. The relative power of a surface to emit heat by radiation, or the ratio of the radiant energy emitted by a surface to that emitted by a blackbody—considered to have almost perfect heat absorption and therefore very low emissivity—when the given surface and the blackbody are at the same temperature.

emittance. The ratio of the radiant energy emitted from a surface at a given temperature to the radiant energy that would be emitted by a perfect blackbody at that same temperature. It should specify surface conditions and then refer to measured quantities. (See **emissivity**.) May be used to refer to angular [watts/m^2/steradian] or spectral [watts/m^2/micrometer] properties as well as total hemispherical radiation.

employment opportunities. Jobs and careers in solar energy can be found in both new and traditional fields. For instance, the skills found throughout the entire building industry, including those of the electrical and plumbing trades, have a role to play in solar applications. Architecture, engineering, research, and information science are other professions that play vital roles in the emerging solar industry. In addition to these traditional work categories, solar will require new specialists such as solar technicians, solar installers, and solar designers. Many institutions offer programs in which solar is treated as an area of specialization within a traditional academic or vocational discipline. However, solar-specific training programs are becoming more common.

endergonic reaction. Chemical reaction with absorption of Gibbs free energy.

endothermic. Heat absorbing. An endothermic reaction is one in which heat must be supplied to further the reaction.

end-use demand. The amount of energy used by final consumers, often given by type of end use. Sometimes measured in primary energy equivalents, which can include conversion losses.

energy. The capacity to do work. It takes such forms as potential, kinetic, heat, chemical, electrical, and radiant energy. Potential energy arises by virtue of the position or configuration of matter. Kinetic energy is energy of motion. Heat energy is the kinetic energy of molecules. Chemical energy arises out of the capacity of atoms to evolve heat as they combine or separate. Electrical energy arises out of the capacity of moving electrons to evolve heat, electromagnetic radiation, and magnetic fields. Radiant energy is energy in transit through space; it is emitted by electrons as they change orbit and by atomic nuclei during fission and fusion; on striking matter, such energy appears ultimately as heat. Only radiant heat can exist alone; all other forms require the presence of matter. Some forms of energy can be converted into other forms, and all forms are ultimately converted into heat. Energy is measured in ergs.

energy audit. An accounting of the forms of energy used during a designated period, such as monthly.

energy consumption. The amount of energy consumed in the form in which it is acquired by the user. The term excludes electrical generation and distribution losses. Also called net energy consumption.

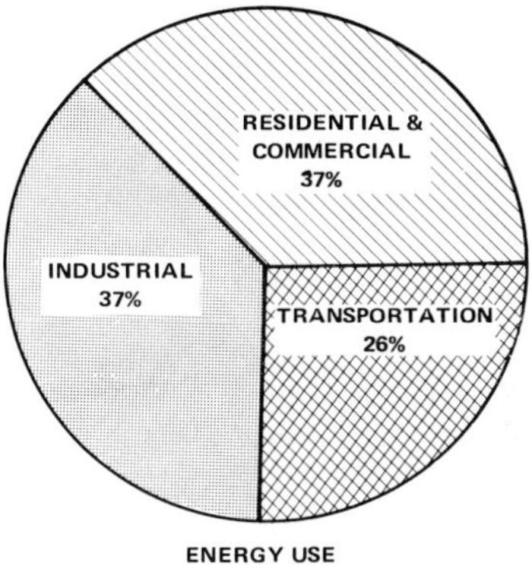

Courtesy DOE

Pie chart illustrating energy consumption by sector.

energy conversion. Transformations of energy from one form to another.

energy crops. Includes such agricultural crops as corn and sugarcane; also non-food crops such as poplar trees. See **biomass**.

energy density. A ratio of energy per pound, usually used to compare different batteries.

energy dissipation. The transformation of mechanical energy into heat energy.

energy efficiency ratio. The amount of useful work or product divided by the fuel or energy input. Example: in electrical generation, it is the amount of electricity produced per unit of fuel consumed; for an air conditioner, it is the amount of cooling provided per unit of electricity used.

energy farm. A concept involving the farming of selected plants for the purpose of providing biomass that can be used as fuel or converted into other fuels or energy products.

energy gap. Range of energies in a semiconductor forbidden to electrons; photons are readily absorbed if their energy exceeds the gap. The electron states below the gap (in energy) constitute the valence band (normally full); above the gap is the conduction band, normally empty, into which valence band electrons can be excited by light. Impurities and atomic displacements produce allowed states in the otherwise forbidden gap.

energy intensity. Describes the relative extent to which energy is required for a process, e.g., manufacturing, farming, or a society.

energy pattern factor. The ratio of the mean cube of the wind speed to the cube of the mean wind speed, the cube of the cube factor.

energy payback time. The time required for any energy-producing system or device to produce as much useful energy as was consumed in its manufacture and construction.

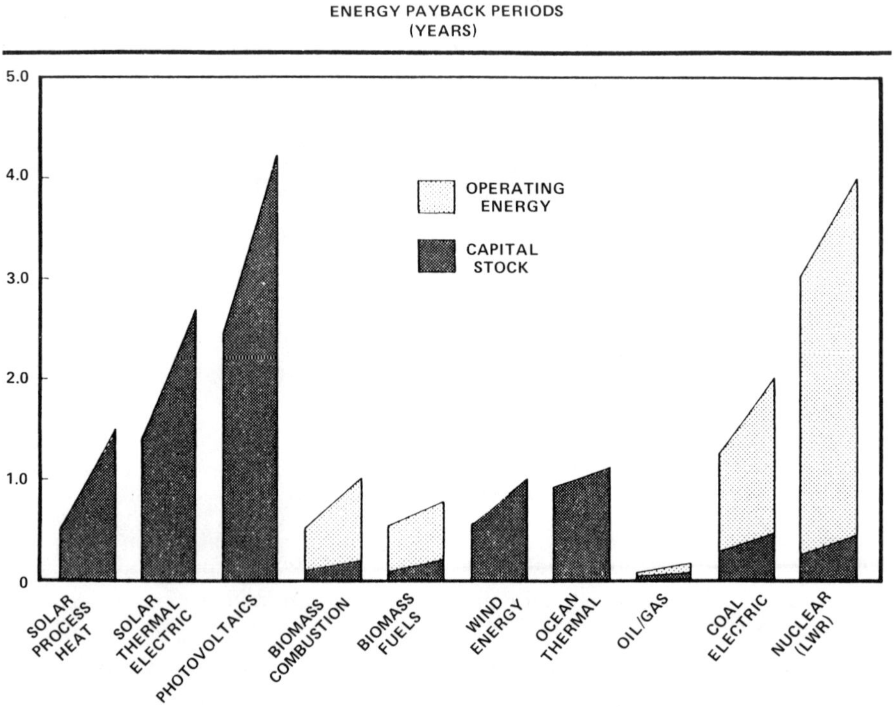

Courtesy Mitre Corporation

Energy payback periods in years are shown for solar and conventional technologies.

energy plantations. The growing of plant material for its rich fuel value; a renewable source of energy-rich, fixed carbon produced by photosynthesis.

energy quality. A loose, qualitative term describing the usefulness of an energy source and normally related to temperature (as of heat energy), energy density, ease of conversion, and entropy.

energy recovery factor. The ratio of the average load on a Wind-Energy Conversion System (WECS) for the period of time considered to the power which would have been reproduced by a WECS with a specified constant power coefficient.

energy requirement. The total yearly energy used by a building to maintain the selected inside design conditions under the dynamic impact of a typical year's climate. It includes raw fossil fuel consumed in the building and all electricity used for lighting and power. Efficiencies of utilization are applied, and all energy is expressed in the common unit of Btu.

energy reserves. Refers to the bank of natural resources such as natural gas, natural gas liquids, petroleum, coal, lignite, and energy available from water power and solar and geothermal energy.

energy rose. A diagram which presents wind-energy measurements from a site analysis in relation to the direction from which the wind occurs at the site. See **wind rose**.

Energy Security Act (P.L. 96-294). June 30, 1980 legislation authorizing a U.S. biomass and alcohol fuel program; established independent Office of Alcohol Fuels within DOE, and authorized program including loan guarantees, price guarantees, and purchase agreements with producers of fuel alcohol.

energy sharing. Collecting solar energy on one building or portion of a building and distributing it to other areas which have poor solar access.

energy storage. The ability to convert energy into other forms, such as heat or a chemical reaction, so that it can be retrieved for later use; also the development, design, construction, and operation of advanced devices for storing energy until needed. The technology includes devices such as batteries, pumped storage for hydroelectric generation, flywheels, and compressed gas.

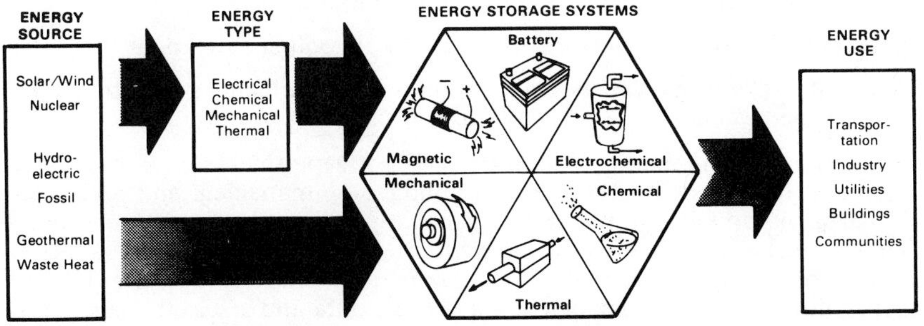

Courtesy DOE

Block diagram of energy storage concepts.

energy supply. The total amount of primary energy resources used. For solar sources, energy supply data are often expressed in fossil equivalents. See **end-use demand**.

engine. A device for turning energy of different kinds into kinetic energy, usually with the intention of putting it to work. An electric engine or motor uses the power of electric current to create a magnetic field. An internal-combustion engine works by converting chemical energy in gasoline or diesel-fuel molecules into heat which is then used to expand gases in a cylinder. The expanded gases drive a piston (as in a steam engine); the piston, in turn, drives a crankshaft.

enrichment. The addition of nitrogen, phosphorus, and carbon compounds or other nutrients into a lake or other waterway that greatly increases the growth potential for algae and other aquatic plants. Most frequently, enrichment results from the inflow of sewage effluent or from agricultural runoff.

ensilage. Immature green forage crops and grains which are preserved by alcohol formed by an anaerobic fermentation process.

enthalpy. For the purpose of air conditioning, enthalpy is the total heat content of air above a datum, usually expressed in units of Btu per pound. It is the sum of sensible and latent heat and ignores internal energy changes due to pressure change.

entrainer. Azeotropic distillation is accomplished by the addition of a compound called an azeotrope former, entrainer, additive, or extractive solvent. When the additive used is taken off as part of the distillate, it is known as the entrainer. One of the methods of getting anhydrous alcohol is to use benzene as an entrainer. The benzene, water, and alcohol form a ternary azeotrope which is distilled out the top of the still. The benzene is fully recovered and returned to the azeotropic column to be used again. Only water and anhydrous alcohol are withdrawn from the distillation columns.

entropy. A measure of the unavailable energy in a system; energy that cannot be converted into another form of energy. The change in entropy, dS, of any system is equal to the increment of heat added divided by its absolute temperature, so that:

$$dS = \frac{dQ \text{ (joules)}}{T \text{ (kelvin)}}$$

The total entropy of the universe is continually increasing.

envelope-dominated. A building with no internal-cooling-only zones.

environment. The sum of all external conditions and influences affecting the life, the development, and, ultimately, the survival of an organism.

environmental development plan (EDP). The Department of Energy (DOE) management document for the planning, budgeting, managing, and reviewing of the broad environmental implications of each energy technology alternative for each major DOE research and development program. The EDP is designed to identify environmental issues, problems, and concerns as early as possible during the program's development; to analyze the available data and assess the current state of knowledge related to each issue, problem, and concern; to set forth strategies to resolve these; to set forth the processes by which the public is involved in identification and resolution of these issues, problems, and concerns; and to designate significant milestones for resolution of these issues, problems, and concerns. The timing of the EDP's milestones reflects the sequencing of the technology development. EDPs, once complete, are made available to the public.

environmental impact assessment (EIA). A report, prepared by the Department of Energy (DOE) program office, which evaluates the environmental impacts of proposed DOE actions to ensure that environmental values are considered at the earliest meaningful point in the decisionmaking process and which, based upon the evaluation, determines whether or not an environmental impact statement should be prepared. The EIA is intended to be a brief, factual, and objective document describing the proposed action, the environment which may be impacted, the potential environmental impacts during construction, operation, and site restoration, potential conflicts with federal, state, regional, or local plans, and the environmental implications of alternatives.

environmental impact statement (EIS). A report prepared at the earliest meaningful point in the decisionmaking process which analyzes the anticipated environmental impacts of proposed Department of Energy (DOE) actions and of reasonably available alternatives, and which reflects responsible public and governmental views and concerns. An EIS is prepared in response to plans in the program's EDP or after the review of an EIA which identifies potentially significant impacts. The EIS goes through a specific preparation process involving agency and public review. The EIS goes through four steps during its preparation. The preliminary draft is reviewed within DOE, the draft is distributed to the public for review and comment, the preliminary final incorporating comments submitted to DOE in response to the draft is reviewed within DOE, and the final EIS is issued reflecting the agency's final review and deliberations. This final EIS is then distributed to the public. Except in special cases, no DOE action subject to EIS preparation can be taken sooner than 30 days after the final EIS has been issued. Required by the National Environmental Policy Act (NEPA), Section 102 (2) (c).

enzyme. A protein biocatalyst produced by living cells of plants and animals. Enzymes act as catalysts for the chemical reactions of biological processes by increasing the chemical reaction time of the substrate.

eolation. The process by which wind modifies land surfaces, both directly by transportation of dust and sand, and by the work of sandblasts, and indirectly by wave action on shores.

eolian. Of, relating to, formed by, or deposited from the wind or currents of air. Eolian was formerly spelled aeolian.

equation of time. A correction in the conversion of standard time to solar time which takes into account perturbations in the earth's orbit and rate of rotation.

equilibrium. When using a column still in alcohol fuel production, it is desirable to have the column still in equilibrium. In an enclosed system, equilibrium is achieved when the interchange of liquid-to-vapor and vapor-to-liquid reaches the point where the number of molecules that leaves the liquid as vapor is equal to the number returning as liquid from the vapor.

A distillation column, however, is not an enclosed system, since new material is constantly being introduced, and the purified product being taken out. Therefore, equilibrium can only be achieved when the phases, or portions of the solution, in the column are changing at the same rate. (Each plate in a perforated plate column, for example, would constitute a phase.)

Since the ratio of the alcohol/water composition determines the boiling-point temperature, and each plate will have its own ratio of this mixture, each phase should therefore have its respective boiling point. In order to get the desired purity of alcohol, these ratios of composition and temperature must be obtained. When each plate has its correct ratio and temperature, the column will be in equilibrium, and the desired purity produced.

equinox. Either of the two times each year when the sun crosses the equator and night and day are everywhere of equal length. The spring equinox occurs about March 21 and the fall equinox about September 23.

equity capital. That portion of the total debt of a corporation in which stock is given in return for invested capital.

equivalent direct radiation. Heat expressed in terms of a square foot of steam radiator surface emitting 240 Btu per hour; in other words, Btu per hour divided by 240.

equivalent evaporation. The quantity of water which would be evaporated by a given apparatus if the water is received by the apparatus at 212°F (100°C) and vaporized at that temperature under atmospheric pressure. It is expressed in pounds per hour.

equivalent length. The resistance of a duct or pipe elbow, valve, damper, orifice, bend, fitting, or other obstruction to flow expressed in the number of feet of straight duct or pipe of the same diameter which would have the same resistance.

equivalent sphere illumination. That illumination which would fall upon a tank covered by an imaginary transparent hemisphere which passes light of the same intensity through each unit area.

equivalent temperature. Also British equivalent temperature (BET). A composite mean radiant temperature (MRT), and air temperature, also defined as the mean temperature of the environment effective in controlling the rate of sensible heat loss from a blackbody in still air when the surface temperature and size of the blackbody are comparable to those of the human body. Where the enclosure surface (MRT) and air temperatures are equal, this temperature is also the BET; when not equal the BET is that temperature at which a body with an 80°F (27°C) surface temperature will lose sensible heat at the same rate as in the given environment.

erosion. The physical and chemical processes by which the earth or rock material is loosened or dissolved and removed from any part of the earth's surface. It includes the processes of weathering.

escalation rate. A number which defines the annual increase in monetary value of a specified quantity.

ester. A compound formed by the reaction of an alcohol and an acid, usually with elimination of water.

estuaries. Areas where the freshwater meets salt water. For example, bays, mouths of rivers, salt marshes, and lagoons. Estuaries are delicate ecosystems; they serve as nurseries, spawning and feeding grounds for a large group of marine life and provide shelter and food for birds and wildlife.

ethanol (ethyl alcohol, C_2H_5OH). The alcohol product of fermentation that is used in alcoholic beverages and for industrial purposes; blended with gasoline to make gasohol; also known as grain alcohol. It is miscible with water, methyl alcohol, ether, chloroform, and acetone. Pure ethanol is a colorless liquid (boiling point, 172.98°F [78.32°C] at a pressure of 760 mm) with an agreeable odor and a burning taste. It is very hygroscopic and is miscible with water in all proportions. Ethanol forms with water a constant-boiling-point mixture [azeotrope (q.v.)], which boils at 172.65°F (78.14°C). See also **specially denatured alcohol**.

ethylene. Colorless, flammable, unsaturated gas with characteristic sweet odor and taste. It is contained in illuminating gas.

eupatheoscope. A black hollow copper cylinder heated electrically to the surface temperature of a human being. The heat loss from the cylinder in a given environment is equivalent to that lost by a human being.

Euphorbia lathyris. A weed that can be grown on semiarid land to produce biomass-generated synthetic petroleum products. A member of the poinsetta family.

Courtesy DOE

Euphorbia lathyris is a weed that can be grown in semi-arid land and be used to produce a synthetic fuel.

eutectic. A mixture of substances which has a melting point lower than that of any mixture of the same substances in other proportions.

eutectic salts. A mixture of two salts with a melting point lower than that of any other combination of the same components. Attractive in solar applications at both low and high temperatures because of the large amount of heat which can be stored and recovered in the phase change in a relatively small volume. At a given temperature, salts melt, absorbing large amounts of heat which will be released as the salts freeze; for example, Glauber's salts. The melt–freeze temperatures vary with different salts, and some occur at convenient temperatures for thermal storage such as in the range of 80° to 120°F (27° to 49°C).

eutrophication. The normally slow aging process by which a lake evolves into a bog or marsh and ultimately assumes a completely terrestrial state and disappears. During eutrophication the lake becomes so rich in nutritive compounds, especially nitrogen and phosphorus, that algae and other microscopic plant life become superabundant, thereby "choking" the lake, and causing it eventually to dry up. Eutrophication may be accelerated by many human activities.

eutrophic lakes. Shallow lakes, weed-choked at the edges and very rich in nutrients. The water is characterized by large amounts of algae, low water transparency, low dissolved oxygen, and high BOD.

Eva-Adam. A proposed method of transporting energy wherein methane and water are catalytically reacted with a high-temperature source of energy to carbon monoxide and hydrogen. These products are then transmitted to a site where they catalytically react to form methane and water, releasing energy. The methane and water are then returned in a second pipe (a chemical heat pipe) to the source of energy to repeat the cycle.

evacuated tube collector. A collector manufactured from concentric glass tubes with an evacuated space between the tubes. They are very efficient at relatively high temperature differences [about 212°F (100°C)].

evaporate. Changing the phase of a material from liquid phase to gaseous phase; for example, water evaporates to form water vapor (steam).

evaporation. Conversion of a liquid to the vapor state by the addition of latent heat of vaporization; usually refers to vaporization into the atmosphere.

evaporation ponds. Shallow, artificial ponds where sewage sludge is pumped, permitted to dry, and either removed or buried by more sludge.

evaporative cooling. Evaporating water cools and humidifies surrounding air; house air is circulated over water as a technique to cool indoor air in dry-climate areas.

evaporative cooling system. A space-conditioning system which accomplishes space cooling by evaporating water as it is circulated through the collectors during those hours which have the appropriate conditions for cooling (usually night and early morning).

evaporative equilibrium. Utilizing a sling psychrometer, it is the point when the wet-bulb temperature (q.v.) finally reaches a stable point.

evaporator. A component of a heating and cooling system in which a working fluid undergoes a change from liquid to gas, extracting heat and producing a cooling effect.

evergreen/coniferous trees. Trees which do not shed their leaves at the end of the growing season.

excise tax (gasoline). A tax collected at the pump to support the construction and maintenance of highways. Gasohol is exempt through 1992 from the $0.04 federal excise tax and in some states from the state excise tax.

exhaust fan. A fan which sucks used air from an area and thereby causes fresh air to enter.

exothermic reaction. Chemical reaction with evolution of heat.

expansion joint. A joint whose primary purpose is not to join pipe but to absorb the longitudinal expansion in the pipeline due to heat.

expansion loop. Either a bend like the letter "U" or a coil in a line of pipe to provide for expansion and contraction.

expansion tank. In a closed-loop system, a tank that provides space for the expanded volume of the heat-transfer liquid as it is heated in the pressurized collector loop.

expansion valve. A special valve used in refrigerating systems through which the liquid refrigerant (under high pressure) is allowed to escape into a lower pressure and thus expand into a gas.

exponential growth. Growth which occurs at a rate which is a constant percentage of the whole.

external combustion engines. An engine in which the fuel is burned outside the cylinders.

external radiation. Ionizing radiation reaching the body from sources outside the body.

external receiver. A solar-energy receiver where the solar radiation is absorbed on the external surfaces.

extraterrestrial radiation. The solar radiation which would be received on a horizontal surface if there were no atmosphere.

F

fabric filters. A device for removing dust and particulate matter from industrial emissions much like a home vacuum cleaner bag. The most common use of fabric filters is the baghouse.

face cord. Wood cut 12, 16, 24 inches (0.31, 0.41, or 0.61 meters) long and stacked 4 feet (1.2 meters) high and 8 feet long (2.4 meters). This is not to be confused with a "cord."

factor of safety. Ratio of a normal working condition to the ultimate condition, such as, in strength of materials, ratio of working stress to ultimate strength. Loosely used as any allowance over and above the estimated required load or condition.

facultative. Having ability to live under severe environmental conditions.

facultative (anaerobe). A microorganism that grows equally well under aerobic and anaerobic conditions.

Fade-shield®. A glazing product of Modico Inc. which is a fairly transparent (T = 50 percent) gray-bronze sheet.

fahrenheit scale. A temperature scale in which the boiling point of water is 212°F and its freezing point 32°F. To convert fahrenheit to centigrade, subtract 32, multiply by 5, and divide the product by 9. See also **centigrade (or Celsius) scale.**

failure (structural). Any one of the following: (1) sudden, locally increased curvature, major spalling, or structural collapse; (2) inability of the structure to resist a further increase in load; (3) structural deflections under design loads that cause significant performance degradation of the component or subsystem.

fall wind. A strong, cold, downslope wind.

fan. A device which blows air in an air-heating solar system or in a forced-air heating system. A fan moves heat by transport heat flow.

fan coil. A unit consisting of a fan and a heat exchanger which transfers heat from liquid to air (or vice versa); usually located in a duct.

fantail. A propeller-type wind turbine mounted sideways on a larger wind machine (horizontal-axis-type) to keep that machine aimed into the wind.

fan velocity pressure. The pressure corresponding to the fan outlet velocity. It is the kinetic energy per unit volume of flowing air.

farm tractor fuel. Any petroleum product, exclusive of gasoline, diesel fuel, and liquefied petroleum gas, which is used for the generation of power for the operation of farm implements.

***f*-chart procedure.** A solar system performance and sizing procedure, developed by the University of Wisconsin, expressed in terms of a universal "*f*-chart" family of curves.

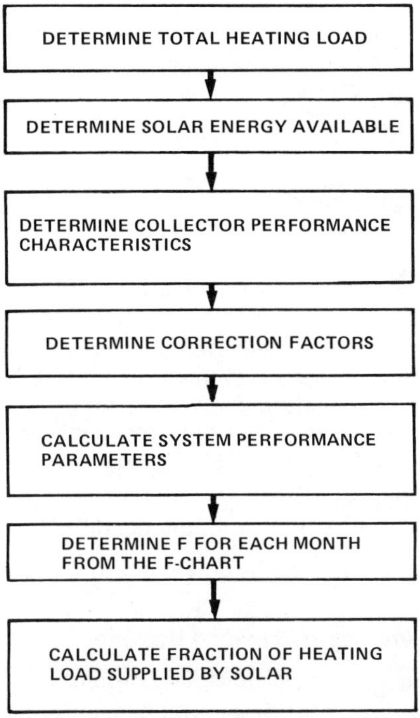

Courtesy SERI

Block diagram flow chart shows steps in calculating fraction of heating load supplied by solar.

feather. To pitch each blade of a wind-energy conversion system rotor to a zero lift condition. (Normally used as a method of wind-system shutdown.)

feathering. The pitch of the blades of a wind machine is altered so that the net lift on the blades is zero to avoid unnecessary stress from severe gusts. (See diagram p. 134.)

fecal coliform bacteria. A group of organisms common to the intestinal tracts of humans and of animals. The presence of fecal coliform bacteria in water is an indicator of pollution and of potentially dangerous bacterial contamination.

feedback. The coupling of an effect or product back to its source to continuously regulate (increase or decrease) the latter's activity.

feedlot. A relatively small, confined land area for raising cattle. Although an

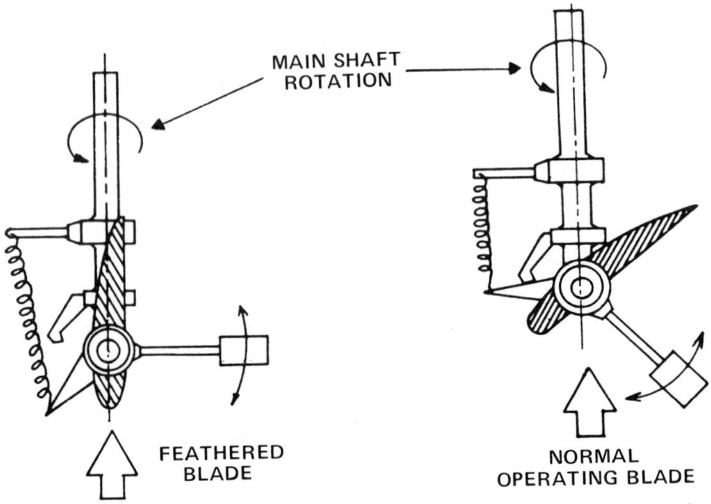

Courtesy DOE

Feathering controls move wind-energy conversion system blades to prevent system damage in high winds.

economical method of fattening beef, feedlots concentrate a large amount of animal wastes in a small area. This excrement cannot be handled by the soil as it could be if the cattle were scattered on open range. In addition, runoff from feedlots contributes excessive quantities of nitrogen, phosphorus, and potassium to nearby waterways, thus contributing to eutrophication.

feed plate position. The theoretical position in a distillation column above which enrichment occurs and below which stripping occurs.

feedstock. A raw material that can be converted to one or more end-products, for example, biomass.

fen. A low-lying land area partly covered by water.

fenestration. Arrangements, proportions, and design of windows in a building.

fermentable sugar. Sugar (usually glucose) derived from starch and cellulose that can be converted to ethanol (also known as reducing sugar or monosaccharide).

fermentation. Decomposition of organic material to alcohol, methane, etc., or organisms, especially bacteria and yeasts, usually in the absence of oxygen. (See diagram p. 135.)

fermentation ethanol. Ethyl alcohol produced from the enzymatic transformation of organic substances.

fermentation lock. A glass tube partially filled with water which allows CO_2 to escape from fermenting mash but prevents organisms from contaminating the mash. (See diagram p. 136.)

Fiberfill®. A window-insulating product of the Du Pont Co. which consists of a polyester quilting material.

Fiberglass®. A fiberglass insulating material by Owens-Corning. Typically pink in color.

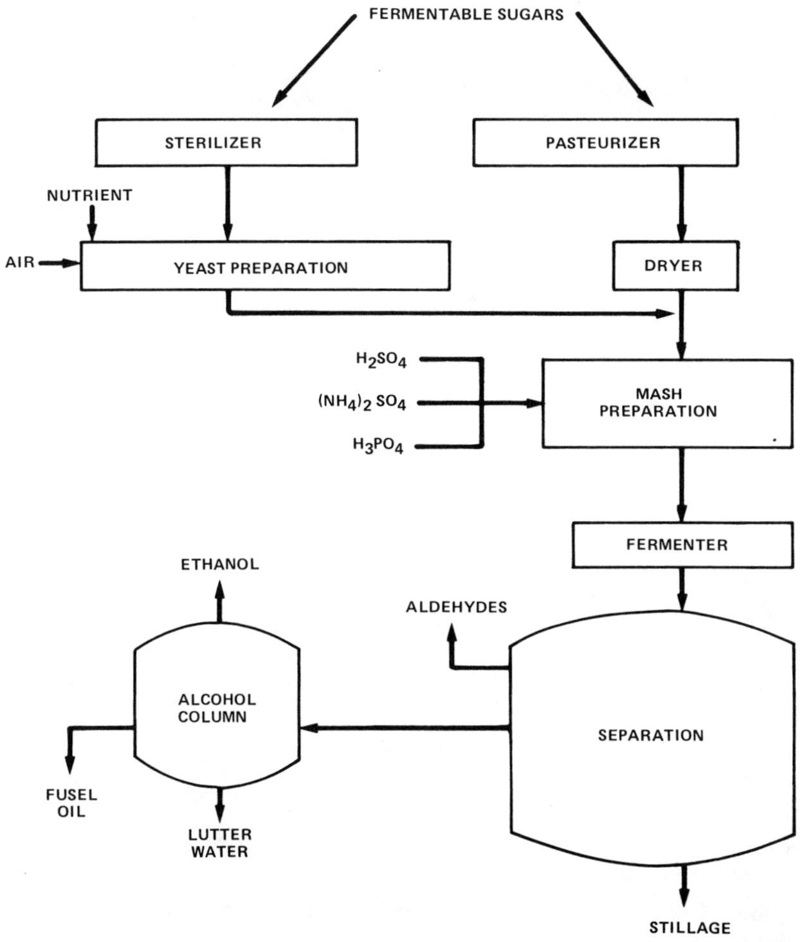

Courtesy EPRI

Block diagram of fermentation process to produce ethanol for power alcohol or gasohol.

fiberglass batting. An insulator made from loosely packed glass fibers, coming in thick layers which can be stapled to a wall or layed between ceiling rafters. Heat flows by conduction through batting since the fibers prevent convection.

fill factor. The ratio of the maximum power a photovoltaic cell can produce to the theoretical limit if both voltage and current were simultaneously at their maximum. A key characteristic in evaluating cell performance.

film coefficient. The heat transferred by convection per unit area per degree temperature difference between the surface and the fluid. Also called unit convection conductance or surface coefficient.

Filon®. A glazing product of Vistron Corporation which is a translucent sheet made of fiberglass and polyester, sometimes with a Tedlar® coating.

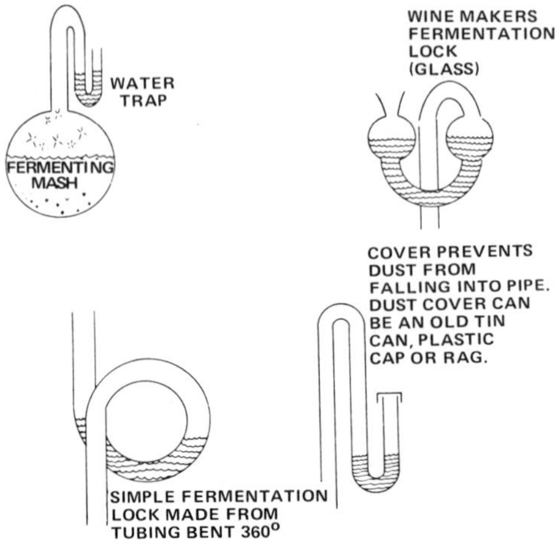

Courtesy Larry W. Carley, How to Make Your Own Alcohol Fuels, *TAB Books Inc.*

Fermentation lock prevents mash contamination by allowing CO_2 to escape while preventing air from getting to mash.

filter. A device for purifying a fluid, such as air or water, that removes suspended matter and or gases or dissolved matter.

filter cake. The compacted solid or semisolid material separated from a liquid and remaining on a filter after filtration.

filtration. In wastewater treatment, the mechanical process that removes particulate matter by separating water from solid material usually by passing it through sand.

fin. A thin sheet of metal in a heat exchanger that conducts heat to the surrounding gas or liquid.

finned tube. Heat-transfer tube or pipe with extended surface in the form of fins, discs, or ribs.

fire brick. Heat-resistant ceramic material formed into bricks and used to line the fireboxes of boilers, furnaces, or other combustion chambers.

fire classification. The National Fire Protection Association classifications are as follows: Class A fires are defined as those in ordinary solid, combustible materials such as coal, wood, rubber, textiles, paper, and rubbish. Class B fires are defined as those in flammable liquids such as fuel or lubricating oils, grease, paint, varnish, and lacquer. Class C fires are defined as those in (live) electric equipment such as oil-filled transformers, generators, motors, switch panels, circuit breakers, insulated electrical conductors, and other electrical devices.

fireplace heat exchanger. The arched chamber added to a box stove to increase efficiency.

firm power. Power intended to have assured availability to consumers to meet all, or any agreed-on portion, of their load requirements.

first cost. The amount you pay for a solar system when you first buy it; sometimes called the initial cost.

first law of thermodynamics. States that energy can neither be created nor destroyed. Also called the law of conservation of energy.

fiscal year (FY). Government's 12-month financial year, from October through September of the following calendar year; for example, FY 1982 extends from October 1981 through September 1982.

fixed charge rate. Share of initial investment cost required as level lifetime payment to recover "first-cost" investments plus annual charges (operation and maintenance, later investments, etc.). Includes tax factor, omits inflation rate.

fixed coning. A wind-energy conversion system rotor which has no articulation in coning.

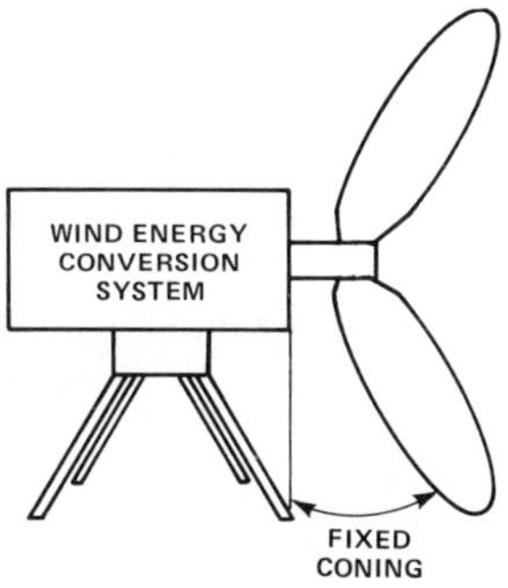

Courtesy DOE

Fixed coning provides a standard angle between wind-energy nacelle and blade of wind-energy conversion system.

fixed mirror/distributed focus (FMDF). A number of hemispherical reflectors concentrate sunlight on receivers, one of which is suspended in each bowl. The receiver is actually a steam boiler, and the steam is used to drive a conventional turbogenerator to yield electricity. The bowllike concentrators are fixed, but the receivers track the sun, swinging through a daily arc and automatically adjusting to the sun's varying position with the seasons. This dual adjustment is made possible by a two-axis tracking mount. In this concept, an energy-storage means is used, and the electrical power output is so regulated that it can interface with an existing power transmission grid.

fixture branch. In plumbing, the branch from the water-distributing pipe in the building to the fixture supply pipe.

fixture supply pipe. The pipe connecting the fixture with the fixture branch at the floor or wall line.

fixture unit. A base figure or weight in which each type of water-using fixture is assigned a number representing its maximum water demand relative to other fixtures.

flame. An ordinarily visible condition resulting from the rapid oxidation of a fuel and producing self-evident heat, light, or both.

flange. In pipework, a ring-shaped plate on the end of a pipe at right angles to the end of the pipe, and provided with holes for bolts to allow fastening the pipe to a similarly equipped adjoining pipe. The resulting joint is a flanged joint.

flange faces. Pipe flanges which have the entire face of the flange faced straight across, and use either a full face or ring gasket, are commonly employed for pressures less than 125 pounds on steam and water lines.

flapping. Wind energy conversion system blade motion in the out-of-plane direction. This motion may be centered at a flap, articulation, hinge, or distributed along the blade. Distributed flexibility.

flap valve. A check valve that uses a flap to block the liquid flow in one direction, but allows flow in the other direction.

flaring. Burning of gas for the purpose of safe disposal.

flash heating. Very rapid heating of material by exposure of small fractions to high temperature and using high flow rates.

flashing. Sheet metal used in waterproofing roof valleys or ridges. A thin impervious material placed in mortar joints and through air spaces in masonry to prevent water penetration or to provide water drainage.

flash point. The temperature at which a combustible liquid will ignite when a flame is introduced, anhydrous ethanol will flash at 51°F (11°C), 90-proof ethanol will flash at 78°F (26°C).

Fuel	Flash Point (°F)	Ignition Temperature (°F)	Flammability Limits in Air (% Vol.)	Saturation Volume (at 100°F)
Ethanol	55	793	3.3-19	15%
Methanol	52	835	6.0-36	31%
Gasoline	—45	700	1.4-7.6	68%

Courtesy U. S. National Alcohol Fuels Commission

Table shows key flash point characteristics of ethanol, methanol, and gasoline.

flash trap. The discharge from a flash trap is intermittent. This type of trap is used only if a pressure differential of 5 psi (34.5 kilopascals) or more exists between the steam supply and condensate return. Flash traps may be used with unit heaters, steam heating coils, steam lines, and other similar equipment.

flat. In the micrometeorological sense terrain may be considered flat only if topographical protrusions or slopes have no significant influence on the statistics of interest (e.g., wind speed or wind shear) throughout the region of interest (e.g., the height of a wind turbine or the dimensions of a cluster). The criteria for flatness may be a function of stability.

flat black paint. Nonglossy paint with a relatively high absorptance.

flat plate collector. A solar collection device for gathering the sun's heat, consisting of a shallow metal box with a glass or plastic transparent lid, where either air or liquid is circulated through the cavity of the box. The basic heat collection

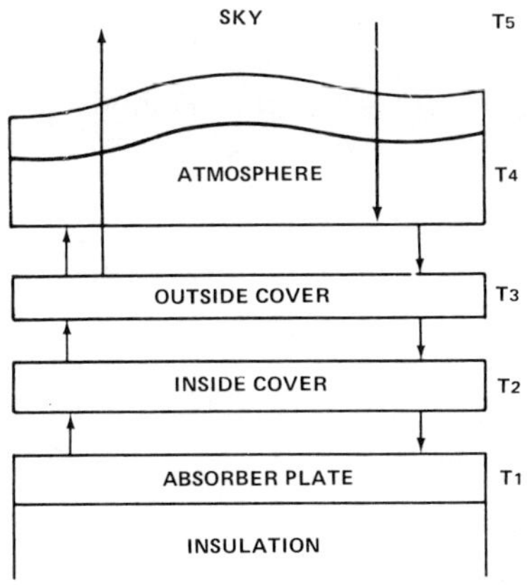

Courtesy SERI

Geometry of a solar flat-plate collector system.

device used in solar heating systems; consists of a "black" plate, insulated on the bottom and edges, and covered by one or more transparent covers. The sunlight striking the black surface is converted to heat without being concentrated.

float and thermostatic trap. A trap is used to drain condensate from heaters, steam heating coils, unit heaters and other apparatus. This combination trap is used where there is a large volume of condensate which would not permit proper operation of a thermostatic trap. Float and thermostatic traps are used in low-pressure heating systems up to a maximum of 15 psi (103.4 kilopascals).

For medium- and high-pressure systems, the same limitations as outlined for thermostatic traps apply.

float trap. The discharge from the float trap is generally continuous. This type is used for draining condensate from steam headers, steam-heating coils, and similar equipment. When a float trap is used for draining a low-pressure system, it should be equipped with a thermostatic air vent.

floc. A clump of solids formed in sewage by biological or chemical action.

flocculation. The process of separating suspended solids by chemical creation of clumps or flocs.

flooded. A situation when the liquid in a vessel or fixture rises to the flood level.

flooded system. In refrigeration, a system where the refrigerant passing the evaporator is only partially evaporated.

flood level. That level at which water begins to overflow the rim of the fixture.

flood tide. The flow, or rising toward the shore, is called the flood tide, and the falling away, ebb tide; a rising tide.

flow condition. State obtained when the heat-transfer fluid is flowing through the collector array under normal operation conditions.

flowmeter. In wastewater treatment, a meter that indicates the rate at which wastewater flows through the plant.

flow rate. The pounds of heat-transfer fluid which pass over or through an absorber plate per hour.

flue gas. A mixture of gases resulting from combustion and emerging from a chimney. Flue gas includes nitrogen oxides, carbon oxides, water vapor, and often sulfur oxides or particulates.

fluid resistance. Impediment or resistance to fluid flow. A high fluid resistance means fluid flows with difficulty, a low fluid resistance means fluid flows easily.

fluidized bed. A gasifier design in which the feedstock is kept in suspension by a rising column of gas.

flume. An inclined channel, usually of wood and often supported on a trestle, for conveying water from a distance to be used for power and transportation.

flutter. The dynamic instability of an elastic body in an airstream.

fluvial. Of or pertaining to streams and rivers; produced by stream or river action, as a fluvial plane.

flux density. The radiant flux incident per unit area.

flux (radiant). The amount of solar energy per unit of time (or power) flowing across a given area. Often imprecisely used to refer to irradiance (W/m^2) or total incident insolation (J/m^2 or kWh/m^2).

flywheel. A heavy wheel used in a rotating system to reduce surges of power input on demand.

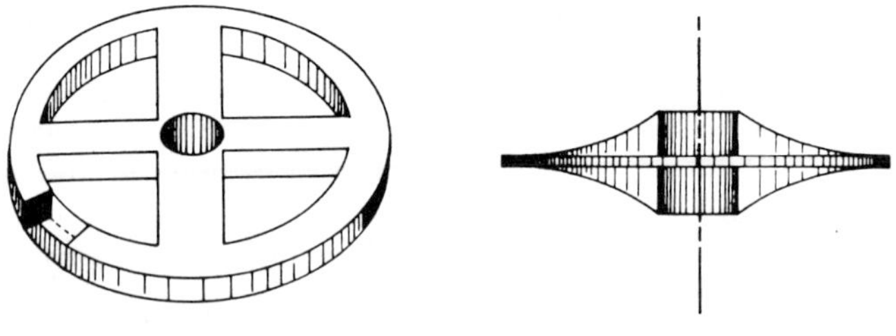

Courtesy DOE

Flywheels can be used for solar energy storage. A conventional and a super flywheel are illustrated.

flywheel effect. The damping of interior temperature fluctuations by massive construction. See **diurnal.**

Foamglass®. A product of Pittsburg-Corning which is a glass foam used for insulation.

foaming. In a boiler, the carryover of slugs of water into the piping due to dirty water.

focusing collectors. Solar-energy collector device that gathers and concentrates the incoming radiation by employing semicircular reflectors made of aluminum to focus sunlight into fluid-bearing copper piping. A gear motor pivots the reflectors to change their angle according to the sun's azimuth. (See diagram p. 141.)

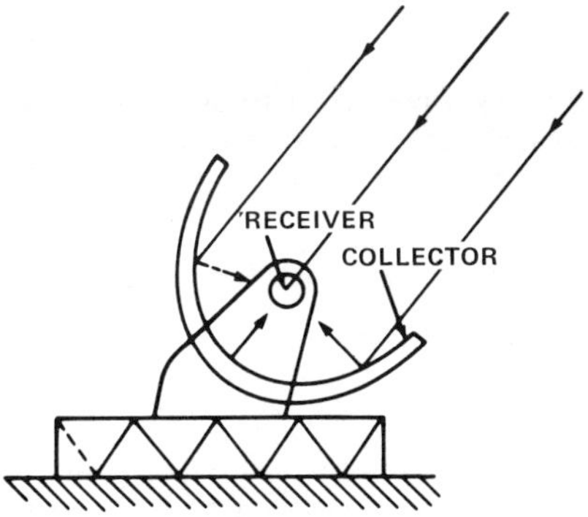

Courtesy DOE

Solar energy is focused on receiver by the collector.

foehn. A warm dry wind on the lee side of a mountain range. The warmth and dryness of the air is due to adiabatic compression upon descending the slopes. It is always associated with a strong wind blowing over a mountain ridge or chain.

fog. Liquid particles formed by condensation of vaporized liquids.

Fome-cor®. A product of Monsanto Corp. used for window insulation, consists of a thin layer of polystyrene foam between cardboard sheets.

food chain. An arrangement of the organisms of an ecological community according to the order of predation in which each uses the next, usually lower, member as a food source.

food waste. Animal and vegetable waste resulting from the handling, storage, sale, preparation, cooking and serving of foods, commonly called garbage.

foot-candle. A point reading of surface illumination. One foot-candle is the illumination at a point 1 foot from a candle.

footing. Structure buried in the ground to support a building or tower.

force. The push or pull that alters the motion of a moving body or moves a stationary body. The unit of force is the dyne or the poundal:

$$\text{Force} = \frac{\text{Mass} \times \text{Velocity}}{\text{Time}}$$

forced-air heat. A conventional heating distribution system which uses a blower to circulate heated air.

forced-air systems. A solar space-conditioning system which uses mechanical equipment (pumps or blowers) to transfer solar heat from collectors to living space or to storage and/or living space.

forced-circulation air heating coil. Heat exchanger, with or without fins, through which either hot water or steam is circulated for the purpose of heating a forced-circulation air stream.

forced-circulation water heater systems. A solar water-heating system that uses mechanical equipment (pumps) to transfer solar-heated water from the collector to storage and, depending upon system design, to the space requiring additional heat.

forced convection. Convection by gas or liquid forced over a surface; for example, a pump forcing water through a pipe or a fan blowing air over a brick—in contrast to natural convection where an object's temperature alone causes flow past its surface.

forced outage rate. The annual amount of unscheduled out-of-service time for power generation units.

forced vibration. Vibration of a structure, generally caused by engines or machines, sometimes by wind.

forest management. The practical application of scientific, economic, and social principles to the administration and working of a forest for specified objectives such as biomass feedstock development.

fossil fuel. Any naturally occurring fuel of an organic nature which originated in a past geologic age (such as coal, crude oil, or natural gas).

foundation wall. That portion of a load-bearing wall below the level of the adjacent grade or below the first floor beams or joists.

fractional distillation. A process of separating alcohol and water (or other mixtures) by boiling and drawing off vapors from different levels of the distilling column.

fractionation. The separation of a substance from a mixture; for example, the separation of one isotope from another of the same element.

fractions. Refiner's term for the portions of oils containing a number of hydrocarbon compounds within certain boiling ranges, separated from other portions in fractional distillation. They are distinguished from pure compounds which have specified boiling temperatures, rather than a range.

Francis turbine. A water turbine operating on a low and medium head, often installed in large hydroelectric schemes. Water enters the turbine radially and leaves axially.

free area. The total area of open space in a grille through which air can pass.

free coning. A wind-energy conversion system rotor which is articulated to allow blade motion in the out-of-plane direction.

free on board (FOB). An economic convention for specifying the cost of hardware at the factory, but not including transportation charges.

freestanding tower. A tower that does not use guy wires.

freeze protection. Protecting a solar hydronic system at night to prevent water from freezing in its collector.

Freidel Crafts catalyst. The catalyst employed in the synthesis of benzene hydrocarbons by the action of alkyl halides on aromatics in the presence of anhydrous aluminum chloride.

French drain. A covered ditch containing a layer of fitted or loose stone or other pervious material.

Freon. A highly stable gas that does not combine readily with other chemical elements and, for that purpose, is used in heat pumps and refrigeration systems.

frequency. The frequency of a function periodic in time is the reciprocal of the period. The unit is the cycle-per-unit time and must be specified. The unit cycle per second is commonly called hertz (Hz).

Fresnel collector. Type of concentrating solar collector which uses a Fresnel lens to focus solar radiation.

Fresnel equation. An equation for the reflection of nonpolarized radiation on passing from one medium to another.

Fresnel lens. An optical device that focuses light like a magnifying glass; concentric rings are faced at slightly different angles so that light falling on any ring is focused to the same point. Fresnel lenses are flat rather than thick in the center, and can be stamped out in a mold.

friction. A force that slows down movement and causes heat.

friction head. The head or energy lost as the result of the disturbances set up by the contact between a moving stream of water and its containing conduit. Friction head in pipes is the additional pressure that the pump must develop to overcome the frictional resistance offered by the pipe, by bends or turns in the pipeline, by changes in the pipe diameter, by valves, and by couplings.

front-end combustion. Combustion air enters the stove at the front and burns the wood from front to back.

fructose ($C_6H_{12}O_6$). A fermentable monosaccharide (simple) sugar. Fructose is a ketohexose.

fuel. Any substance that can be burned to produce heat.

Solar Energy to Fuels

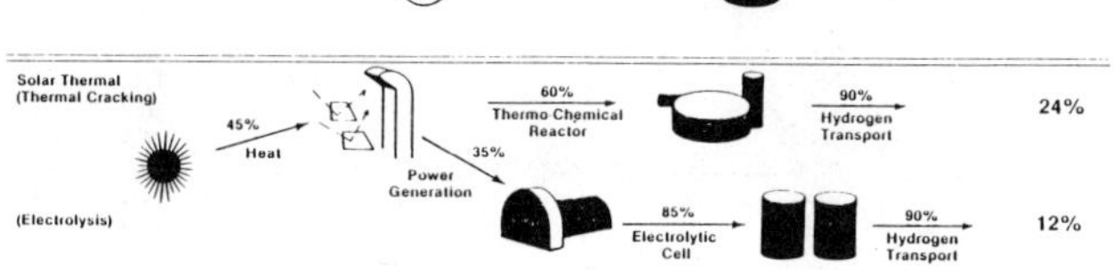

Courtesy Mitre Corporation

Illustration of efficiency in converting solar energy to fuels.

fuel cell. A device for combining hydrogen and oxygen in an electrochemical reaction to generate electricity. Chemical energy is converted directly into electric energy without combustion. (See diagram p. 144.)

fuel cell battery. A chemical battery, with sodium, chlorine, and mercury among the possible ingredients, which will have self-generating tendencies.

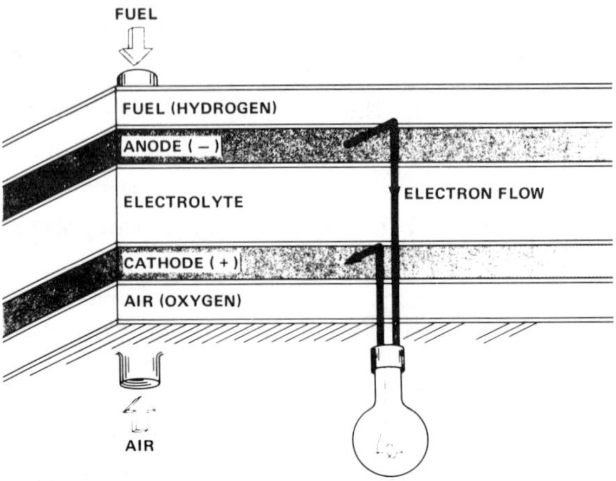

Courtesy EPRI

Diagram of fuel cell operation concept.

fuel depot. A bulk storage installation composed of storage tanks and related facilities such as docks, loading racks, and pumping units.

fuel efficiency. The ratio of the heat produced by a fuel for doing work to the available heat of the fuel. This efficiency is determined by the non-heat-forming materials in the fuel and the non-work-producing heat which is developed by the fuel.

fuel gas. Synthetic gas used in heating and cooling.

fuel grade alcohol. Usually refers to ethanol of 160 to 200 proof.

ALCOHOL		PROOF
0%		0
5%		10
10%		20
15%		30
20%		40
25%	↑ ↑	50
30%		60
35%		70
40%	BEVERAGE GRADE ALCOHOL	80
50%		100
55%		110
60%		120
65%		130
70%		140
75%		150
80%	FUEL GRADE ALCOHOL	160
85%		170
90%		180
95%		190
100%	↓ ↓	200

Courtesy Larry W. Carley, How to Make Your Own Alcohol Fuel, *TAB Books Inc.*

Fuel grade alcohol is 160 proof or higher. Gasohol is almost 200 proof.

fuel oil. Any liquid or liquefiable petroleum product burned for the generation of heat in a furnace or firebox or for the generation of power in an engine, exclusive of oils with a flash point below 100°F (38°C), tag closed-cup tester, and oils burned in cotton or wool-wick burners.

fuel rate. The amount of fuel needed to generate 1 kilowatt-hour (kWh) of electricity. In 1969 the rates were 0.88 pound (0.4 kilogram) of coal, 0.076 gallon (0.29 liter) of oil, or 10.4 cubic feet (0.3 m^3) of natural gas.

fuel saver. A solar device used solely to save fuel at conventional fossil fuel-burning facilities. The conventional systems provide the needed system reliability.

fuel savings. The money you make each year on a solar heater because it eliminates some of the fuel costs you would otherwise have.

full sun. The full sun condition is the amount of power density received at the surface of the earth at noon on a clear day—about 100 mW/cm^2. Lower levels of sunlight are often expressed as 0.5 sun or 0.1 sun. A figure of 0.5 sun means that the power density of the sunlight is one-half of that of a full sun.

fume. Tiny solid particles commonly formed by the condensation of vapors of solid matter.

fumigant. A pesticide that is burned or evaporated to form a gas or vapor that destroys pests. Fumigants are often used in buildings or greenhouses.

fundamental particle. Ultimate units occurring in the atom or formed by impact of high-energy particles from an outside source. In chemistry, the most important fundamental particles are the proton, the neutron, the electron, and the photon. Besides these, there are mesons of various designations, such as neutrinos and positrons, which are of more direct concern to the physicist than to the chemist.

fungi. Small, often microscopic plants without chlorophyll. Some fungi infect and cause disease in plants or animals; other fungi are useful in stabilizing sewage or in breaking down wastes for compost. Fungi are important elements of bioconversion of waste molecules for new energy sources.

fungicide. A pesticide chemical that kills fungi or prevents them from causing diseases, usually on plants of economic importance. See **pesticide**.

furling. Stopping the rotors of a wind generator so that they cannot turn.

furling speed. The wind speed at which the wind machine must be shut down to prevent high wind damage. See **cut-out speed.**

furnace. When used in a central heating system, a self-contained appliance for heating air by transfer of heat of combustion through metal to the air.

furnace burner unit. Warm-air furnace sold integrally with the burner.

furnace volume. The cubic contents of the combustion space of a boiler bounded by the grates, direct heating surface, and tube sheet.

fusel oil. A clear, colorless, poisonous liquid mixture of alcohols obtained as a byproduct of grain fermentation; major constituents are amyl, isoamyl, propyl, isopropyl, butyl, and isobutyl alcohols.

fusible plug. An insert of metal with low melting point placed in boilers, sprinklers, and other devices to melt when the temperature becomes dangerously high, so that the melting will relieve pressure, allow water flow, or otherwise tend to alleviate the dangerous condition.

fusion weld. Joining metals by fusion, using oxyacetylene or electric arc.

G

gallium (Ga). A chemical element, atomic number 31, metallic in nature, used in making certain kinds of solar cells.

gallium arsenide (GaAs). A compound used in making photovoltaic cells.

gallon. A unit of measure. A U.S. gallon contains 231 cubic inches (in.^{3}), 0.133 cubic feet (ft.^{3}), or 3.785 liters. It is 0.83 times the imperial gallon.

galvanic cell. A cell in which chemical change is the source of electrical energy. It usually consists of either two dissimilar conductors in contact with each other and with an electrolyte, or two similar conductors in contact with each other and with dissimilar electrolytes.

galvanic corrosion. Deterioration of metallic parts occurring when a heat-transfer fluid that conducts electricity (water and antifreeze solution do; silicone and hydrocarbon oils do not) contacts two dissimilar metals that are not properly isolated electrically.

galvanic coupling. Contact between dissimilar metals immersed in an electrically conductive solution (such as water-based heat-transfer media) results in a potential (voltage) difference, causing the less noble material metal to corrode. The galvanic series, which tabulates the potential differences between metals under corrosive conditions in a given environment or electrolyte, form the basis for predicting corrosion tendencies.

galvanized pipe. Steel pipe coated with zinc to resist corrosion.

garbage grinding. A method of grinding food waste by a household disposal, for example, and washing it into the sewer system. Ground garbage then must be disposed of as sewage sludge.

gas. That state of matter which has neither independent shape nor volume. It expands to fill the entire container in which it is held. It is one of the three forms of matter: solid, liquid, and gas.

gas conditioning. The removal of objectionable constituents and addition of desirable constituents.

gas constant. The constant *R* in the perfect gas equation $PV = RT$. In English units, this constant is 1545 divided by the molecular weight of gas.

gas holder. A gas-tight receptacle or container in which gas is stored for future use. There are two general ways of storing gas; at approximately constant pressure (low-pressure containers), in which case the volume of the container changes; and in containers of constant volume (usually high-pressure containers), in which case the quantity of gas stored varies with the pressure.

gasification. Any chemical or heat process used to convert a substance to gas.

gasifier. Device which generates from organic residues a low-energy gas which can substitute for gas and oil in existing equipment.

gasket. A piece of solid material, usually metal, rubber, or plastic, placed between two pieces of pipe or between an automotive cylinder head and the cylinder block, for example, to make the metal-to-metal structure fluid-tight.

Gasohol. A program established by the State of Nebraska to aid in the development of an alternative automotive fuel containing a blend of 10 percent agriculturally derived ethyl alcohol and 90 percent unleaded gasoline. Also a trademark registered by the Agricultural Products Industrial Utilization Committee (APIUC), a Nebraska state agency established to administer their Gasohol program.

gasoline. A volatile flammable liquid obtained from petroleum which has a boiling range of approximately 84° to 421°F (29° to 216°C), and is used as fuel for spark ignition internal-combustion engines.

gas turbine. A prime mover in which gas under pressure, formed by combustion, is directed against a series of turbine blades; the energy in the expanding gas is converted into mechanical energy supplying power at the shaft.

gas turbine combustion engine. A new generation engine in which a continuous flow of heated, compressed air is directed at high velocity against a bladed turbine which in turn drives the transmission and vehicle wheels. Any type of burnable liquid can be used to provide the heat.

gas turbine generating station. An electric-generating station in which the prime mover is a gas-turbine engine.

gate valve. A valve employing a gate, often wedge-shaped, allowing fluid to flow when the gate is lifted from the seat. Such valves have less resistance to flow than globe valves.

gearbox. Converts mechanical rotary motion to vertical motion. The gearbox in a wind-energy conversion system can also contain a spring mechanism which allows the tail to fold parallel to the wheel in a high wind.

gear ratio. A ratio of speeds (rpm) between the rotor power shaft and the pump, generator, or other device power shaft; applies both to speed-increasing and speed-decreasing transmissions.

gelatinization. The rupture of starch granules by heat to form a gel of soluble starch and dextrins.

gel ponds. A polymer gel or cover over the surface of a solar pond. The gel is lighter than the brine and therefore floats on top of the solar pond. (See diagram p. 148.)

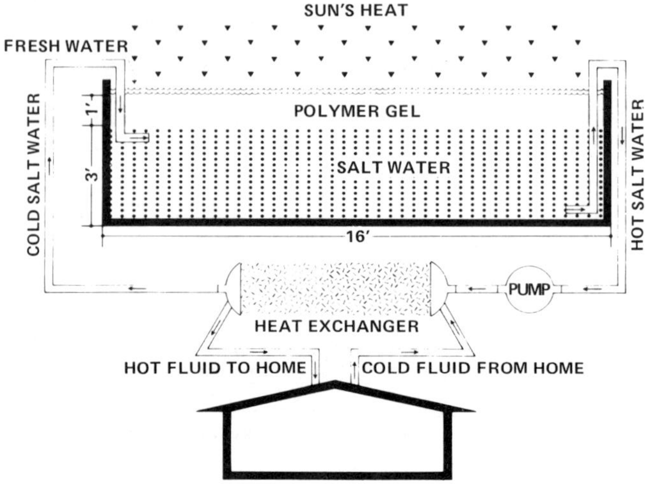

Courtesy Popular Science, *April, 1981*

Process diagram of polymer gel on surface of solar pond.

generator. A machine or subsystem that transforms mechanical energy into electrical energy.

generic system. A system that is designated a "typical" system, when there are numerous variations of the system being studied.

genetic selection. The applied science and art of systematic genetic improvement of a species or population, often through breeding. Used to improve tree production for biomass fuels.

geophysics. A study of subsurface geological conditions of structure or material through the interpretation of measurement variations in density, magnetics, elasticity, electrical conductivity, temperature, and/or radioactivity.

geostrophic motion. Motion which is unaccelerated and frictionless.

geostrophic wind. That horizontal wind velocity for which the Coriolis acceleration exactly balances the horizontal pressure force.

geostrophic wind level. The lowest level, observed to be between 3900 and 5220 feet (1.2 and 1.6 kilometers), at which the wind becomes geostrophic. The geostrophic wind level may be considered to be the base of the free atmosphere.

geothermal energy. Energy produced by the exploitation of the internal heat of the earth.

geothermal energy resources. Thermal energy resources found in the earth; these include hydrothermal convective systems, pressurized water reserves, hot dry rocks, manual gradients, and magma.

geothermal generating station. An electric-generating station in which the prime mover is a steam turbine, and the source of the steam is generated in the earth by heat from the earth's magma.

geothermal gradient. The change in temperature of the earth with depth, expressed either in degrees per unit depth, or in units of depth per degree.

geothermal heat. The natural migration of heat outward from the earth's molten core. In unbroken portions of the earth's crust, heat is transferred by conduction, a slow and diffuse process revealed by gentle temperature gradients that decrease toward the earth's surface. But where that surface is weakened, as by the fracture and faults at crustal plate boundaries, magma intrudes a hot mass transported from below. Geothermal heat is thus concentrated in geologically active regions. Hydrothermal convection cells transfer the heat up, where it vents naturally in hot springs, fumaroles, or geysers—or to a level that is accessible for extraction through wells.

geothermal power. Power plants using hot water or steam that is stored in the earth from volcanic activity. Sources of geothermal energy are currently under development in this country and New Zealand. Geysers in California presently produce electricity in the United States. In a few places natural steam is available. In many places, hot water can be tapped as a usable energy source. Also, there are areas of intensely hot underground rock that can be used by fracturing the rock and forcing cold water down to it. The heated water can be returned to the surface to produce steam power.

geothermal steam. Steam drawn from deep within the earth. There are about 90 known places in the continental United States where geothermal steam could be harnessed for power. These are located in California, Idaho, Nevada, and Oregon.

germicide. A chemical or agent that kills microorganisms such as bacteria and prevents them from causing disease. Such compounds must be registered as pesticides with the Environmental Protection Agency.

germination of malting. Germination is the awakening of the dormant seed by the proper amount of heat, air, and moisture, to begin the process of renewed life. During germination, several complex changes take place. The visible signs are the sprouting of the acrospire (the incipient stalk) and the rootlet. Inside the seed, enzymes are activated, particularly along the scutellum, a membrane separating the embryo from the endosperm (starchy part of the kernel). These enzymes partially break down this cell wall and enter the endosperm, converting the hard, flinty starch into a softer and more soluble form. Some of this starch is converted to maltose and glucose, then passed through the scutellum and reconverted to starch, to begin building the rootlet. The acrospire is also formed and grows upward from the embryo along the kernel on the inside of the hull. This will eventually erupt at the top of the seed and become the stalk of the plant.

getters. A column or cartridge containing an active metal which will be sacrificed to protect some other metal in the system against galvanic corrosion.

geyser. A volcano in miniature from which hot water and steam (instead of lava and ashes) are erupted periodically during the waning phase of volcanic activity.

gigawatt (GW). Power unit equal to 1 billion (10^9) watts (W), 1000 megawatts (MW), or 1 million (10^6) kilowatts (kW).

gin pole. A pipe, board, or tower used to improve leverage while raising a tower.

gland bushing. A metal bushing installed between the packing nut and the packing to transmit the force exerted by the packing nut against the packing.

Glauber's salt. A term for sodium sulfate decahydrate, which melts at 90°F (32°C); a component of eutectic salts.

glazed area (or glazing). For solar collection, glazing refers to all materials which are translucent or transparent to shortwave solar radiation, including glass, plexiglass, Kalwall®, etc.

glazing. A covering of transparent or translucent material (glass or plastic) used for admitting light. Glazing retards heat losses from retardation and convection. Examples: windows, skylights, greenhouse, and collector coverings.

glazing materials. Glass, plastic films, and sheets used on solar collectors. They admit as much solar radiation as possible and reduce the upward loss of heat to the lowest attainable value.

global insolation. The total insolation on a horizontal surface on the earth, averaged over some specified period of time.

global radiation. Sum of diffuse and direct insolation (W/m^2). See **total insolation**.

globe thermometer. A thermometer in a hollow spherical black globe, readings from which show a higher value, due to radiation, than a conventional thermometer, so that the globe device measures the effective radiation temperature.

globe valve. One with a somewhat globe-shaped body with a manually raised or lowered disc which, when closed, rests on a seat so as to prevent passage of a fluid.

glow. The incandescence of a heated substance, or the light from such a substance; white or red heat.

glucose ($C_6H_{12}O_6$). Glucose is a monosaccharide (simple sugar). This is one of the three sugars (glucose, fructose, and mannose) that the yeast *Saccharomyces cerevisiae* will internalize and ferment. If the raw material is a starch, this must be hydrolyzed to glucose before it can be fermented. Glucose is an aldohexose.

glucosidase. An enzyme that hydrolyzes polymers of glucose monomers (glucoside); specific flucosidases must be used to hydrolyze specific glucosides; e.g, β-glucosidases are used by hydrolyze cellulose; α-glucosidases are used by hydrolyze starch.

glycol. An antifreeze additive to water.

governor. A device for regulating the speed of an engine or motor under varying conditions of load and pressure; also a device that automatically controls the speed of the rotors in a wind generator by changing their angle. In a high wind, the governor turns the blades into the wind (called feathering) to prevent damage to the machine.

grade. In piping, the slope of a line compared to horizontal, usually expressed in inches fall per foot of pipe.

gradient wind. Any horizontal wind velocity that is tangent to the contour line of a constant-pressure surface.

grain. A unit of weight equivalent to 65 milligrams (mg) or 2/1000 of an ounce.

grain alcohol. See **ethanol**.

grain loading. The rate of emission of particulate matter from a polluting source. Measurement is made in grains of particulate matter per cubic foot of gas emitted.

grate area. Area of surface on which fuel is burned, ordinarily measured at the top surface of the grates.

gravitational energy. The energy of attraction between two material bodies. It is directly proportional to the product of the masses of the two bodies and inversely proportional to the square of the distance between their centers.

gravity. The force by which substances are attracted to each other, or fall to earth.

gravity convection. The natural movement of heat that occurs when a warm fluid rises and a cool fluid sinks under the influence of gravity. See **convection.**

gravity wave. A wave in a fluid medium in which restoring forces are primarily due to gravity buoyancy rather than by compression.

green belts. Certain areas restricted from being used for buildings and houses; they often serve as separating buffers between pollution sources and concentrations of population.

greenhouse. A solar greenhouse is a popular means of passive solar heating. It can be incorporated into new construction or attached to existing houses which have southern exposure. The greenhouse uses south-facing glass (or other glazing material) to collect solar energy. This incoming energy is then absorbed by walls and/or floors made of materials that store heat, such as masonry. This heat provides both a suitable environment for growing plants year-round as well as a source of supplemental heat for the rest of the home.

Solar greenhouses have certain distinct advantages. They can be planned and built by do-it-yourselfers for as little as several hundred dollars. Professional units usually begin at about $2,500. In addition, greenhouses provide attractive living space and the potential for growing some of your own food supply. Greenhouses are also quite adaptable in design; they can be as large or small, complex or simple as the designer or owner wants them to be. Consumer options range from prepackaged building modules to do-it-yourself projects to custom designs by professionals.

greenhouse effect. The trapping of energy with a consequential rise in temperature due to a multiplicity of causes, such as reduced convection or reflection, increased absorption, etc. Used to describe the basis for the intentional rise in solar collectors, as well as the earth's possible temperature rise due to increasing carbon dioxide content in the atmosphere.

green wood. Wood that is still standing or just freshly cut and has the highest moisture content. This is not to be confused with wet wood.

grid. Network to transmit and distribute electrical energy from point of production to point of end-use. Usually alternating current being tightly controlled to 50 or 60 Hz.

grill. A covering over an inlet or outlet with openings through which a fluid passes.

gross area. The total frontal area of a collector, including framing and structural supports.

gross collector area. The maximum projected area that encompasses the collector and its associated manifolds, and other elements normally part of the complete collector module.

gross cross-sectional area. The overall or outside area of a flat-plate collector. It is usually slightly larger than the absorber area since it includes the framework required to hold the absorber.

gross national product (GNP). The total market value of the goods and services produced in a national economy, during a given year, for final consumption, capital investment, and governmental use. (Note that GNP does not include the value of intermediate goods and services sold to producers and used in the production process itself.) A national economic measure often related to energy consumption with implications for solar policy.

ground cover. Grasses or other plants grown to keep soil from being blown or washed away.

ground joint. One where the parts to be joined are precisely finished and then ground in so that the seal is tight.

ground reflection. Solar radiation reflected from the ground (e.g., by snow) into a solar collector; more heat can be captured by a collector with ground reflection.

groundwater. Water below the ground surface, such as the supply of freshwater under the earth's surface, in an aquifer (q.v.) or soil that forms the natural reservoir for man's use.

groundwater runoff. Groundwater discharged into a stream channel as spring or seepage water.

grouting. The filling or finishing of cracks with a thin mortar or caulking compound or any similar material which will mend the cracks and make the surface water-resistant.

growing season. The season which is warm enough for the growth of plants. On the whole, the growing season is confined to that period of the year when the daily means are above 43°F (6°C).

gust. A positive or negative short-term wind variation from the mean wind speed.

gust factor. Ratio of wind speed averaged over a short period to the mean wind speed.

H

habitat. The sum total of environmental conditions of a specific place that is occupied by an organism, a population, or a community.

Hall effect. When a conductor or semiconductor carrying a current is placed in a magnetic field that has a component normal to the direction of the current density, the distribution of the current density over the cross-section of the conductor or semiconductor is changed, and a potential gradient is set up in a direction normal to the directions of both the current density and the field.

hammer mill. A broad category of high-speed equipment that uses pivoted or fixed hammers or cutters to crush, grind, chip, or shred solid wastes.

hammer mill shredder. A cylindrical machine which is lined with spike-shaped projections which are utilized to tear and break up organic waste material.

hard water. Water containing dissolved minerals such as calcium, iron, and magnesium. The most notable characteristic of hard water is its inability to lather soap.

hazardous air pollutant. According to law, a pollutant to which no ambient air-quality standard is applicable, and that may cause or contribute to an increase in mortality or in serious illness. For example, asbestos, beryllium, and mercury have been declared hazardous air pollutants.

head. A unit of pressure intensity usually given in inches or feet of a column of the fluid under consideration. Thus, 1 foot of water head is the pressure from a column of water 1 ft high. A measure of height a pump must lift water.

header. The pipe that runs across the edge of an array of solar collectors, gathering (or distributing) the heat-transfer fluid from (or to) the risers in the individual collectors. This ensures that equal flow rates and pressure are maintained. See **risers.**

head loss. A measure of friction loss caused from water flow through pipes.

head pressure. The discharge pressure at the refrigeration compressor; also the condenser pressure.

heads. The first vapors to boil off in alcohol fuel distillation. The heads contain the highest portion of alcohol while those toward the end of distillation will be mostly water.

heat. A form of kinetic energy that flows from one body to another because of a temperature difference between them. It may be generated or transferred by combustion, chemical reaction, mechanical means, or passage of electricity. Heat is usually measured in calories (cal) or British thermal units (Btu).

heat actuated cooling. Use of thermal energy rather than mechanical energy to transfer heat from indoors to a (warmer) outdoor environment.

heat balance. The accounting of the energy output from a system to equal the energy input.

heat budget. The total amount of the sun's heat received on the earth during any one year must equal exactly the total amount which is lost from the earth by reflection and radiation into space.

heat capacity (specific heat). The quantity of heat required to raise the temperature of a given mass of a substance 1°F.

heat conduction. The oceans receive heat by conduction through the sea bottom. Since the amount is very small (50 to 80 gram calories per square centimeter per year), it is neglected when considering the heat budget.

heat content. The rate of change in heat content or enthalpy of air per unit time, measured in British thermal units per hour.

heat distribution method. The release of radiant heat from a storage mass within the habitable areas of the structure, or convective heating from a storage mass through airflow paths provided by openings or by ducts (with or without the assistance of a fan or pump generally having a horsepower rating of less than 1 horsepower) in the storage mass, to habitable areas of structure.

heat engine. A device for producing mechanical (motion) energy directly from two heat reservoirs of different temperatures. A machine which converts thermal energy to useful, mechanical work, e.g., a steam turbine.

heat exchanger. A device which is designed to transfer heat easily. Heat transfer can be between a liquid and gas or between liquid and liquid; analogous to the tap on the rainwater-collecting system. (See diagram p. 155.)

heat flow. The movement of heat from one place to another due to temperature differences; analogous to volume flow.

heat-flow path. Path which heat takes as it moves by conduction through a material from a high temperature to a lower one. Length of flow path is how far the heat travels; area of the flow path is how much material the heat flows through at each point along the path.

heat gain. An increase in the amount of heat contained in a space, resulting from direct solar radiation and the heat given off by people, lights, equipment, machinery, and other sources.

heating capacity. The quantity of heat necessary to raise the temperature of a system or substance by 1 degree of temperature, expressed in calories per degree centigrade (cal/°C); also the quantity of heat in British thermal units (Btu) that a room air conditioner is capable of adding to a room in 1 hour's time.

heating degree-day. A measure of the coldness of the weather experienced, based on the extent to which the daily mean temperature falls below a reference temperature, usually 18°C. For example, on a day when the mean outdoor dry-bulb

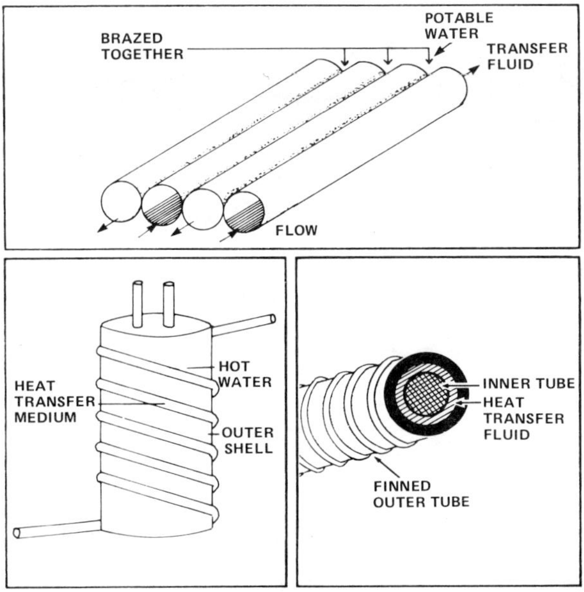

Courtesy HUD/SERI

Several different double-walled heat exchangers for solar applications.

temperature is 2°C, there would be −1 degree-days experienced. A daily mean temperature usually represents the sum of the high and low readings divided by two.

heating effect. Concerns the ratio of heat output of a radiator or convector in the living zone of a room with the total output measured by steam condensation or water temperature drop.

heating load. The amount of heat and humidity that must be added to a building to maintain occupant comfort.

heating medium. A fluid used for conveying heat from a heat source to heat dissipating devices; includes air, water, and steam.

heating season. Portion of the year (usually October to May in the northern hemisphere) when outdoor cold weather makes indoor heating necessary to maintain comfort. (See diagram p. 156.)

heating surface. Broadly, the area intended for transferring heat.

heating value. The amount of heat produced by the complete combustion of a unit quantity of fuel. The gross or higher heating value is that which is obtained when all of the products of combustion are cooled to the temperature existing before combustion, the water vapor formed during combustion is condensed, and all the necessary corrections have been made. The net or lower heating value is obtained by subtracting the latent heat of vaporization of the water vapor formed by the combustion of the hydrogen in the fuel from the gross, or higher, heating value.

heating, ventilating, air-conditioning (HVAC) system. System that provides

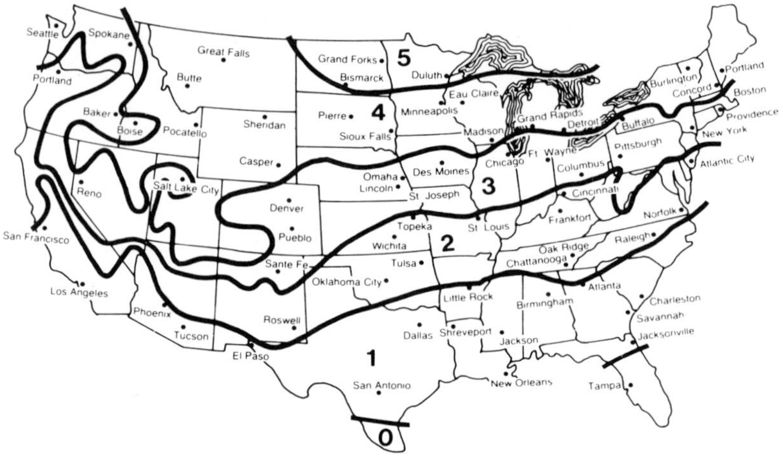

HEATING ZONE	RECOMMENDED FOR		
	CEILING	WALL	FLOOR
0,1	R-26	R-13	R-11
2	R-26	R-19	R-13
3	R-30	R-19	R-19
4	R-33	R-19	R-22
5	R-38	R-19	R-22

Courtesy HUD

Recommended insulation for various heating seasons in the United States.

heating, ventilating and/or air conditioning within or associated with a building.

heat island effect. An air-circulation problem peculiar to cities. Tall buildings, heat from pavements, and concentrations of pollutants create a haze dome that prevents rising hot air from being cooled at its normal rate. A self-contained circulation system is put in motion that can be broken by relatively strong winds. If such winds are absent, the heat island can trap high concentrations of pollutants and present a serious health problem.

heat liberation rate. The amount of heat which is liberated per unit times per cubic foot of combustion space.

heat loss. The sum cooling effect of the building structure when the outdoor temperature is lower than the desired indoor temperature. It represents the amount of heat that must be provided to a space to maintain indoor comfort conditions.

Heat-mirror®. A product of Suntak Research Associates and Thermofilm Corporation which is a transparent insulating sheet that includes a far-infrared-reflecting film.

heat of combustion. The heat released when a substance is completely burned in oxygen.

heat of condensation. The same as the heat of vaporization, except that the heat is given up as the vapor condenses to a liquid at its boiling point.

heat of fusion. The heat released when a liquid freezes. It is equal to the amount of heat absorbed when the solid melts.

heat of liquid. The heat necessary to raise the temperature of the liquid, as distinguished from the heat of vaporization.

heat of vaporization. Alcohol fuel has a greater latent heat of vaporization than does gasoline. This means that it takes more heat to vaporize alcohol than it does to vaporize gasoline. The amount of heat necessary to vaporize ethanol is 361 Btu per pound compared to 140 Btu per pound for gasoline.

heat pipe. A heat-transfer device consisting of a cylinder or tube which absorbs heat by vaporization of a liquid at one end and releases heat by condensation of vapor at the other end.

heat pump. An electrically operated machine for heating and cooling. When heating, it transfers heat from one medium at a lower temperature (called the heat source) to a medium at a higher temperature (called the heat sink), thereby cooling the source (outside air), and warming the sink (the house). When cooling, the heat pump functions much like an air conditioner—taking unwanted heat from the heat source (a building) and dumping it to the heat sink (the outside).

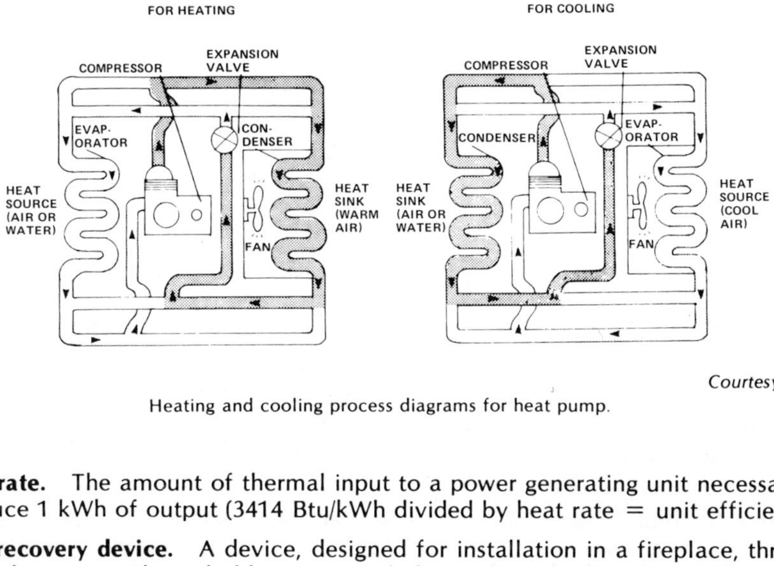

Courtesy DOE

Heating and cooling process diagrams for heat pump.

heat rate. The amount of thermal input to a power generating unit necessary to produce 1 kWh of output (3414 Btu/kWh divided by heat rate = unit efficiency).

heat-recovery device. A device, designed for installation in a fireplace, through which house air or household water is cycled to reclaim the fire's heat before it can escape up the chimney.

heat register. The grilled opening into a room by which the amount of warm air from a furnace can be directed or controlled; it includes a damper assembly.

heat regulation device. Shading or venting mechanism to control the amount of solar heat admitted through solar collection areas. Nighttime insulation or its equivalent to control the amount of heat permitted to escape from the interior of a structure.

heat sink. A structure designed to absorb excess heat. Hot components are mounted on a heat sink, which is usually designed with cooling fins to carry away the excess heat by convection.

heat source. A medium from which heat is extracted. A heat pump uses air or water outside a structure as a heat source during its heating cycle.

heat storage. A device or medium that absorbs collected solar heat and stores it for use during periods of inclement or cold weather.

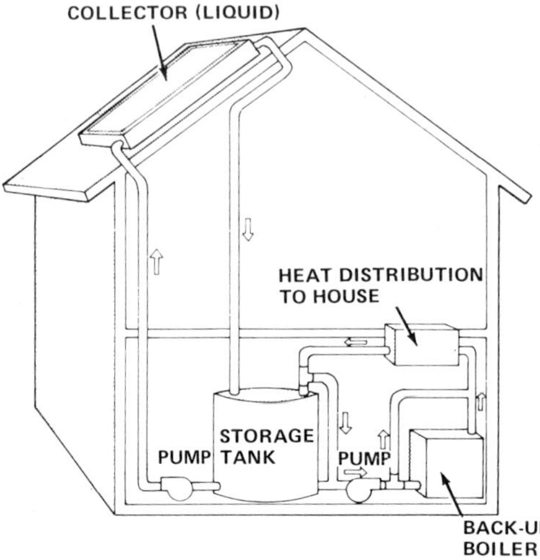

Courtesy HUD/DOE

Water tank storage for solar heat is shown in this system diagram.

heat storage capacity. The amount of heat which can be stored by a material.

heat tax. Applies to the heat energy that becomes unavailable for further use whenever energy is converted from one form to another.

heat transfer. The methods (radiation, convection, and conduction) by which heat may be propagated or conveyed from one place to another. This term is sometimes used to mean rate of heat transfer.

heat-transfer fluid. Fluid that carries heat from solar collector to storage. In direct systems, this is plain water; in indirect systems, it may be treated water, air, or a special oil. Since it reduces collector temperature by carrying away heat, this fluid is sometimes referred to as "collector coolant."

heat transmission coefficient. Any one of a number of coefficients used in the calculation of heat transmission by conduction, convection, or radiation through various materials and structures.

heat transport fluid. The fluid used for transporting or transferring thermal energy from one area to another within the system.

heat treatment. Heating and cooling a solid metal or alloy in such a way as to obtain desired conditions or properties of the metal to improve its functions.

HEAT-TRANSFER FLUID

Medium	Specific gravity	Viscosity (centipoise)	Heat capacity (Btu/lb°F)	Freezing point (°F)	Anticorrosion protection	High-temperature reactions	Toxicity requires double wall?
Aqueous fluids							
Water	1.000	0.5-0.9	1	+32	None for copper, nitrites, or chromates and a pH buffer in a closed loop for steel or aluminum	Boils at 212°F	No (yes, if chromate or nitrite inhibitors are added)
50 wt.% Water-ethylene glycol	1.050	1.2-4.4	0.83	-33	Silicates, pH buffer	Boils at 230°F; degrades slowly to formic acid above 300°F	Yes
50% Water-propylene glycol	1.022	1.4-7.0	0.87	-24	Phosphates, pH buffer	Boils at 230°F; not recommended above 300°F	Yes
Organic fluids				(pour point)			
Paraffinic oils (Exxon Caloria HT43 or Dowtherm HP)	0.821	11	0.51	+20	Not needed	Flash point 375°F; oxidizes slowly if exposed to air above 230°F	Yes
Aromatic oils (Dowtherm J or Monsanto Therminol 55)	0.830	0.6	0.46	-95	Not needed	Flash point 145°F; oxidizes slowly on exposure to air above 300°F	Yes
Silicone oils (GE SF-96 or Dow Corning Q2-1132)	0.970 0.960	50 15	0.36 0.38	-120 -185	Not needed	No known problems, flash point is above 500°F	Yes

heat unit. A unit of quantity of heat, such as the British thermal unit (Btu) or the calorie (cal).

heat value. The amount of heat obtainable from a fuel and expressed in British thermal units per pound (Btu/lb).

height-diameter ratio. Ratio of wind-energy conversion system height from the ground of the area centroid of the rotor-swept area to the rotor diameter.

helical screw expander. A spiral-shaped machine for driving a generator through which hot water and steam expand.

heliochemical process. The process by which solar energy is utilized through photosynthesis.

heliochronometry. Telling of time by means of the sun.

heliodon. A solar angle simulator useful in conducting shading assessments on buildings or solar-collector arrays.

helioelectrical process. The process by which energy is utilized by photovoltaic converters.

heliostat. A combination of mirrors, support structure, drive mechanism, and mounting foundation that tracks in two axes of motion to continuously reflect the sun's rays onto a fixed receiver. An instrument consisting of a mirror mounted on an axis moved by clockwork; the heliostat reflects sunbeams in one direction, usually to a central absorber located in a tower. (See photograph p. 161.)

heliostat packing density. The ratio of total reflective surface area to the total land area used by the heliostats (collector subsystem).

heliothermal. Any process that uses the sun's radiation to produce heat.

heliothermic. Site planning that takes into account natural solar heating and cooling processes and their relationships to building shape and orientation.

heliothermometer. An instrument that measures the sun's heat.

heliotropic. Any device (or plant) that follows the sun's apparent motion across the sky.

heliotropic wind. A diurnal component of the wind velocity leading to diurnal shift of the wind or turning of the wind with the sun, produced by the east-to-west progression of daytime surface heating.

hemicellulose. Long-chain molecules made up of five-carbon sugars.

hemispherical bowl technology. The hemispherical bowl concentrator technology (also referred to as FMDF, Fixed Mirror Distributed Focus) develops and characterizes the concept of a stationary hemispherical concentrator. This concept will focus solar radiation on a radial focal area which will be tracked by a linear receiver during the daily rotation of the earth. The reflective surface may be built in larger units than can tracking collectors, as the mirrored element is stationary, and as a result, the tracking system has to move only the receiver. The FMDF is capable of operating temperatures of up to 1000°F (538°C). Potential markets include: electrical generation for small communities, total energy, irrigation, and industrial process heat applications.

The Crosbyton Hemispherical Bowl Project located at Crosbyton, TX, is the lead project for FMDF research and development. A recent achievement of the FMDF

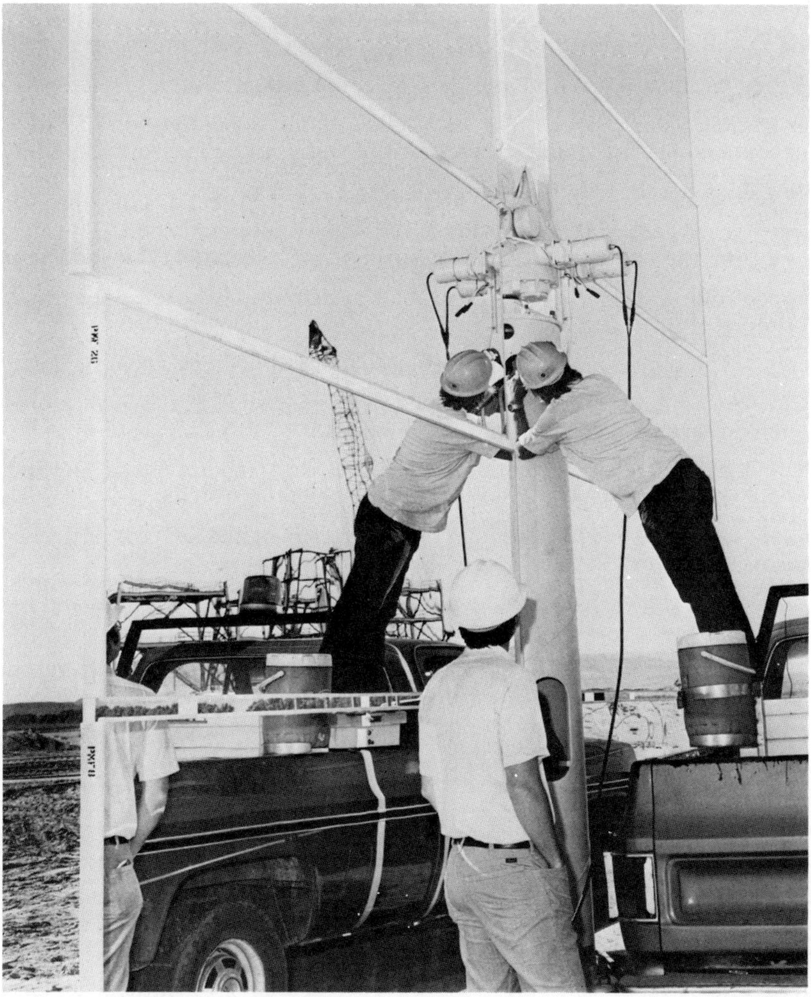

Courtesy DOE

One of 1818 heliostats (mirrors) which reflect the sun's energy to a receiving "power tower" at Solar One.

project has been the installation of a 65-foot (19.8-meter), bowl at Crosbyton which is currently undergoing performance testing.

herbaceous. Nonwoody, lacking a large proportion of lignin.

herbivore. An organism that feeds on vegetation.

hertz (Hz). A unit of frequency used in electrical and electronic measurements and equal to one cycle or one wavelength of electrical energy per second.

heterojunction. Zone of electrical contact between two dissimilar materials. See **homojunction.**

heterotrophic organism. An organism dependent on organic matter for food.

hexode. Any of various simple sugars that have six carbon atoms per molecule.

high head. Refers to generation of hydroelectric power using large dams.

higher heating value (HHV). The heat released during combustion of fuel if all products are cooled to room temperature and water is condensed to liquid.

high-low emitter cell. Normal solar cell with extra doping.

highside. That portion of a mechanical refrigerating system under relatively high pressure from the compressor to the expansion valve, including the condenser.

high-speed rotor. A wind-energy conversion system rotor whose tip speed ratio is five or more.

hingepin. The pin upon which the disc or clapper of a check valve swings.

HISPER. A computer program developed by the Marshall Space Flight Center to define and/or evaluate solar-system performance.

HITEC. A eutectic combination of potassium nitrate, sodium nitrite, and sodium nitrate, used as a working fluid in solar thermal electric systems.

hi-volume sampler. A device used in the measurement and analysis of suspended particulate pollution. Also called a Hi-Vol.

hog(ger). A machine which reduces wood to coarse chips, e.g., for fuel and for the manufacture of wood pulp and wood chipboard.

hole. A vacancy where an electron would normally be in a perfect crystalline structure.

homojunction. The zone of contact between the *n*-layer and the *p*-layer in a single material, the two layers having been created by doping the basic crystal with other substances. See **heterojunction.**

homopolar generator. A solid rotor machine without commutators or permanent magnets.

hood. A canopy or collecting structure over a process area or piece of equipment to collect fumes, gases, vapors, or odors which are then exhausted or treated in some manner.

hopper. A vessel into which materials are fed. It is usually constructed in the form of an inverted pyramid or cone terminating in an opening through which the materials are discharged (not primarily intended for storage).

horizontal axis wind turbine. A wind-energy conversion system whose axis of rotation is parallel with the ground and the mean wind velocity. (See photo p. 163.)

horizontal branch. In plumbing, the drain line from the fixture drain to the waste stack.

horsepower. A unit of power equivalent to 33,000 foot-pounds per minute (ft-lb/min), or 550 foot-pounds per second (ft-lb/sec) (mechanical horsepower), or 746 watts (W).

horsepower hour. One horsepower expended for 1 hour, or the horsepower multiplied by the number of hours. One horsepower hour equals 1,980,000 foot-pounds (ft-lb), 0.745 kilowatt-hours (kWh), 2545 British thermal units (Btu) (mean).

Courtesy DOE

Dunlite® 12-foot (3.7-meter) diameter small wind-energy conversion system rated at 2000 W at 25 mph windspeed.

hose coupler. A metal or plastic device for joining two pieces of hose (or one hose and a piece of rigid piping), by means of screw threads in the coupler which mate with threaded connections on the hoses or pipes.

hot/cold tank storage. A thermal-energy storage system utilizing separate tanks for the charged (hot) and uncharged (cold) storage medium.

hot deck. A hot-air chamber forming part of a ventilating unit.

hot dry rock. Evidence suggests that the largest potential geothermal energy resource may be hot rock that has not come into contact with sufficient underground water to be categorized as hydrothermal. Before this resource can be developed, a technology must be perfected to extract heat from deposits of dry rock. One promising method under study employs a hydrofracturing technique comparable to that used in petroleum recovery. Through this method, high-pressure water is used to create a large crack in hot rock, such as granite, which overlays an area of high heat flow from the earth's interior. Pressurized water is then circulated through the crack to extract the heat. Another method of extracting heat from hot dry rock may be to use explosives, instead of hydrofracturing, to form cavities with increased surface area for water contact. Both methods—explosion and hydrofracture—require extensive research programs before being put into use.

hot-rock reservoir. A potential source of geothermal power. The "hot-rock" system requires drilling deep enough to reach heated rock then fracturing it to create a reservoir into which water can be pumped.

hot-water heater. Used to heat water for home use. More than 400 kilowatt hours per year (kWh/yr) can be saved by lowering the thermostat setting on the electric hot-water heater from 150°F to 130°F (66°C to 54°C). Insulating hot-water tanks and bare pipes further reduce heat loss.

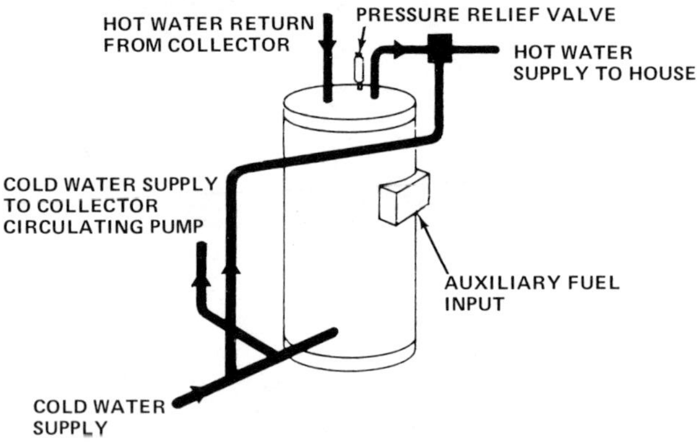

Courtesy National Copper Association

Single domestic hot water heater and controls.

hot-water heating system. A system in which water is the heating medium.

hot water sizing. The size of a solar domestic water preheat system depends on the number of gallons of hot water used by a household. This is estimated to be 15 to 20 gallons (57 to 76 liters) of hot water per person per day. On that basis, for every person in the house, you need 15 to 20 square feet (1.4 to 1.9 square meters) of collector area. The amount of storage required is the same as for an active space heating system: 1 to 2.5 gallons (3.8 to 9.5 liters) of liquid for every square foot (0.1 square meter) of collector.

hot-water wells. Geothermal wells in the Geysers' region and elsewhere which produce dry steam free of water, and more commonly a mixture of steam and hot brines; a wet steam, liquid-dominated, hydrothermal convection system which has a temperature range from below 190°F (88°C) to greater than 660°F (349°C) and can produce a mixture of vapor and liquid, or liquid alone. The thermal energy from these hot-water wells is used to generate electricity and, more notably, to provide heating.

hour angle. Each hour equals 15° of longitude with solar noon being zero, mornings positive, and afternoons negative.

hours of storage. The number of hours a plant can produce electricity or provide power to an industrial process when operating exclusively from an initially fully charged storage subsystem.

house heating. See **space heating.**

hub height. The height above ground of the axis of a horizontal axis wind turbine. It is frequently assumed that hub height winds may be used to construct an adequate representation of the winds affecting the entire wind turbine. For a vertical axis wind turbine such a height would be midway between the top and bottom of the rotor.

humid heat. Ratio of the increase in total heat per pound of dry air to the rise in temperature, with constant pressure and humidity ratio.

humidifier. A mechanical means of increasing the relative humidity by injecting water or water vapor into the air.

humidifying effect. The quantity of water evaporated per unit of time (usually 1 hour) times the latent heat of vaporization at the evaporating temperature.

humidistat. A regulating device, actuated by changes in humidity, used for the automatic control of relative humidity.

humidity (relative). Ratio of the actual water vapor content of the air to the water vapor that would be found in the same volume of saturated air at the same temperature and pressure.

humus. Decomposed organic material.

hybrid. A power-generating plant comprised of a solar energy collection subsystem and a nonsolar energy subsystem.

hybrid solar energy system. A system which incorporates a major passive aspect, where at least one of the significant thermal energy flows is by natural means and at least one is by forced means.

hybrid system. A combination of a solar technology with a conventional technology to provide the controlled availability needed for everyday use.

hydrate. A solid material resulting from the combination of a gas with water under pressure. The greater the pressure in the equipment, the higher the temperature at which the hydrate will form.

hydration. The chemical process of combination or union of water with another substance.

hydraulic. Strictly, having to do with water in motion, but the term is extended to cover all liquids which convey, store, or transfer pressure energy to reactants.

hydraulic turbine. An enclosed rotary type of prime mover which produces mechanical energy by the force of water directed against blades fastened to a vertical or horizontal shaft.

hydro. Meaning water or the presence of hydrogen; a prefix used to identify a type of generating station in which the prime mover is operated by water power.

hydrocarbon. Any of a large class of organic compounds composed solely of carbon and hydrogen. The compounds having a small number of carbon and hydrogen atoms in their molecule are usually gaseous; those with a larger number of atoms are liquid and the compounds with the largest number of atoms are solid.

hydrodynamics. The branch of science that deals with the cause and effect of fluid migration; that branch of hydraulics which relates to the flow of liquids over weirs, or through pipes, channels, and openings.

hydroelectricity. System in which the potential energy of natural water is used after harnessing it by impounding it in dams, or by releasing it through turbogenerators. The conversion of the kinetic energy in moving water (generally first held behind a dam) to mechanical (rotary) energy, and then to electricity by a generator. Sometimes considered a solar technology.

hydroelectric plant. An electric power plant in which the energy of falling water is converted into electricity by turning a turbine generator.

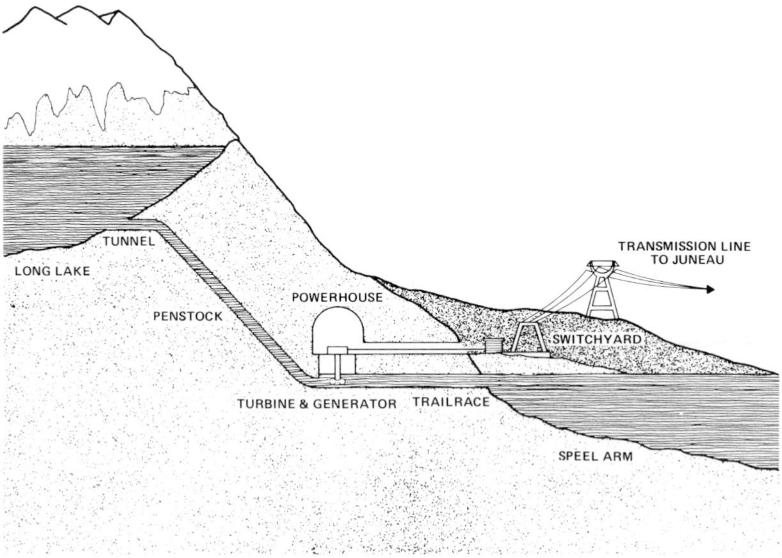

Courtesy DOE

Pictorial diagram of basic system elements of a hydroelectric facility.

hydrogenase. An enzyme of various microorganisms that promotes the formation of gaseous hydrogen.

hydrogenation. A process whereby hydrogen atoms are added to an organic molecule to form a new compound; such reactions usually require heat and pressure in the presence of a catalyst.

hydrogen-chlorine battery. An advanced battery system which uses gaseous hydrogen and chlorine as reactants.

hydrogen economy. Use of hydrogen (which has been separated from water electrically at central production centers) as an energy source to be piped to local storage or usage areas. The hydrogen would be burned to produce heat, used as a transportation fuel, in industrial processes, or in fuel cells to generate electricity.

hydrogen fuel. Use of hydrogen gas as a fuel. Hydrogen is virtually an ideal fuel since it burns in air to form nonpolluting water vapor. Hydrogen would be easily transported in existing natural gas lines and readily stored near where it is needed for power generation.

hydrogen sulfide (H_2S). A malodorous gas made up of hydrogen and sulfur with the characteristic odor of rotten eggs. It is emitted in the natural decomposition of organic matter and is also the natural accompaniment of advanced stages of eutrophication.

hydrologic cycle. The complete cycle of phenomena through which water passes, commencing as atmospheric water vapor, passing into liquid and solid form as precipitation, thence along or into the ground surface, and finally again returning to the form of atmospheric water vapor through evaporation and transpiration. Also called the water cycle.

hydrology. The science dealing with the properties, distribution, and circulation of water and snow.

hydrolysis. Decomposition of a chemical compound by reaction with water. For example, starch and cellulose can by hydrolyzed by acids and enzymes to produce simple sugars which can be fermented to produce ethanol.

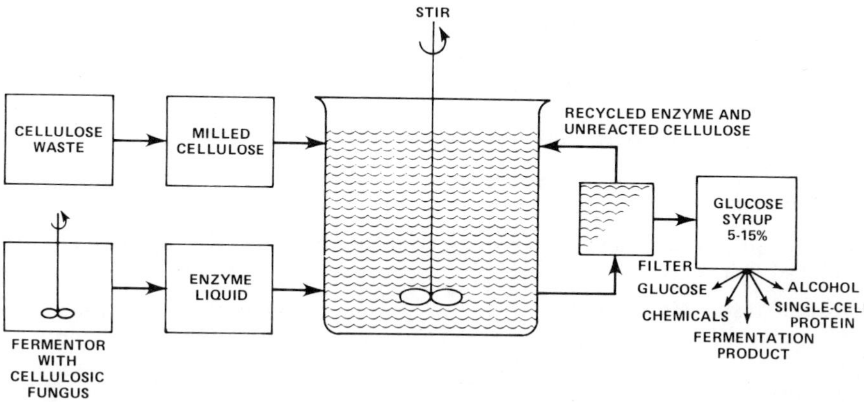

Reprinted with permission from Russell E. Anderson, Biological Paths to Self-Reliance, *Van Nostrand Reinhold*

Process block diagram of enzymatic cellulose hydrolysis.

hydrometer. A long-stemmed glass tube with a weighted bottom. It floats at different levels depending on the relative weight (specific gravity) of the liquid; the specific gravity or other information is read where the calibrated stem emerges from the liquid. (See diagram p. 168.)

hydronic. A solar heating system that heats water in its collector and stores heat in a water tank; a water-based heating system.

hydropower. Power produced by falling water. See **hydroelectricity.**

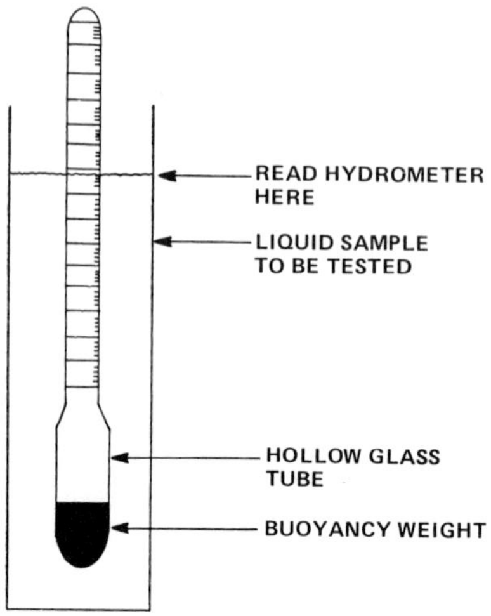

Courtesy DOE

The buoyancy weight at the bottom determines where the hydrometer will float. The hollow glass tube shows the specific gravity, degrees balling (percent sugar), or proof.

hydrosphere. The aqueous envelope of the earth including the ocean, all lakes, streams, and underground waters, and the water vapor in the atmosphere.

hydrostatic design stress. The estimated maximum tensile stress due to internal hydrostatic pressure in the wall of the pipe in the circumferential orientation that can be applied continuously with a high degree of certainty that failure of the pipe will not occur.

hydrostatic head. The pressure created by the weight of a column of water.

hydrostatic pressure. The pressure of, or corresponding to, the weight of a column of water at rest.

hydrostatic test. A strength test of equipment (pipe) in which the item is filled with liquid, subjected to suitable pressure, and then shut in and the pressure monitored; a test to determine whether a container will hold a certain pressure; also a test to locate leaks or to prove that there are no leaks.

hydrostorage. Technique to store power utilizing a dam by pumping water into a reservoir, to be drawn out when power needs to be generated.

hydrothermal reservoir. A body of porous, permeable rock, gravel, or soil containing natural steam or naturally heated water at a temperature significantly above the average temperature at the earth's surface.

hygrometer. A device for measuring moisture in the air.

hygroscopic. When a substance has the ability to absorb moisture from the air, it is hygroscopic. Ethanol is hygroscopic. If 200-proof ethanol is left open to moisture-laden air, it will quickly absorb water and dilute itself back to 192-proof. However, lower proof alcohol absorbs moisture slowly, so that a straight alcohol fuel (which usually is no higher than 180-proof) will not, if properly stored, have a hygroscopic problem. (Alcohol fuel can be adequately stored by the same methods used for storing gasoline.)

hygrostat. A device sensitive to humidity changes and arranged to actuate other equipments when a predetermined humidity is attained.

hysteresis. The time lag exhibited by a body in reacting to changes in the forces affecting it.

I

IBC cell. Interdigitated back contact photovoltaic cell.

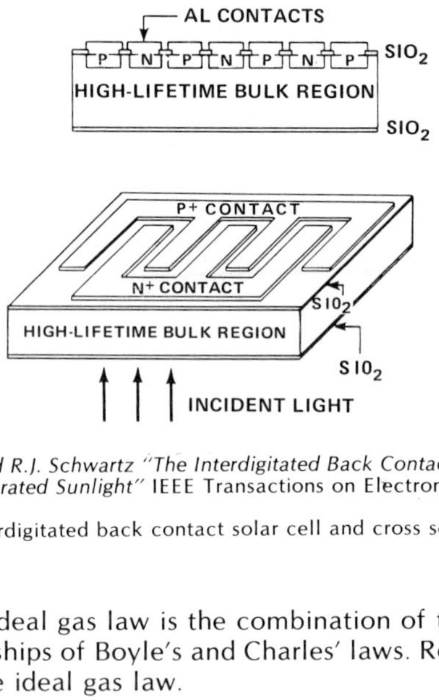

Courtesy M.D. Lammert, and R.J. Schwartz "The Interdigitated Back Contact Solar Cell: A Silicon Solar Cell for Use in Concentrated Sunlight" IEEE Transactions on Electron Devices, *vol.ED-24(1977) p.337*

An interdigitated back contact solar cell and cross section view.

ideal gas law. The ideal gas law is the combination of the volume, temperature, and pressure relationships of Boyle's and Charles' laws. Real gases deviate by varying amounts from the ideal gas law.

ignition control. Gasoline additives to reduce deposits in the combustion chamber.

ignition point. Of solids and liquids, the minimum temperature at which combustion can occur, but at which it is not necessarily continuous; of combustible gases, the flash point.

ignition temperature. The lowest temperature at which, in the presence of oxygen, combustion will be independent of a supply of external heat.

illuminants. The group of unsaturated or heavy hydrocarbons in a manufactured gas, such as ethylene and benzene, which burns with a luminous flame.

image furnace. Apparatus for the production of a very high temperature in a small space by focusing the radiation from the sun (solar furnace), or from an electric arc (arc-image furnace), by means of mirrors and/or lenses.

immersion-formed layer. Layer formed by immersion of a solid in a liquid, with the layer formation usually occurring because of electrolytic action.

immiscible. Pertaining to substances that are not capable of mixing; for example, oil and water.

impeller. Rotating member of centrifugal pump, which receives inflowing water at or near its center and accelerates it radially to the periphery, where it is discharged with the kinetic energy (initial + added) needed to carry it through the pumping system.

impermeable. Having a texture that does not permit water to move through it perceptibly under the head difference ordinarily found in subsurface water. Synonym for impervious.

impermeable dry rock. Geothermal regions where the heat is contained almost entirely in impermeable rock of very low porosity; a source of thermal energy.

impoundment. A body of water, such as a pond, confined by a dam, dike, floodgate, or other barrier.

impulse trap. Under normal loads, the discharge from this trap is intermittent. When the load is heavy, however, the discharge is continuous. This type of trap may be used on any equipment where the pressure at the trap outlet does not exceed 25 percent of the inlet pressure.

impulse turbine. A turbine driven by high-velocity jets of water or steam which impinge on some kind of vane or bucket attached to a wheel. The high-velocity jets are produced by forcing the water and steam through a nozzle.

inch of mercury. A pressure unit representing the pressure required to support a column of mercury 1 inch high at a specified temperature; 2.036 inches of mercury (at 32°F [0°C] and standard gravity of 32.174 feet per second per second [ft sec^2]) is equal to 1 pound per square inch (psi)(6.9 kilopascals).

inch of water. A pressure unit representing the pressure required to support a column of water 1 inch high. Usually reported as inches W. C. (water column); 27.68 inches of water is equal to 1 pound per square inch (psi)(6.9 kilopascals).

incident. In reference to solar radiation, those light rays falling on or striking a surface.

incident angle. The angle between the sun's rays and a line normal to the irradiated surface. In the case of tubular collectors, the angle between the normal to the tube centerline and the sun's rays.

incident light. The amount of light reaching an object.

incident radiation. The amount of radiant energy incident on a surface per unit time and unit area.

incident sunlight. Direct sunlight.

incineration. The controlled process by which solid, liquid, or gaseous combustible wastes are burned and changed into gases; the residue produced contains little or no combustible material.

incomplete combustion. A term applied to combustion in which all of the fuel is not burned; for example, leaving unburned carbon in ashes.

index of refraction. The ratio of the velocity of light, or of other radiation, in the first of two media to its velocity in the second as it passes from one medium into the other, the first medium usually being taken to be a vacuum or air. Also called refractive index.

indicated horsepower. The horsepower determined from the pressure-volume indicator diagram. This is the power developed within the cylinder of the engine and is more than the power delivered at the driving shaft, by the amount of mechanical friction.

indirect energy. Any form of energy used in the manufacture of goods or provision of services which themselves may or may not use energy.

indirect gain solar. A type of passive solar heating system in which the storage is interposed between the collecting and the distributing surfaces (e.g., trombe wall, water wall, or roof pond).

Courtesy HUD

Sun passes through solar glazing and is stored in trombe wall. Vents provide warm air circulation flow.

indirect-gain system. Passive solar heating system in which the sun directly warms a heat storage element in one area of house, and heat is then distributed from that element to the rest of house by natural convection, conduction, or radiation.

indirect heating surface. Any type of heating surface which does not come in direct contact with the original source of heat.

indirect system. See **closed-loop system.**

Indolene. A standard mixture of chemicals used in comparative performance tests of automotive fuels.

induced ventilation. The thermally assisted movement of fresh air through a building.

induction. The production of magnetization or electrification in a body by the proximity of magnetized or electrified bodies, or of an electric current in a conductor by the variation of the magnetic field in its vicinity.

induction generator. A type of electric generator which runs above synchronous speed and draws all of its excitation (reactive power) from the system it is feeding into. (It generates at a constant power factor.)

induction meter. An instrument for measuring alternating current circuits that will measure both power and energy. It is the type used for household and other consumer electricity meters.

industrial air conditioning. Air conditioning in industrial plants where the objective is the furtherance of a manufacturing process rather than the comfort of persons.

industrial alcohol. Ethyl alcohol produced and sold for other than beverage purposes; depending on the use, may or may not be denatured.

industrial degree-day. A degree-day unit based on 45°F (7°C) or 55°F (13°C) mean daily temperature so as to be applicable to industrial buildings maintained at relatively low temperatures.

industrial process heat. The end-use applications of thermal energy in industry covering a wide range of temperatures. The term is also used to indicate that portion of the national solar program which deals with this end-use.

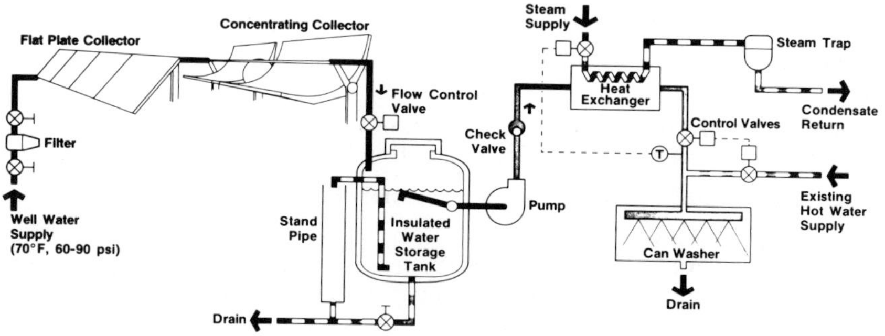

Courtesy DOE

Process flow diagram of use of solar energy for industrial process hot water.

industrial revenue bonds (IDR bonds). Debt incurred through industrial revenue authorities in numerous states; since such authorities are permitted to issue bonds for sale to the public which are payable in tax-exempt interest rates, IDR or tax-exempt bonds normally reflect an interest rate substantially below the prime rate.

inert. Not acted upon chemically by the surrounding environment. Nitrogen and carbon dioxide are examples of inert constituents of natural gases; they dilute the gas and do not burn and thus add no heating value.

inert gas. A gas that does not react with other substances under ordinary conditions.

inertial force. The fictitious force acting on a body as a result of using a noninertial frame of reference; examples are centrifugal and Coriolis forces that appear in the rotating-earth coordinate system.

inertial separator. An air-pollution control device that uses the principle of inertia to remove particulate matter from a stream of air or gas.

infiltration. Inward air leakage through cracks and joints in any building element, and around or through windows and doors of a building; it is caused by the pressure, effects of wind, and/or the effect of differences in the indoor and outdoor temperatures.

influent. A flow into a building or system.

infrared. Invisible long wavelength radiation (heat); also capable of producing a thermal or photovoltaic effect, though less effective than visible light.

infrared radiant heater. A self-contained, vented or unvented heater used to convert the flame energy to radiant energy, a substantial portion of which is in the infrared spectrum, for the purpose of direct heat transfer.

infrared radiation. Radiant energy of wavelengths exceeding 7600 angstroms and not visible to the human eye. Sunlight contains some 60 percent of such rays.

infusible. Not transformable by heating from solid to liquid state under specified conditions of pressure, temperature, and time.

initial cost. The amount you pay for a solar system when you first buy it; sometimes called the first cost.

inlet. Where the air or water enters a solar collector or other device.

inoculum. A small amount of bacteria produced from a pure culture which is used to start a new culture.

inorganic. All substances that do not contain carbon as a constituent, such as metals, rocks, minerals, and a variety of earths.

in-plane. Denotes wind-energy conversion system blade motion in the direction of the rotor plane.

inside coil. Portion of a heat pump that is located inside the building and functions as the heat-transfer point for warming or cooling inside air.

In-sider®. A glazing product of Plaskolete Inc. which is a transparent 0.040 inch (1.02 mm) semirigid sheet of vinyl.

in situ. In the natural or original position; applied to a rock, soil, or fossil when occurring in the situation in which it was originally formed or deposited.

insolation. The solar power density incident on a surface of stated orientation, usually measured in W/m^2 or $Btu/ft^2/hr$. This implies total solar insolation which is the sum of diffuse and direct components on a unit surface in unit time.

installation cost. The cost of getting a solar system from the crates in which it is delivered to a working solar system. Installation costs are a part of the first cost.

installed capacity. The maximum load for which a machine, apparatus, device, plant, or system is designed or constructed, not limited by existing service conditions.

instantaneous efficiency. Pertaining to a solar collector, as the amount of energy removed by the transfer fluid per unit of transparent frontal area over a given 15-min period divided by the total incident solar radiation onto the collector per unit area for the 15-min period.

instantaneous peak demand. The maximum demand at the instant of greatest load.

insulate. To separate or to shield (a conductor) from conducting bodies by means of nonconductors, so as to prevent transfer of electricity or heat.

insulated-plate collector. An unglazed solar collector with an insulated metal plate but no means for extracting solar heat.

insulation. General measure of thermal resistance. Insulation resists the flow of heat.

insulator. A material that prevents heat from flowing easily by conduction. High insulation corresponds to high thermal resistance; opposite to a conductor.

insurance. Research on potential hazards of solar heating and cooling has begun in several areas. The National Association of Mutual Insurance Companies has compiled guidelines on insuring new systems, as well as processing claims. To date, the types of claims filed include water damage and freezing, and damage to ground-mounted structures. In general, no claims have been denied on the basis of their deriving from a solar-energy system. There have been no reports of added costs due to solar or of premium penalties due to the inclusion of a solar energy system.

integral collector/storage system. An integrated solar water-heating system that combines the collection and storage of solar energy in a water tank(s). A typical breadbox system consists of an insulated box containing water tanks painted black with solar glazing and insulating lids.

integrated average insolation. The total energy per unit area received by a surface for a specified time period divided by the time period (W/m^2, $Btu/h/ft^2$).

integrated demand. The demand averaged over a specified period, usually determined by an integrating demand meter or by the integration of a load curve. It is the average of the instantaneous demands during a specified demand interval.

integrated farm system. A farm designed to produce, process, and use agricultural biomass for energy production.

integrated pest control. A system of managing pests by using biological, cultural, and chemical means.

integration. The combination of isolated components (e.g., of technology) into an interwoven system.

intensity of radiation. At a given place, the energy per unit time entering a sphere of unit cross-sectional area centered at that place. The unit of intensity is the erg per square centimeter second ($erg/cm^2/sec$) or the watt per square centimeter (W/cm^2).

intensive forestry. A practice of forestry which results in a high level of volume and quality of output per unit area, through the application of the best techniques of silviculture and management.

interconnected system. A wind-energy conversion system which is electrically connected to the utility grid in a manner that allows it to feed power into that grid.

intercooling. Extraction of heat from a compressed gas between two stages of compression.

intercropping. Planting different crops in alternate rows to increase the effective land area.

interfacial energy. Tension at interfaces between the various phases of a system which may include solid, liquid, and gas interfaces, of varying combinations and qualities.

intermediate plant. An electrical plant which is used to meet daily or seasonal variations in electrical load. The annual average use of this type of plant is less than that of baseload plants, but more than peak-load plants.

intermittent stream. A stream that flows only in direct response to precipitation, such as a rainstorm or melting snow. It receives little water from springs and is dry for a large part of the year.

internal combustion. Pertaining to any engine in which the heat source pressure necessary to produce power is developed in the engine cylinder by the burning of a mixture of air and fuel, and converted into mechanical work by means of a piston. The automobile engine is a common example.

internal gain. Combined heat release from all heat sources in a building other than its heating system. Major sources include lighting, people, machinery, and combustion.

internal mass. Massive materials with heat-storage potential contained within the building as walls, floors, or freestanding elements.

internal rate of return. Expected rate of return if a solar system is viewed as an investment.

International Solar Energy Society (ISES). The principal worldwide organization for exchanging technical information on all forms of solar energy. U.S. headquarters are located at Killeen, TX. The international headquarters' mailing address is P. O. Box 52, Parkville, Victoria, Australia 3052.

interruptible service. Low-priority service offered to customers under schedules or contracts which anticipate and permit interruption on short notice, generally in peak-load seasons, by reason of the claim of firm-service customers, and higher priority users. Unlike off-peak service, gas is available at any time of the year if the supply is sufficient.

interstate waters. According to law, waters defined as: (1) rivers, lakes, and other waters that flow across or form a part of state or international boundaries; (2) waters of the Great Lakes; (3) coastal waters—whose scope has been defined to include ocean waters seaward to the territorial limits and waters along the coastline (including inland streams) influenced by the tide.

intrinsic layer. A layer of semiconductor material whose properties are essentially those of the pure undoped material.

inulin. A polymeric carbohydrate comprised of fructose monomers found in the roots of many plants, particularly Jerusalem artichokes.

inventory. The supplies of all factors of production which must be maintained in storage, i.e., grain, wood, enzymes, etc.

inversion. An atmospheric condition where a layer of cool air is trapped by a layer of warm air so that it cannot rise. Inversions spread polluted air horizontally rather than vertically so that contaminating substances cannot be widely dispersed. An inversion of several days can cause an air-pollution episode.

inverted bucket trap. The discharge from the inverted bucket trap is intermittent and requires a differential pressure between the inlet and discharge of the trap to lift the condensate from the heating coils. Inverted bucket traps are well suited for draining condensate from steam lines or equipment where an abnormal amount of air is to be discharged and where dirt may drain into the trap.

inverter. A device which converts direct current to alternating current and generates its own frequency and voltage references. See **synchronous inverter.**

in vitro. Isolated from a living organism and artificially maintained.

iodine test. Iodine reacts with starch to form a blue to black coloration, but it does not react with sugar. The iodine test is used to determine if all the starch in alcohol mash has been converted to sugar.

ion. An atom or group of molecularly bound atoms that carries a positive or negative electric charge as a result of having lost or gained one or more electrons. When a salt dissolves in water, it dissociates, i.e., splits up into two charged ions (a positive and a negative), e.g., common salt, sodium chloride, NaCl, in water becomes $Na^{+}Cl$.

ion exchange. A chemical process in which ions from two different molecules are exchanged.

ionization. The removal of some or all electrons from an atom, leaving the atom with a positive charge, or the addition of one or more electrons, resulting in a negative charge.

ionizing radiation. Any radiation capable of displacing electrons from atoms or molecules, thereby producing ions. Examples: alpha, beta, and gamma radiation; shortwave ultraviolet light.

ion pair. A closely associated positive ion and negative ion (usually an electron) having charges of the same magnitude and formed from a neutral atom or molecule by radiation.

irradiance. Radiant flux density. The amount of radiant flux power per unit area that flows across or onto a unit area of surface.

irradiation. Solar radiation, direct, diffuse, and reflected, that strikes a surface. The radiant energy incident per unit area. The product of flux density (irradiance) and its duration.

isallobaric wind. The wind velocity whose Coriolis force exactly balances a locally accelerating geostrophic wind.

isobar. An imaginary line, or a line on a map or chart, connecting or marking places on the surface of the earth where the height of the barometer reduced to sea level is the same either at a given time or for a certain period.

isolated solar gain. A type of passive solar heating system in which heat is collected in one area to be used in another (e.g., greenhouse or attic collector).

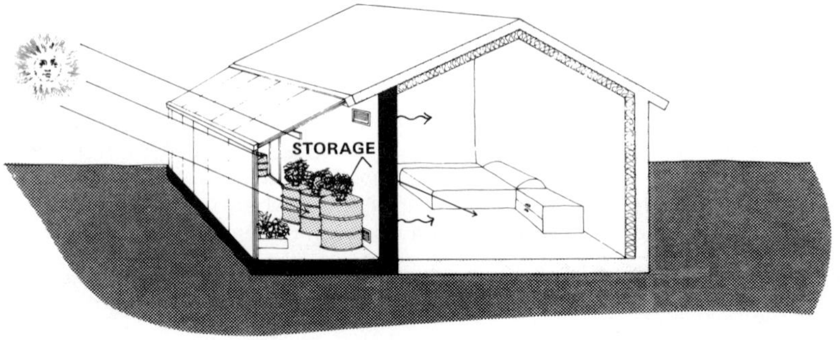

Courtesy HUD

Solar energy is stored in water-filled tanks in greenhouse. Secondary or isolated solar gain occurs when air flows through.

isolation. The solar power density incident on a surface of stated orientation, usually measured in W/m^2 or Btu/ft^2/hr. This implies total isolation, which is the sum of diffuse and direct components.

isopsychric. Constant relative humidity.

isotherm. A line on a map or chart of the earth's surface connecting points having the same temperature at a given time or the same mean temperature for a given period.

isothermal. A change taking place at a constant temperature

isotropic distribution. Approximation (when referring to diffuse radiant energy fluxes from the sky) in which the radiance of the sky is taken to be independent of position in the sky.

I-V **curve.** A graphical presentation of the current versus the voltage from a photovoltaic cell as the load is increased from the short circuit (no load) condition to the open circuit (maximum voltage) condition. The shape of the curve characterizes cell performance.

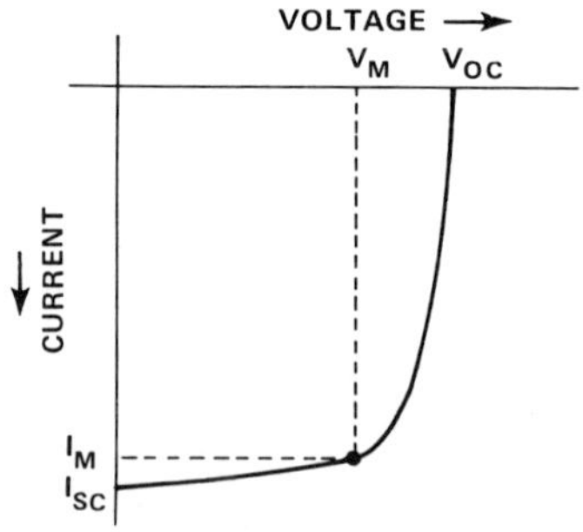

Courtesy DOE

Solar cell current (*I*)-voltage (*V*) characteristic curve.

jacket. The enclosure on an appliance such as a water heater, furnace, or boiler; the space surrounding a cylinder of an engine through which a cooling liquid flows or which provides structural support.

jobber price. The price at which a petroleum jobber purchases a refined product from a refiner or terminal operator.

joint compounds. Materials to be used on pipe joints, primarily to lubricate the threads, and secondarily to prevent leakage.

joist. One of a series of evenly spaced horizontal structural beams, usually made of lumber, steel, or concrete, for supporting floors and ceilings.

joule. The absolute meter-kilogram-second unit of work or energy that equals 10^7 ergs or approximately 0.7375 foot-pound (ft-lb) or 0.2390 gram calorie (g cal).

Joule's law. (1) The rate at which heat is produced by a steady current in any part of an electric circuit is jointly proportional to the resistance and to the square of the current. (2) The internal energy of an ideal gas depends only on its temperature and is irrespective of volume and pressure.

katabatic wind. Any wind blowing down an incline. If the wind is warm, it is called a foehn; if cold, it may be a fall wind, such as the bora.

Kata thermometer. A thermometer developed by Heberden, Hill, and Griffiths which is heated before being exposed to the environment to be measured and the rate of fall noted. The rate of fall of the dry-bulb is a measure of the radiation and convection; of the wet-bulb, a measure of evaporation plus radiation and convection. If the dry-bulb is cooled in a vacuum bottle, the rate is a measure of radiation alone.

ketone. Any of various organic compounds containing a carbonyl group (CO) and a hydrocarbon group such as CH_3. The carbonyl group is linked to the hydrocarbon groups in the middle of the chain resulting in at least one hydrocarbon group on each side of the carbonyl group.

kilogram (kg). Formerly a force unit under the older Metric System. Under the SI Metric System, force is now designated by the newton, symbol N. In the SI Metric System, the kilogram is used strictly in reference to mass.

kilovolt-ampere (kVA). A unit of electric measurement equal to the product of a volt (V) and an ampere (A); 1000 volt-amperes.

kilowatt (kW). A unit of power equal to 1000 watts (W) or to energy consumption at a rate of 1000 joules per second (J/s). It is usually used for electrical power. An electric motor rated at 1 horsepower (hp) uses electrical energy at a rate of 776 W, or 0.776 kW.

kilowatt-hour (kWh). The unit of energy equal to that expended in 1 hour at a rate of 1 kilowatt (kW) or 3413 British thermal units (Btu). Domestic water heating for a family of four requires 400 kWh per month on the average.

kinetic energy. The energy associated with an object's motion; varies directly in proportion with the object's mass and with the square of its velocity.

King-Lux®. A solar product of Kingston Industries Corporation which is an insulating sheet of preanodized aluminum 0.010 to 0.040 inch (0.25 to 1.02 mm) thick.

kite anemometer. A kite which has been calibrated to give quantitative wind-speed data.

knock. The sound of "ping" associated with the autoignition in the combustion chamber of an automobile engine of a portion of the fuel-air mixture ahead of the advancing flame front.

knot. Measure of wind speed; equal to 1 nautical mile per hour.

kool motor. A blend of 16 percent alcohol and 84 percent gasoline used in British vehicles in the 1930s.

Krebs cycle. Cyclic series of reactions that occur in living organisms, forming a phase of the metabolic function in which acetic acid (or acetyl equivalent) is oxidized through intermediate acids to CO and H_2O, thus providing energy for storage in the form of energy-rich phosphate bonds.

K-value. Btu/hr/ft²/°F per inch. See **conductivity.**

Kynar®. A solar glazing product of Westlake Plastics Co. which is a thin sheet of polyringlidene flouride.

L

labor-intensive. Indicates that a relatively large percentage of costs are labor-related.

lactic acid ($C_3H_6O_3$). The acid formed from milk sugar (lactose) and produced as a result of fermentation of carbohydrates by bacteria called *Lactobacillus*.

lactose ($C_{12}H_{22}O_{11} \cdot H_2O$). Milk sugar, saccharum lactis. This is a sugar derived from whey. Cow's milk contains about 5 percent lactose. Lactose is fermented by the yeast *Saccharomyces fragilis*.

lagoon. In wastewater treatment, a shallow pond usually man made where sunlight, bacterial action, and oxygen interact to restore wastewater to a reasonable state of purity.

laminar flow. This describes a streamline flow of the alcohol/water channeling down the inside of a packed column. This condition is undesirable, as the alcohol has no chance to be vaporized. Usually, laminar flow results from a poor choice of packing, such as lathe turnings, etc.

laminated glass. Two or more sheets of glass held together by an intervening layer or layers of plastic material.

langley. A measure of solar radiation; it equals 1 calorie per square centimeter (cal/cm^2), or 3.69 Btu per square foot (Btu/ft^2).

lapped joint. A pipe joint made by using loose flanges on lengths of pipe whose ends are turned over or lapped over to produce a bearing surface for a gasket or metal-to-metal joint.

lap weld pipe. Pipe made by welding along a scarfed longitudinal seam in which one part is overlapped by the other.

latent heat. The change in heat content that occurs with a change in phase and without change in temperature; the heat stored in the material during melting or vaporization. Latent heat is recovered by freezing a liquid or by condensing a gas.

latent heat-of-fusion storage. The use of phase change material, such as Glauber's salt, for thermal storage. The change of phase in a suitable temperature range releases or absorbs the latent heat of fusion.

latent heat of vaporization. The quantity of heat required to change a unit weight of liquid to vapor with no change in temperature.

latitude. How far a point on the earth's surface is from the equator. A solar collector near the equator (low latitude) usually has a smaller tilt angle than one nearer to the poles (high altitude.)

law of thermodynamics. The First Law of Thermodynamics states that energy can neither be created nor destroyed. The Second Law of Thermodynamics states that when a free exchange of heat takes place between two bodies, the heat is always transferred from the warmer to the cooler body. See **first law of thermodynamics, second law of thermodynamics.**

law of the cube. This formula states that at any given moment the power content of the wind is proportional to the cube of the wind velocity. When the wind speed doubles, the power output from a wind machine will increase eight times.

leachate. Water which flows through the soil or ground and ends up in a water course. If the water is contaminated initially and does not receive enough purification as it passes through the soil, the leachate can cause pollution. Similarly if rain, etc., falls into a highly organic sludge, it can carry polluting material with it. Liquid that has percolated through solid waste or other mediums and has extracted dissolved or suspended materials from it.

leaching. The process by which soluble materials in the soil, such as nutrients, pesticide chemicals, or contaminants, are washed into a lower layer of soil or are dissolved and carried away by water.

lead-acid battery. Conventional automobile battery; uses lead and lead oxide for electrodes and sulfuric acid for the electrolyte; secondary or storage battery which can be recharged electrically, as opposed to a primary or dry-cell battery which is used until the chemical compounds that produce the electricity have been exhausted and it is then discarded.

leader. In warm-air heating, the supply duct running horizontally from the furnace to the riser, or stack. In plumbing, a water conductor from the roof to the storm drain.

leading edge. In wind-energy conversion systems the leading edge is the point on an airfoil which is the farthest from the trailing edge; area of blade surface which is the first to come in contact with the air flow.

Courtesy DOE

Front edge of wind-energy conversion system blade.

leafy plants. Deciduous plants which block the sun in summer but allow solar heating in winter; used for seasonal control of solar radiation.

lean fuel mixture. An excess of air in the air/fuel ratio; gasohol has a leaning effect over gasoline because the alcohol adds oxygen to the system.

learning curve. The ability of manufacturers to reduce the unit cost of a given "item" as total production volume accumulates. A product is said to be on an 80 percent learning curve if the cost of the product drops 20 percent with a doubling of cumulative production.

lead-acid battery. Conventional automobile battery; uses lead and lead oxide for electrodes and sulfuric acid for the electrolyte; secondary or storage battery which can be recharged electrically, as opposed to a primary or dry-cell battery which is used until the chemical compounds that produce the electricity have been exhausted and it is then discarded.

leader. In warm-air heating, the supply duct running horizontally from the furnace to the riser, or stack. In plumbing, a water conductor from the roof to the storm drain.

leasing. A form of financing whereby debt and ownership are retained by a third party who is normally not involved in the management of the operation; allows for the loan of both debt and equity capital to the market.

leclanche cell. A zinc-carbon primary cell whose exciting liquid is a solution of salammoniac.

lee wave. A gravity wave train which is excited by a mountain or ridge and extends downwind, in the lee of the mountain. Wave clouds frequently form at the crest of each wave.

legume. A class of plants whose roots are infected by bacteria which synthesize ammonia for the plant's use, thus reducing the need for nitrogen fertilizers.

lethe. A measure of air purity. The term, coined by Dr. W. D. Wells, University of Pennsylvania, is the equivalent of one complete air change of pure air. The air quantity required for 1 lethe varies with the size of the room but is independent of time.

levelized annual cost. An annual sum, which, if expended each year over a specified time for equipment or services, would be equivalent to the summation of all actual costs during the same period, for fixed and variable charges, including burdens.

levelized busbar energy cost. The constant annual revenue per unit of energy required over the lifetime of a plant to compensate for its fixed and variable costs, interest costs, and shareholder return (mills per kWh).

Lexan®. A glazing product of General Electric Company which is a transparent plate of polycarbonate.

life cycle. The phases, changes, or stages an organism passes through during its lifetime.

life-cycle costing. A method of cost analysis in which operating, maintenance, fuel, and other ownership costs are estimated for predicted lifetime of a device, and considered along with initial cost; often used to compare costs of solar heating or cooling systems and conventionally fueled systems.

lift. The force which "pulls" a wind turbine blade along, as opposed to drag.

lift device. Wind-energy conversion system which depends primarily on lift to extract energy from the wind.

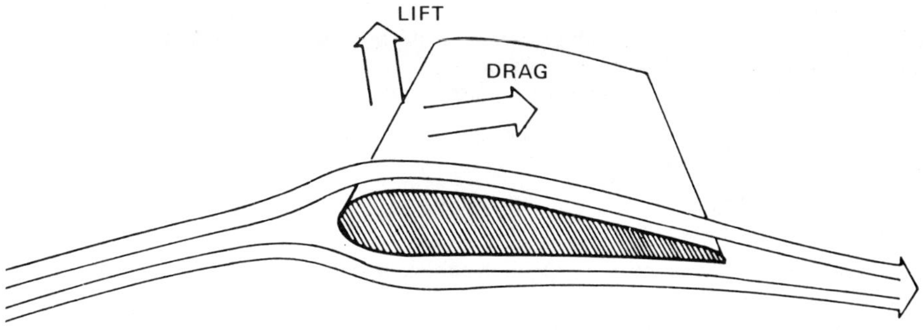

Courtesy DOE

Forces of lift and drag acting on a wind-energy conversion system blade.

light. A form of radiant energy which gives rise to the sensation of sight. The portion of the electromagnetic spectrum considered visible light in wavelengths from approximately 430 millimicrons to 690 millimicrons (1 millimicron = 10^{-9} meter).

lignified cellulose. Cellulose polymer wrapped in a polymeric sheath (ligin) and extremely resistant to hydrolysis because of the strength of its linkages.

lignin. A polymeric, noncarbohydrate constituent of wood that functions as a binder and support for cellulose fibers. Highly resistant long-chain molecules (polymer of aromatic alcohols) which constitute up to 25 percent wood's weight.

lime dehydration. Technique for drying or dehydrating alcohol fuel. The liquid fuel is poured into a tank with lime calcium oxide or CaO. The lime reacts with any water present to form calcium hydroxide, an insoluble compound. The water-free alcohol is drained off the settled lime.

limit switch. A control to limit some function. Examples are pressure limit switches which shut off the fuel burner when the steam pressure reaches a predetermined point; similarly, temperature limit switches for hot water and warm air.

limited partnership. Legal mechanism in which investors can limit their liability by investing in a "managing partner" who operates the company; the managing partner is usually a corporation and limited partners are allocated tax credits and depreciation in a ratio which may exceed their pro rata share; basically used for tax-shelter-type investments, since the limited partners can be allocated depreciation, tax credits/losses which exceed their original investments.

limnology. The study of the physical, chemical, meteorological, and biological aspects of fresh waters.

line drop. Loss in voltage owing to the resistance of conductors conveying electricity from a power station to the consumer.

line transformer. A transformer classified as distribution line equipment, generally having a rated primary voltage of 2,300 to 15,000 volts.

linkage. The bond or chemical connection between constituents of a polymeric molecule.

liquefaction. The change in the phase of a substance to the liquid state; in the case of fermentation, the conversion of water-insoluble carbohydrate to water-soluble carbohydrate. The process of breaking down and dissolving organic solids, by the action of bacterial enzymes outside their cells.

liquid. Liquid storage accommodates heat in proportion to its specific heat capacity and temperature increase.

liquid-based solar heating system. A solar heating system in which liquid, either water or an antifreeze solution, is heated in the solar collectors.

liquid-dominated convective hydrothermal resources. Hydrothermal convective systems in the upper part of the earth's crust which transfer heat from a deep igneous source to a depth sufficiently shallow to be tapped by drill holes.

liquid fuel. Any liquid used as fuel which can be poured or pumped. Petroleum or crude oil is a natural liquid fuel, while distilled oil, coal tar, and residual oil are prepared liquid fuels.

liquid-to-air heat exchanger. A device that transfers heat from liquid to a gas, for example, a car radiator or a baseboard heater.

liquid-to-liquid heat exchanger. A device that transfers heat from a liquid to another liquid, for example, a coil in a heat storage tank.

liquid-type collector. A solar heat collector designed to use a liquid as heat-transfer fluid.

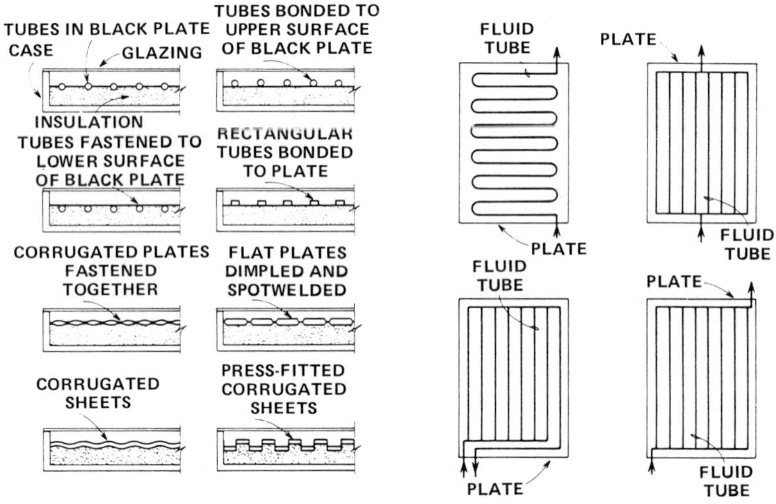

Courtesy - Popular Science Book/Harper and Row, Home Energy How-To *by A.J. Hand*

Flat-plate liquid-type solar collector configurations.

liter. The primary standard of capacity in the metric system, equal to the volume of 1 kilogram (kg) of pure water at maximum density, at approximately 4°C, and under normal atmospheric pressure. 1 liter = 0.264 gallon, 1.05 quarts, or 2.11 pints.

lithium-sulfur battery. An advanced battery, which may be able to store large amounts of energy per unit weight and size. It consists of a series of cells which contain lithium in the negative electrode and a metal sulfide in the positive electrode. The electrolyte is a molten salt. They operate at about 752° to 797°F (400° to 425°C). Present prototype lithium-sulfur cells of 200 to 250 watt-hour size have a storage capacity of about 40 to 50 watt-hours per pound; experimental versions have demonstrated 75 watt-hours per pound (in comparison, present lead-acid battery has a storage capacity of about 12 watt-hours per pound).

live steam. Steam direct from the boiler and under full pressure; distinguished from exhaust steam, which has been deprived of its available energy.

load. The energy tapped from any power source; in water quality use, the quantity of material carried by flowing water generally expressed as pounds per day; in the electric industry, the amount of electric power delivered or required at any specified point or points on a system. Load originates primarily at the power-consuming equipment of the customers.

load center. A point at which the load of a given area is assumed to be concentrated.

load duration. The time during which the load (utility power demand) exceeds a given magnitude. Usually summed for time periods of particular interest.

load duration curve. A graph made by plotting data in order of magnitude, against time intervals, for a specified period. The ordinate may be an absolute quantity or percentage.

load factor. The average power output of an energy system divided by its rated power output.

load-frequency control. The regulation of the power output of electric generators within a prescribed area in response to changes in system frequency or tie-line loading.

load growth. The growth in energy and power demands by a utility's customers.

load leveling. Deferment of certain loads to limit electrical power demand to a predetermined level.

load match. The phase relationship between the load on the utility network and the wind-generated power; i.e., wind-generated power peaking in the afternoon with maximum load occurring in the afternoon is a good load match. Load matching also applies to seasonal wind and wind variations.

load power. The load power of an energy load is the average rate of flow of energy through the terminals of that load when connected to a specified source.

load profile. Time distribution of building heating, cooling, and electrical loads.

local mean time. Hour angle of mean sun. Local apparent noon is the transit time at which the upper limb of the sun crosses the local meridian. When making sun observations, this discrepancy must be corrected by use of the equation of time from the *Nautical Almanac.*

local solar time. System of astronomical time in which the sun always crosses the true north-south meridian at 12 noon. This system of time differs from local time according to longitude, time zone, and equation of time.

local winds. Winds that differ, over a small area, from those appropriate to the general pressure distribution.

logarithmic velocity profile. The variation of the mean wind speed with height in the surface boundary layer.

log law. Log law refers to a wind profile in which the wind speed increases with the logarithm of height, or to wind profile descriptions that modify the logarithmic description to account for stability effects.

long-wave radiation. Infrared or radiant heat.

loss factor. Parameter characterizing solar collector operating conditions.

loss of head. The decrease of energy head between two points such as that caused by friction.

loss stop option. Option for photovoltaic manufacturing firm to halt photovoltaic production under government long-term agreement, when production results in losses.

louvers. Slanted overlapping slats arranged to allow the circulation of air.

low Btu gas (LBG). A gas composed of CO_2, O_2, CO, H_2, CH_4, and N_2 with a heating value of 100-200 Btu per standard cubic foot (Btu/scf).

lower heating value. The heat released during combustion of fuel if all products are cooled to room temperature while water remains as steam; the lower heating value is more representative of typical fuel burning than high heating value.

lowest normal tides. A plane of reference lower than mean sea level by half the maximum range. This does not take into account wind or barometric pressure fluctuations.

low head. Refers to generation of hydroelectric power with relatively small dams.

lowside. That portion of a mechanical refrigeration system under relatively low pressure, extending from the expansion valve through the evaporator to, but not including, the compressor.

lubricant. Oil, grease, graphite, and, in general, anything of the sort used to overcome friction and permit a freer action of parts.

Lucite®. A glazing product of Du Pont Co. which is a transparent plate of methyl methacrylate.

Lumar®. A glazing product of Martin Processing Inc. which is a thin sheet of polyester film.

lumen. The quantity of light required to illuminate 1 square foot to an average intensity of 1 foot-candle. By multiplying the mean candlepower by 3.63, the quantity of light expressed in lumens for the solid cone with a spread of 130° is obtained.

luminescence. Emission of light produced by the action of biological or chemical processes, or by radiation, or by any other cause except high temperature (which produces incandescence).

LWECS. Large-scale wind-energy conversion system.

macroencapsulation. A means for containing latent-heat storage material in small easily handled packages, such as cans or plastic chubs.

magma. Magma is the very hot liquid, or partly liquid, rock that is found under the entire surface of the earth, starting at depths of 6 to 9 miles (9.7 to 14.5 km). These depths are too great for economic utilization. In some places, however, the magma exists close to, or even at, the earth's surface—sometimes in large volume. No techniques are known yet for converting this energy into useful forms for commercial exploitation.

magnetic anomaly. Variation of the measured magnetic pattern from a theoretical or empirically smooth magnetic field on the earth's surface.

magnetic field. The field of force that exists around electrons traveling along a conductor or orbiting the atomic nucleus.

magnetic force. The mechanical force exerted by a magnetic field upon a magnetic pole placed in it.

magnetic south. "South" as indicated by a compass; changes markedly from one location to another because of the latitudinal relationship to the earth's magnetic fields.

maintenance. Proper maintenance of a solar energy system is important for good performance. Different kinds of systems require different upkeep tasks. Make sure to check with your contractor about the specific maintenance requirements of your system. For example, the condition of pumps or fans should be checked periodically and collectors should be kept clean. In systems circulating antifreeze mixed with water, proper concentration ratios must be maintained and pH balance tested at least once a year. Some contractors now offer maintenance contracts.

majority carrier. Current carriers (either free electrons or holes) which are in excess in a specific layer of a semiconductor material (electrons in the *n*-layer, holes in the *p*-layer) of a cell.

make-up air. Air brought into a building from the outside to replace that exhausted by a ventilating system.

make-up water. Water supplied to replenish that lost by leaks, evaporation, etc.

malleable iron. Cast iron heat-treated to reduce its brittleness. The process enables the material to stretch to some extent and to stand greater shock.

malt. Generally, malt refers to barley grains that have been artificially germinated. Certain enzymes are activated and others are newly formed when the seed germinates. These enzymes include cystase, phytase, the amylases, phosphorylases, proteases, and all those which the embryo plant will use in breaking down food material stored in the endosperm. The purpose of these versatile catalysts is both to tear apart the seed and to rebuild these molecules into the root and stem structures. For alcohol fuel production, it is the α- and β-amylase enzymes that are desired, and the malting process must be carefully controlled to achieve the optimum production of these.

maltase. An enzyme produced by the yeast cell which hydrolyzes maltose sugar to glucose (glucase, alpha-glucosidase).

maltose (maltobiose; malt sugar; $C_{12}H_{22}O_{11} \cdot H_2O$). The most common reducing disaccharide that will be used in alcohol fuel production. It is composed of two molecules of glucose. When starch is hydrolyzed with β-amylase enzymes, maltose sugar is the result.

manhole. An opening into a tank, boiler, furnace, vault, or other equipment through which a man can enter to service equipment; can be sealed with a removable plate or door.

manifold. A chamber or tube having several inlets and one outlet, or a single inlet and several outlets.

manometer. Any instrument which measures gaseous pressure. Measures pressure or a pressure difference by balancing the applied pressure against the hydrostatic head of a column of liquid of known density.

marginal cost. The cost of one additional unit within a group of like units.

market discount rate. The rate of return that can be expected on the best investment.

market penetration. The amount of a product that will be sold on a yearly basis as it gains consumer acceptability over a specified time.

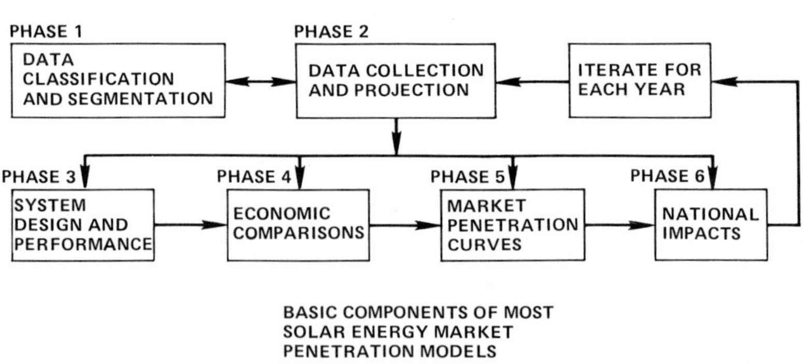

Courtesy SERI

Block diagram of market penetration model basic elements.

markup. Percentage added to direct costs to cover overhead, indirect costs, miscellaneous costs, and profit.

marsh. A low-lying tract of soft, wet land that provides an important ecosystem for a variety of plant and animal life but often is destroyed by dredging and filling.

mash. A mixture of grain and other ingredients with water to prepare wort for brewing operations.

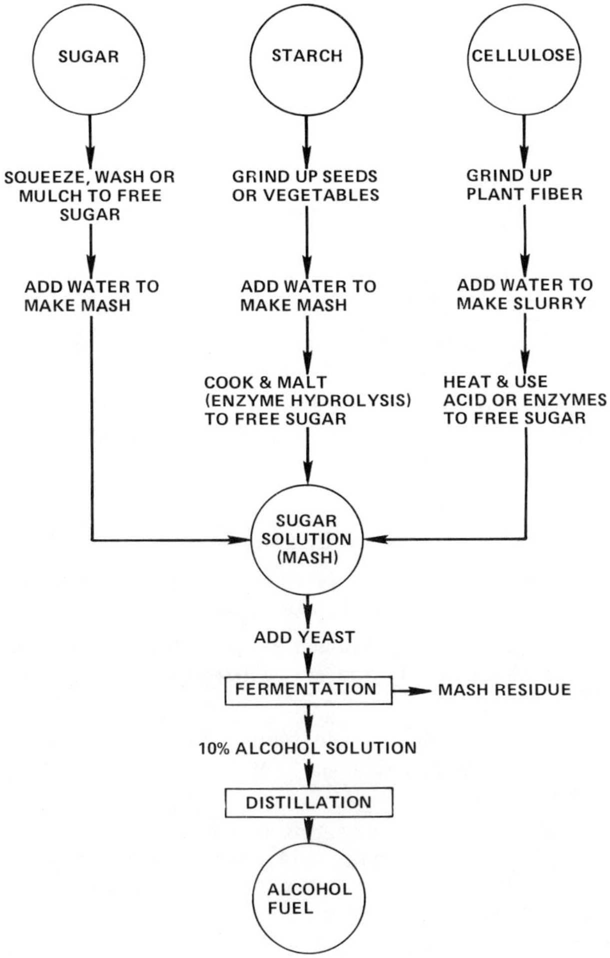

Courtesy DOE

Block diagram of sugar, starch, or cellulose conversion to mash and then alcohol fuel.

masonry wall. A passive solar house that lets solar radiation heat a glass-covered masonry wall. Heat is conducted into the house through the wall.

mass. A basic property of matter, for everyday purposes identical with weight. But even when objects become weightless (for example, in earth orbit), they lose none of their mass—they still have inertia and momentum. For the nuclear physicist, mass and energy are interchangeable. There are two ways of determining the mass of a body. One is to weigh it and apply Newton's Law (force equals mass times acceleration). The other is to apply a force to it and to calculate the mass from the resulting acceleration.

mass materials. Concrete, masonry, or other heavy materials used to store thermal energy for heating or cooling.

matter. The substance of which a physical object is composed. All materials in the universe have the same inner nature, that is, they are composed of atoms, arranged in different ways; the specific atoms and the specific arrangements identify the various materials.

maximum demand. The greatest of all the demands under consideration during a specified period of time.

maximum designed wind speed. The highest wind speed which a turbine is designed to withstand intact.

maximum flow temperture. Maximum temperature in a component when the heat-transfer fluid is flowing through the system.

maximum no-flow temperature. Maximum temperature in a component when the heat-transfer fluid is not flowing through the system.

maximum service temperature. Maximum temperature to which a component will be exposed in actual service, either with or without the flow of the heat-transfer fluid.

Mbh. One thousand British thermal units per hour.

meal. A granular substance produced by grinding.

mean daily temperature. Average of the minimum and maximum daily temperatures used to determine the number of degree days.

mean power output. The average power output of a wind-energy conversion system (WECS) at a given mean wind speed based on the Rayleigh frequency distribution.

mean radiant temperature. That single temperature of all enclosing surfaces which would result in the same heat emission as the same surfaces with various different temperatures.

mean temperature difference. In a heat exchanger, the temperatures of the two fluids change throughout. The mean or average of the temperature differences between the two fluids is the mean temperature difference. In general use, it is taken as the log mean temperature difference.

mean wind. The definition varies with context. If no averaging interval is specified, such as annual mean wind, the reference is probably indicative of an averaging interval just long enough to filter out the turbulence in the wind, e.g., 15 minutes to 1 hour.

mean wind speed. Arithmetic average wind speed over the specified time period and at specified height above ground level (the international [World Meteorological

Organizational] standard 10-m [32.8-ft] level may be implied; however, the majority of U.S. National Weather Service anemometers are at a height of 6.1 m [20 ft]).

mechanical equivalent of heat. The conversion factor for transforming heat units into mechanical units of work. One British thermal unit (Btu) equals 778 foot-pounds (ft-lb).

mechanical turbulence. The erratic movement of air caused by local obstructions such as buildings.

medium Btu gas (MBG). A gas composed of CO_2, O_2, CO, H_2, CH_4, and N_2 but with a higher proportion of H_2 and CH_4 than low Btu gas and a heating value of 250-500 Btu/scf.

median wind speed. The 50 percentile (i.e., 50 percent probable) wind-speed value.

medium pressure. When applied to valves and fittings, implies they are suitable for a working pressure of from 125 to 175 pounds per square inch (psi) (862 to 1207 kPa).

mega. A prefix that multiplies a basic unit by 1,000,000 (10^6).

megawatt (MW). The unit by which the capacity of production of electricity is usually measured. A megawatt is 1 million watts (W) or 1000 kilowatts (kW).

melting. Changing phase from a solid to a liquid. Most phase-change materials (like wax) require extra (or latent) heat to melt.

melting point. That temperature at which a single, pure solid phase changes to a liquid or to a liquid plus another solid phase, upon the addition of heat at a specific pressure. Unless otherwise specified, melting points are usually stated in terms of 1 atmosphere pressure.

membrane. A sheet polymer which separates components of solutions by permitting passage of certain substances but preventing passage of others.

merchantable volume. Traditionally the volume of wood contained in tree stems 5 inches (127 mm) or greater in diameter at breast height, measured to a 4-inch (102 mm) top diameter.

mercury switch. A glass tube employing mercury to establish electrical contact between circuits when the tube is tilted so that the mercury bridges the gap between contacts, and conversely.

meridional wind. The wind or wind component along the local meridian, as distinguished from the zonal wind (along the local parallel of latitude).

met. A convenient and approximate unit of human heat production equivalent to the average metabolic heat produced by resting man, about 18.50 British thermal units per square foot per hour (Btu/ft^2/hr).

metal insulator semiconductor. A barrier type of photovoltaic cell designed for lower cost; with successive layers of semiconductor, insulator, and metal.

methane (CH_4). A colorless, odorless, flammable gas that is the product of decomposition of organic matter. It forms the major portion of marsh gas and natural gas.

methanogenesis. Process of producing methane.

methanogenic bacteria. Microorganisms capable of producing methane.

methanol (methyl alcohol; CH_3OH). A light volatile, flammable, poisonous, liquid alcohol formed in the destructive distillation of wood (hence also known as wood alcohol) or made synthetically and used especially as a fuel, a solvent, an antifreeze, or a denaturant for ethyl alcohol, and in the synthesis of other chemicals. Methanol can be used as fuel for motor vehicles. It is a poisonous liquid, and is the lowest member of the alcohol series.

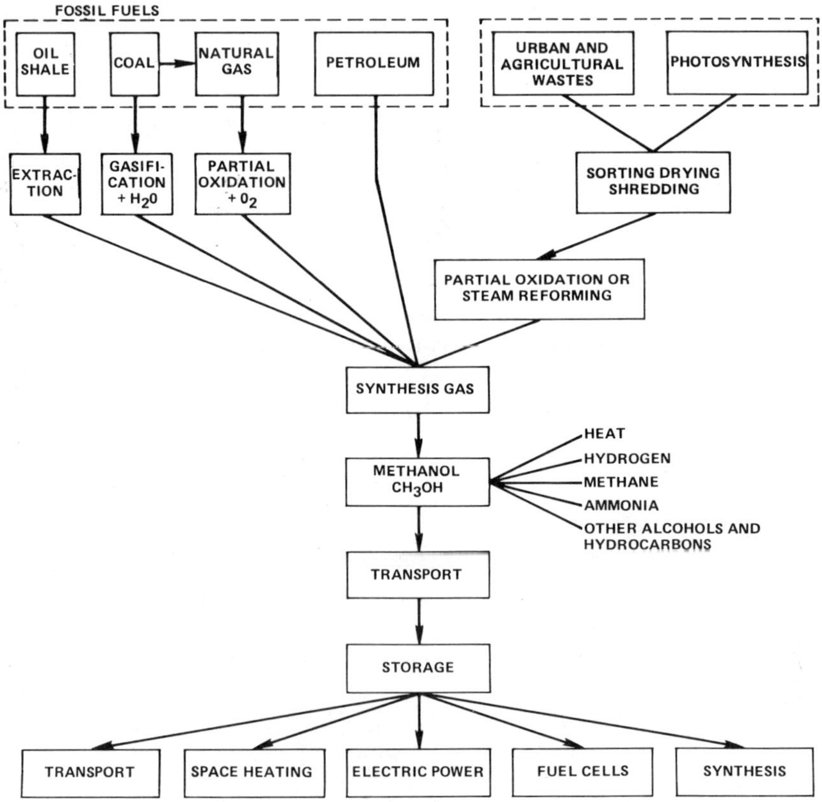

Courtesy Russell E. Anderson, Biological Paths to Self-Reliance, *Van Nostrand Reinhold*
Block diagram of methanol fuel sources and applications.

methogens. The bacteria which produce methane.

methylation. Replacement of one or more hydrogen atoms with a methyl group.

metric ton. A unit of mass and weight that equals 1000 kilograms (kg) or 2204.6 avoirdupois pounds.

MeV. One million (10^6) electron volts. Also written as Mev.

microbes. Minute plant or animal life. Some disease-causing microbes exist in sewage.

microencapsulation. A type of containment for latent-heat storage material which involves dispersion of the substance throughout a "host" material, such as ceiling tile, wallboard, or concrete block. The materials of construction then become effective heat storage devices.

microclimate. The local climate of a specific place or habitat, compared with the climate of the entire area of which it is a part. The microclimate is especially influenced by the local terrain.

micron. One thousandth (10^{-3}) of a millimeter.

Mid-Temperature Solar Systems Test Facility. The Mid-Temperature Solar Systems Test Facility in Albuquerque, NM, serves as a national engineering center for midtemperature components, subsystems, and systems. The facility actually functions as two test installations: (1) the Collector Module Test Facility, which provides support to research and development projects with collectors and receivers; and (2) the Systems Test Facility, which contributes in the areas of equipment characterization, operation, and maintenance experience and system data collection. This test facility is used by trough manufacturers for testing their collectors.

Courtesy DOE

Parabolic dish collector field at Sandia Laboratories Mid-Temperature Solar Systems Test Facility (MSSTF).

million barrels of oil equivalent (MBOE). A means of expressing a quantity of energy by relating it to the amount of petroleum which must be burned to produce the equivalent thermal energy. (The daily energy consumption for the United States is about 38 MBOE.)

million (10^6) Btu (MBtu). The predominant unit of energy in the United States. Generally, 1 million Btu are worth between $1 and $19. Frequently written MMBtu.

mill length. Also known as random length. Run-of-mill pipe is 16 to 20 ft (4.9 to 6.1 m) in length. Some pipe is made in double length of 30 to 35 ft (9.1 to 10.7 m).

minimum energy dwelling. The minimum energy dwelling uses available energy-saving building techniques, advanced household appliances, and a solar/natural gas central energy system for space heating, cooling, and water heating. It is designed to save energy in building techniques and materials, site selection, and construction.

minimum service temperature. Minimum temperature to which a component will be exposed in actual service, either with or without the flow of the heat-transfer fluid.

MINI-SHAC. A computer program developed by the International Business Machines Corporation for use in the rapid evaluation of solar system performance.

minority carrier. Current carriers (either electrons or holes) which are in the minority in a specific layer of semiconductor material. It is the diffusion of minority carriers through the cell barrier that creates a voltage, and constitutes a current, in a photovoltaic device. The process becomes more efficient with increasing minority carrier diffusion length.

miscible. When one liquid has the ability or tendency to uniformly blend with another, it is termed miscible. Alcohol and water are miscible, while gasoline and water are immiscible (incapable of mixing).

misdirected solar radiation. Concentrated and focused solar radiation resulting from the accidental defocusing of the heliostat field from the central receiver unit can be an invisible but serious danger. This misdirected radiation can cause fires, burns, serious glare problems, and possible eye damage.

mist. Liquid particles in air formed by condensation of vaporized liquids.

mixing valve. Automatic thermostatic valve that can be installed downstream of a solar heated hot-water storage tank to prevent accidental scalding of user. The mixing (or tempering) valve adds cold water to the hot water to maintain a present safe temperature.

MMBtu. One million (10^6) British thermal units.

mobile source. A moving source of air pollution such as an automobile.

model. A simplified description of a system, used as an aid to understanding the system. Models may be verbal or mathematical. Mathematical models are constructed from numerical values given to the components of the system and the relationships among components. If the model is sufficiently accurate to be approximately true, it can be used to predict the effect of changes within the system. Models may be constructed as initially complex models, by starting with a replica and discarding nonessential details, or as initially simple models, which begin with the most essential elements and add others as they are needed. Apart from mathematical models, models may be physical.

modified degree-day procedure. A steady-state procedure for determining the building thermal load which assumes that the heat loss is proportional to the equivalent heat-loss coefficient for the building envelope.

modular. A solar system that is made in individual units or modules, each module having the capability of an entire system; for example, a thermic diode is a modular heating system.

modular integrated utility systems. Small plants within housing developments or communities that provide all utility services.

moisture content. The water content of a substance as measured under specified conditions.

moisture tons. Latent heat load expressed in tons.

molasses. This is the thick liquid left after all crystalline sucrose has been removed from the mother liquor in sugar manufacturing. It contains about 20 percent sucrose (not economically feasible to extract), and 20 percent reducing sugars. With proper techniques, this blackstrap molasses can be fermented into alcohol.

molecular sieve. A column which separates molecules by selective adsorption of molecules on the basis of size.

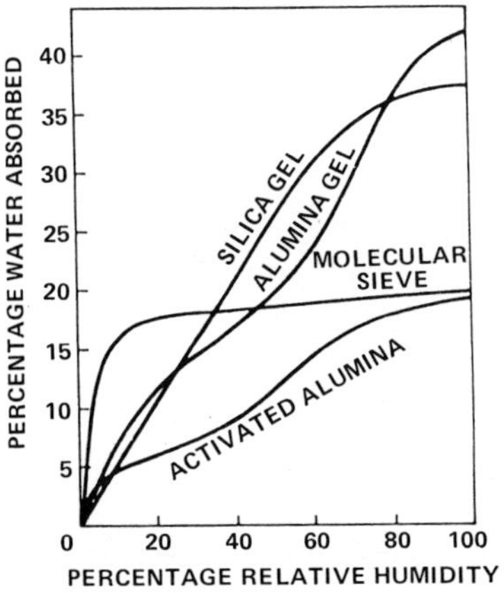

Courtesy DOE

Molecular sieve material can be used to remove water from alcohol.

molecular weight. The sum of the atomic weights of all the atoms in a molecule.

molecule. A group of atoms held together by chemical forces. The atoms in the molecule may be identical as in H_2, S_2, and S_8, or different as in H_2O and CO_2. A molecule is the smallest unit of compound matter which can exist by itself and retain all its chemical properties.

Mollier chart. A form of graph named for its inventor and covering heat properties of a fluid and with entropy as one coordinate and enthalpy as the other.

momentum. The measure of the motion of a body, determined by multiplying its mass by its velocity.

monitoring. Periodic or continuous determination of the amount of pollutants or radioactive contamination present in the environment.

monomer. A simple molecule which is capable of combining with a number of like or unlike molecules to form a polymer.

Monopollin. A mixture of 25 percent alcohol and 75 percent gasoline used in German vehicles between World Wars I and II.

monosaccharide ($C_6H_{12}O_6$). The basic unit of sugars (simple sugars) that can be link together to form other types of sugars or compounds. The most common types of monosaccharides are glucose, fructose, and galactose. Carbohydrates containing more than 10 simple sugar units are classed as polysaccharides.

Monsanto 602®. A glazing product from Monsanto Chemical Co. which is a 0.004-0.006 in. (0.1-0.15 mm) thick polyethylene film used for greenhouse roofs.

monsoon. A name for seasonal winds. The primary cause is the annual variation of temperature over large land areas compared with neighboring ocean surfaces, causing an excess of pressure over the continents in winter and a deficit in summer; but other factors, such as the relief features of the land, have a considerable effect.

motor gasoline production. Total production of motor gasoline by refiners, measured at refinery outlet. Relatively small quantities of motor gasoline are produced at natural gas processing plants, but these quantities are not included.

motor gasoline stocks. Primary motor gasoline stocks held by gasoline producers. Stocks at natural gas processing plants are not included.

motorized flue damper. A device that reduces the amount of heat escaping up the chimney, by opening when the furnace is operating and closing when the furnace shuts off.

mountain and valley winds. A system of diurnal winds, prevailing in calm, clear weather, along the axis of a valley, blowing uphill and upvalley by day (anabatic wind), and downhill and downvalley by night (katabatic wind).

movable insulation. An insulation device which reduces heat loss at night, or during cloud periods, and permits heat gain in sunny periods (e.g., Beadwall®, insulated draperies, automatic shutters); it may also be used to reduce heat gains in summer.

moving annual cost total. In the study of process costs, a series of costs-per-unit observed and recorded at regular intervals (usually in monthly financial summaries cross-referenced to analyzed detail cost). Twelve months are covered and each month the new month's figures are added and those for the corresponding month of the previous year are removed. Therefore, like periods are always compared and seasonal fluctuations are smoothed out.

muck soils. Soils made from decaying plant materials.

mulch. A layer of wood chips, dry leaves, straw, hay, plastic strips, or other material placed on the soil around plants to retain moisture, to prevent weeds from growing and to enrich soil.

multiple-effect evaporator. A series of evaporators in which the vapors removed from each unit are used to supply heat to the next unit in the series.

multiple junction cell. A photovoltaic cell containing two or more cell barriers, each of which is optimized for a particular portion of the solar spectrum to achieve greater overall efficiency in converting sunlight into electricity. See **split spectrum cell; vertical multiple junction cell.**

multiplier. In refrigeration, the ratio of total cooling load to sensible cooling load.

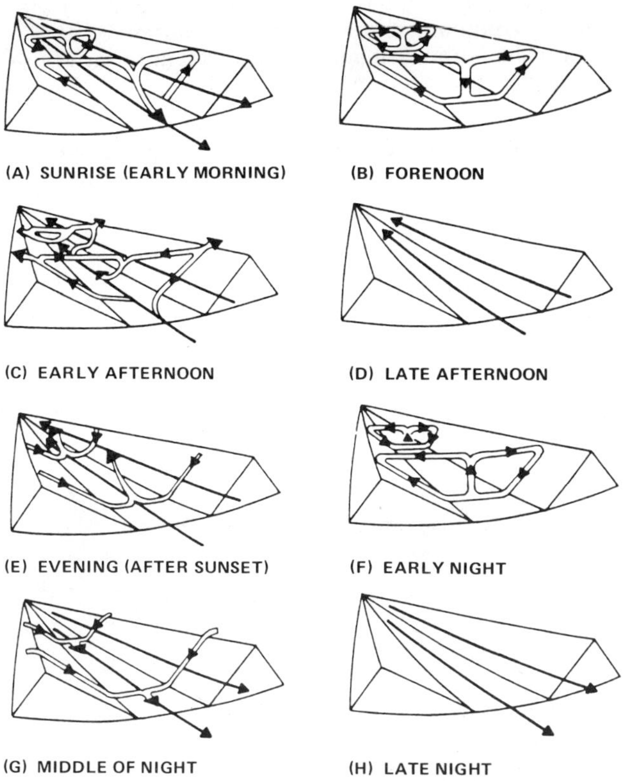

Courtesy DOE

Daily sequence of mountain and valley winds are affected by terrain.

multipurpose transmission line. Employment of a transmission line for more than one function, such as regular transmission, wheeling, reserve capacity, and peak capacity usage.

municipal solid waste. Combined residential and commercial wastes generated within a municipal area and consisting of any materials, including food wastes, that are discarded or rejected as spent, worthless, useless, or in excess and that are not wet enough to be free-flowing.

municipal waste. Defined in the Energy Security Act (P.L. 96-294), 1980, as "any organic matter, including sewage, sewage sludge, and industrial or commercial waste, and mixtures of such matter and inorganic refuse from any publicly or privately operated municipal waste collection or similar disposal system, or from similar waste flows (other than such flows which constitute agricultural wastes or residues, or wood wastes or residues from wood harvesting activities or production of forest products)."

municipal waste energy project. Defined in the Energy Security Act (P.L. 96-294),

1980, as "any facility located in the United States primarily for the production of biomass fuel (and byproducts) from municipal waste; or the combustion of municipal waste for the purpose of generating steam or forms of useful energy, including industrial process heat, mechanical power, or electricity (including cogeneration)."

mutation. Accidental or induced changes in an organism's genetic material.

N

nacelle. A permanent covering placed over the wind-energy conversion system (WECS) generator, drive train, and other elements on top of a WECS tower to protect them from environmental conditions.

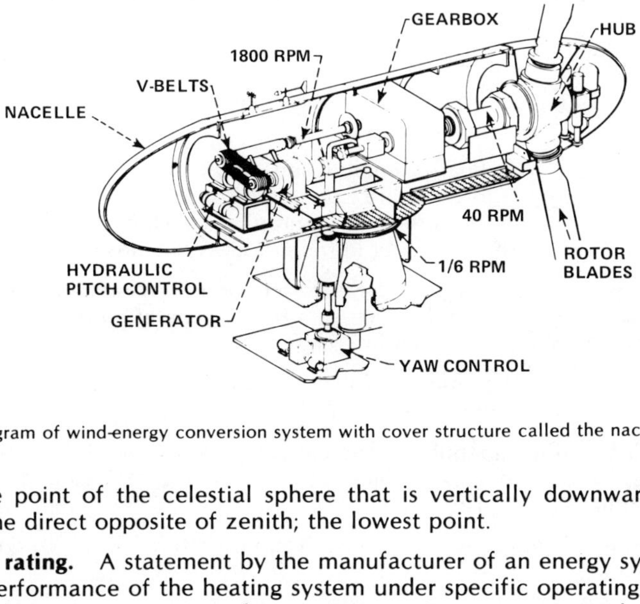

Courtesy DOE

Diagram of wind-energy conversion system with cover structure called the nacelle.

nadir. The point of the celestial sphere that is vertically downward from the observer; the direct opposite of zenith; the lowest point.

name plate rating. A statement by the manufacturer of an energy system which gives the performance of the heating system under specific operating conditions; the full-load continuous rating of a generator, prime mover, or other electrical equipment under specified conditions as designated by the manufacturer.

natural background radiation. The amount of radiation present in the environment which is not the result of man's activities.

natural convection. Convection caused solely by an object's temperature. Gas or liquid near a hot surface rises, near a cold surface falls; differs from forced convection where the gas or liquid is forced over the surface. The motion of a gas or liquid caused by temperature or density difference by which heat is transported.

natural cooling. Space cooling alternatives to energy-consumptive central air-conditioning systems. The five principal means of natural cooling are: shading, ventilation, conduction control, radiation, and evaporation.

natural draft. Draft that is caused by a thermal upset in which temperature differences change the weight (pressure) of air.

natural energy flow. The term "natural energy flow" is not synonymous with "unregulated energy flow." Natural energy flow can, in fact, be highly regulated by mechanically actuated controls, such as dampers or moving insulation. The important point is that the flow comes from nonmechanical sources.

natural frequency. The frequency at which a system will tend to vibrate without the application of a periodic forcing function.

natural gas. Naturally occurring mixture of hydrocarbon gases, found in porous geologic formations beneath the earth's crust. The principal constituent of natural gas is methane.

natural ventilation. The unassisted movement of fresh air through a building.

necrosis. Death of plant cells resulting in a discolored, sunken area or death of the entire plant.

needle valve. A valve provided with a long tapering point in place of the ordinary valve disc. The tapering point permits fine graduation of the opening.

negative gradient. Descriptive of conditions in a layer where the temperature decreases with increasing depth.

net area. Area of a solar collector aperture through which radiation may pass; gross area net of all opaque areas.

net calorific value. Net heat of combustion. In the case of solid fuels and liquid fuels of low volatility, a lower value calculated from gross calorific values as the heat produced by combustion of unit quantity, at constant atmospheric pressure, under conditions such that all water in the products remains in the form of vapor.

net electric system distribution. The kilowatt-hours (kWh) available for total electric system or company load. It is the sum of net generation by the system's own plants, purchased energy, and net interchange.

net energy balance. The amount of energy available from fuel when it is burned, less the amount of energy it takes to produce the fuel.

net generating station capability. The capability of a generating station as determined by test or operating experience less the power generated and used for auxiliaries and other station uses.

net generation. Total generation minus the kilowatt-hours (kWh) consumed for station use.

net reserves. The recoverable quantity of an energy resource that can be produced and delivered.

network. A system of transmission or distribution lines so cross-connected and operated as to permit multiple supply to any principal point on it.

neutral temperature. Ambient temperature at which solar and internal gains cancel the building heat losses so that no net heating or cooling load exists.

new income. The income you actually get from a solar heater each year—annual fuel savings less maintenance and interest.

newton (N). The unit of force in the SI Metric System. It replaces the older metric kilogram-force. The newton is defined as that force which gives to a mass of 1 kilogram (kg) an acceleration of 1 meter per second per second (m/s^2).

nighttime loss. Heat loss from a solar collector at night. A controller of an active system turns off the pump or fan at night to prevent nighttime heat loss. Passive systems use movable insulation and other means.

nipple. A tubular pipe fitting usually threaded on both ends and under 12 inches (0.3 m) in length. Pipe over 12 inches (0.3 m) long is regarded as cut pipe. See also **close nipple; short nipple; shoulder nipple.**

nitinol heat engines. A solar heat engine incorporating as its working fluid intermetallic nickel titanium compound, nitinol.

nitric oxide (NO). A gas formed in great part from atmospheric nitrogen and oxygen when combustion takes place under high temperture and high pressure, as in internal combustion engines. NO is not itself a pollutant; however, in the ambient air, it converts to nitrogen dioxide, a major contributor to photochemical smog.

nitrogen (N). An odorless, colorless, generally inert gas. It comprises 79 percent of the earth's atmosphere in the free state.

nitrogenase. The enzyme system in certain microorganisms which converts nitrogen gas and hydrogen ions (protons) to ammonia.

nitrogen fertilizer. Chemical compounds (e.g., nitrates, urea, ammonia) containing nitrogen in a form suitable for plant uptake and conversion to protein.

nitrogen fixation. The process of converting atmospheric (free) nitrogen to nitrogen compounds for use in either agricultural or industrial operations.

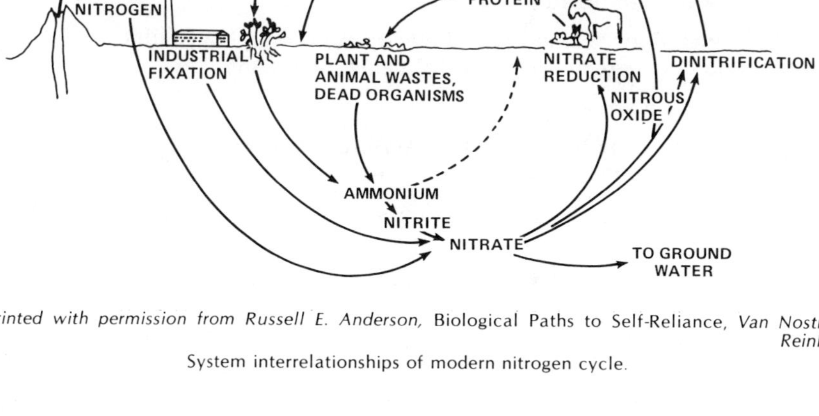

Reprinted with permission from Russell E. Anderson, Biological Paths to Self-Reliance, *Van Nostrand Reinhold*

System interrelationships of modern nitrogen cycle.

nitrogenous wastes. Wastes of animal or plant origin that contain a significant concentration of nitrogen.

nitrogen oxides (NO_X). Compounds of nitrogen and oxygen which may be produced by the burning of fossil fuels. They are very harmful to health and a contributor to the formation of smogs.

nocturnal cooling. A method of cooling through radiation of heat from warm surfaces to a clear night sky.

"no-flow" condition. That condition obtained when the heat transfer fluid is not flowing through the collector array due to shut-down or malfunction and the collector is exposed to the amount of solar radiation that it would receive under normal operating conditions.

nominal capacity. A figure expressed in tons per hour used in the title of the flowsheet and in general descriptions of the plant, applying to the plant as a whole and to the specific project under consideration. It may be taken as representing the approximate tonnage expected to be supplied to the plant during the hour of greatest load.

nominal wall thickness. Specified wall thickness of pipe without adding an allowance to compensate for the underthickness tolerances permitted in approved specifications.

noncoincident demand. The sum of two or more individual maximum demands, regardless of time of occcurrence, within a specified period.

nonconcentrating collector. A nonconcentrating or "flat-plate" collector in which the absorbing surface for solar radiation is essentially flat with no means for concentrating the incoming solar radiation.

nonimaging concentrator. Solar collector which concentrates sunlight onto a relatively small absorber without bringing the solar radiation to a sharp point or line focus.

nonpotable. Water that is not suitable for drinking or cooking purposes.

nonrenewable energy source. A mineral energy source which is in limited supply, such as fossil (gas, oil, and coal) fuels.

nontidal currents. Permanent currents in the general circulatory system of the sea as well as temporary currents arising from meteorologic conditions such as winds, density and water temperature.

nonutility generation. Generation of electric power required in the conduct of the industrial and commercial operations of generating plants.

normal. Direction perpendicular to a surface.

normal operating conditions and temperature. Maximum solar incidence, 100 mW/cm^2 (Air-Mass-One) on a photovoltaic cell at 82°F (28°C).

normal pressure. Usually equal to the weight of a column of mercury 760 millimeters (29.92 in.) in height.

normal radiation. The component of solar radiation which is perpendicular to the absorbing surface.

normal recovery capacity. Amount of water in U.S. gallons raised 100°F (38°C) per hour or per minute when calculated on a thermal efficiency of 70 percent, representing the water heated by a gas input of 1190 British thermal units per gallon (Btu/gal).

normal temperature. Normal temperature and pressure are taken as 32°F (0°C, 273°K) and 30 in. (760 mm) of mercury pressure. Also called standard temperature.

***n*-silicon.** Silicon containing a minute quantity of impurity, or dopant, such as phosphorus, which causes the crystalline structure to contain more electrons than required to exactly complete the crystal structure.There is no electrical imbalance, however.

nucleus. The small, massive, positively charged core of an atom that comprises nearly all of the atomic mass but only a minute part of the atom's volume. The nucleus consists of neutrons and protons, in the common isotype of hydrogen, which consists of one proton only.

nutrients. Elements or compounds essential as raw materials for organism growth and development; for example, carbon, oxygen, nitrogen, and phosphorus.

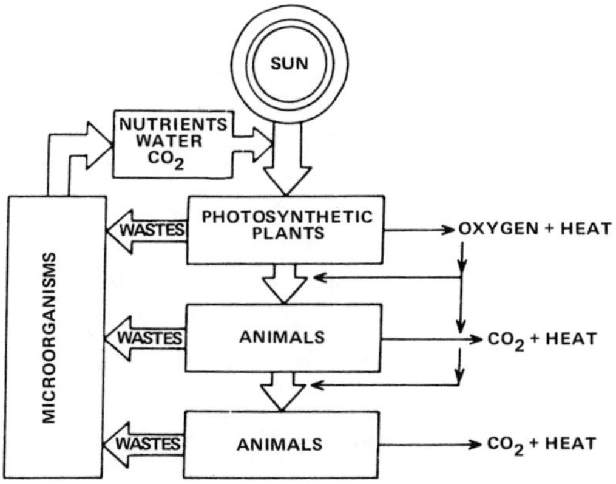

Rerinted with permission from Russell E. Anderson, Biological Paths to Self-Reliance, *Van Nostrand Reinhold*

Block diagram of nutrient and energy cycles in nature.

O

oakum. A caulking filler used to fill extra wide cracks or as a backup for elastomeric caulking.

obstructions. Steam supply mains may be looped over obstructions if a small pipe is run below the obstruction to take care of condensate. The reverse procedure is followed for condensate return mains. The larger pipe is carried under the obstruction.

occupied space. Space within a building or structure or other enclosed space which is normally occupied by people engaged in working, other activities, or normal living.

ocean basin. That part of the floor of the ocean that is more than about 600 feet (183 m) below sea level.

oceanic stratosphere. The cold, deep layers of the ocean consisting of waters of polar or subpolar origin.

oceanography. The broad field of science pertaining to the sea, which includes the studies of the boundaries of the ocean, its bottom topography, the physics and chemistry of sea water, the characteristics of its motion, and marine biology.

ocean temperature. The mean surface temperature of both the Pacific and Atlantic Oceans is 63°F (17°C); that of the Indian Ocean is 64°F (18°C). Maximum temperatures are, respectively, 90°F, 86°F, and 95°F (32°C, 30°C, and 35°C).

ocean thermal energy conversion. Providing power by harnessing the temperature differences between the surface waters and the ocean depths.

Ocean Thermal Energy Conversion power plant. The basic operating principles of an Ocean Thermal Energy Conversion (OTEC) power plant have remained unchanged since d'Arsonval proposed them almost 100 years ago. The ocean is, for all practical purposes, an infinite heat source, converting and storing the incident solar energy from the sun in the form of warm surface water. The warm water is pumped through an evaporator containing a working fluid in a closed Rankine-cycle system. The vaporized working fluid drives a gas turbine which provides the plant's power. Having passed through the turbine, the vapor is condensed by colder water drawn up from deep in the ocean and then pumped back into the evaporator for reuse in the same cycle. No "fuel" of any kind is used; the enclosed working fluid simply is

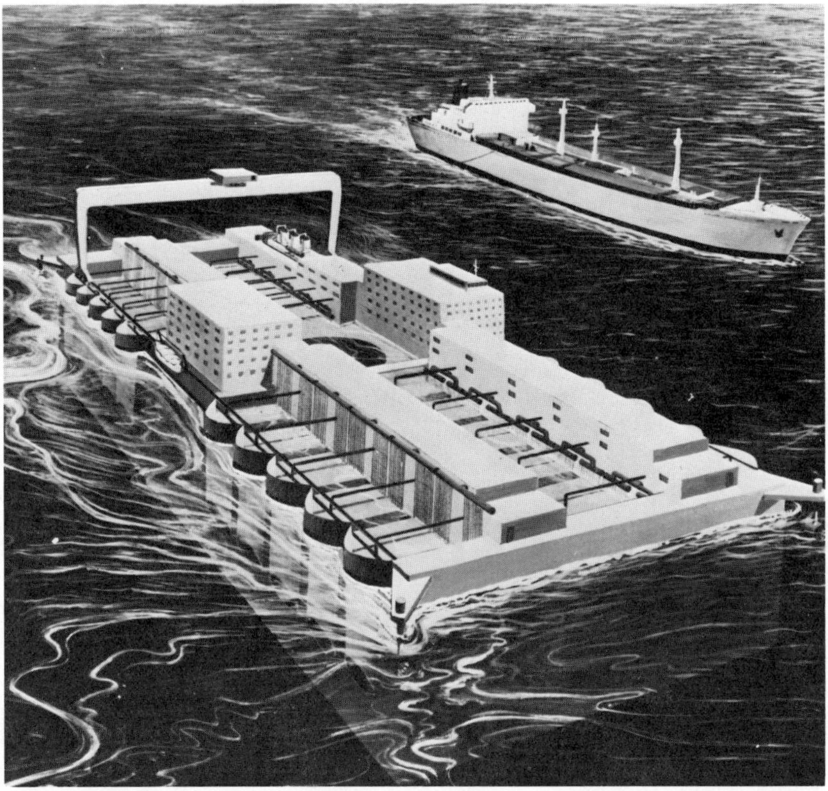

Courtesy DOE

Ocean thermal conversion plant ship; a floating factory which could make 586,000 tons of ammonia a year.

evaporated and condensed over and over by the warm surface and colder deep ocean water. Although OTEC operating principles are simple and well known, both the closed- and open-cycle systems pose complex engineering and cost problems. In both cases, the small temperature differentials [approximately 7°F (4°C) vs 968°F (538°C) in coal-fired boilers] dictate that large quantities of water must be pumped, the pumping power being subtracted from the net power of the system. (See diagram p. 208.)

ocean thermal gradient. Differences in the temperature of layers of the ocean potentially useful for running a heat engine. (See diagram p. 209.)

ocean tidal power. Energy source producing electricity from the mechanical motion created in the tide.

octane. (1) A rating scale used to grade gasoline as to its antiknock properties. (2) Any of several isomeric liquid paraffin hydrocarbons, C_8H_{18}; especially, normal octane, a colorless liquid boiling at 256.3°F (124.6°C) found in petroleum.

octane number. A rating which indicates the tendency to knock when a fuel is used in a standard internal combustion engine under standard conditions.

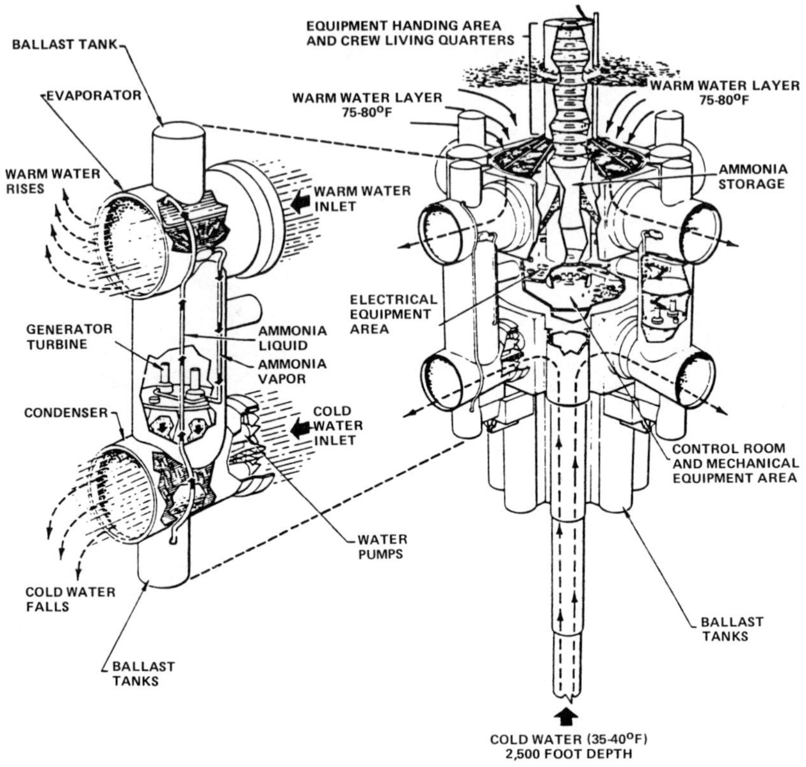

Courtesy DOE

Lockheed's concept of an ocean thermal energy conversion plant.

offering statement. A prospectus of an investment in the form of an offering of stock or partnership in the investment in return for cash invested into the company.

Office of Alcohol Fuels. An independent office within the Department of Energy responsible for administration of all alcohol fuels programs in the department and coordination of related programs in other federal agencies.

Office of Management and Budget. Reports to the President and has final authority in recommendations of all budgets to Congress.

off-peak. Period of low demand, usually during 24-hour period. When applied to capacity, refers to facilities preferably employed during such periods. When applied to customer rates, refers to rates charged during consumption at such periods.

offset. An arrangement of fittings or bends changing the line of piping to another line but parallel to the original line.

off-shore wind-energy conversion system. A wind-energy conversion system which is mounted on an off-shore platform, flotilla, or buoy.

offshore winds. Land breeze; winds blowing seaward from the coast.

ohm(Ω). The practical unit of measurement of electrical resistance that equals the

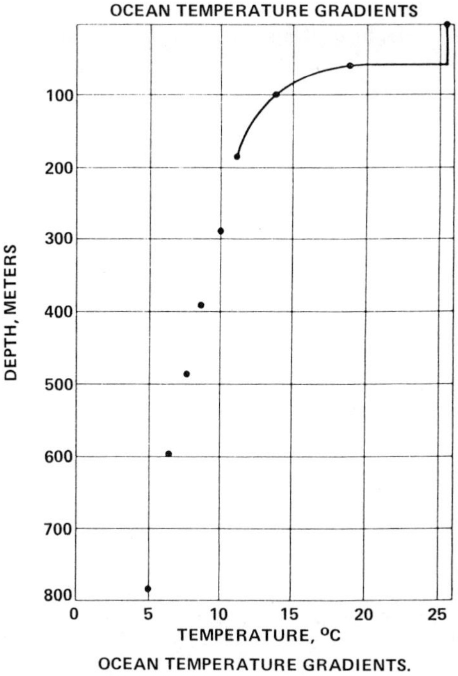

Courtesy DOE

Ocean temperature profile as a function of depth.

resistance of a circuit in which a potential difference of 1 volt (V) produces a current of 1 ampere (A).

ohm-cm. Units of electrical resistivity; an important parameter for semiconductors used in photovoltaic devices.

ohmic contact. Areas comprising the electrical contacts for solar cells for connection to external circuitry.

oilless bearing. Sleeve bearings of porous metal which depend solely on the porosity of the metal for oil storage.

oil valve. A check valve in a thermic diode uses oil in a chamber to block flow in one direction but allows flow in the other direction.

olefin (alkene) (C_nH_{2n}). Group of unsaturated aliphatic hydrocarbons; contains one or more double bonds and, as such, is chemically reactive. Olefins containing one double bond are called alkenes; those containing two double bonds are called alkadienes.

oligosaccharide. Carbohydrates that contain from 2 to 10 monosaccharide units are classed as oligosaccharides. Disaccharides are in this group.

oligotrophic lakes. Deep lakes that have a low supply of nutrients and thus contain little organic matter. Such lakes are characterized by high water transparency and high dissolved oxygen.

one-pipe hot-water system. A system in which the heated and cooled water to and from the radiators are carried in the same "main" but in the same direction, with the hot water on top.

one-pipe steam system. A steam heating system in which steam supplied to radiator travels in the same pipe but in the opposite direction as the cold water or condensate returning from radiator.

one sun. Maximum value of natural solar insolation.

one-tank system. Solar water-heating system using a single tank to store all household hot water, whether heated in the solar collectors or by an auxiliary electric unit.

on-orbit power generation. Large orbital plants using solar cells for energy conversion have been proposed. Transmission of the generated power to earth is accomplished by a microwave beam. The solar cell arrays are oriented by an attitude control system so as to face the sun, while the transmitting antenna constantly faces the ground receiving station. Consequently, the antenna must be mounted on gimbal joints. The frame structure joining the two large array areas is nonmetallic

Courtesy NASA

Artist's rendering of satellite power station photovoltaic cell array in space.

to allow passage of the microwave beam. To reduce weight and cost of cell arrays, plane mirrors would be arranged to effect concentration of incident sunlight. The concentration ratio would be limited to 2 in order to limit cell heating and the consequent reduction in cell effficiency.

on-peak energy. Energy supplied during periods of high system demands.

onshore winds. Sea breeze; winds blowing shoreward from the sea.

on-site generation. Generation of any electrical energy on a customer's property, with or without utilization of recoverable heat.

opacity. Degree of obscuration of light. For example, a window has zero opacity; a wall is 100 percent opaque. The Ringelmann system of evaluating smoke density is based on opacity.

open burning. Uncontrolled burning of wastes in an open area.

open circuit voltage. Photovoltage developed on an open circuit, which is the maximum available at a given irradiance.

open-cycle ocean thermal energy. Energy conversion where the OTEC working fluid is ocean water which itself is vaporized. See **closed-cycle ocean thermal energy conversion.**

Courtesy DOE

Artist's rendering of a 100-MWe open cycle ocean thermal energy power plant.

open-loop (or direct) system. Solar water-heating system in which household water is heated directly in the solar collectors. Circulation may be forced by a pump or it may be by natural means (thermosiphoning and gravity flow).

open system. Some part of the system is vented to the atmosphere, or the system contains fresh or changeable water.

operating water temperature. Systems for heating with hot water are flexible in regard to operating temperature. For instance, if the system is designed for 250°F (121°C) operating temperature it is easy to use a lower temperature in spring and fall by setting the boiler thermostat at a temperature that will satisfy the service water heater, say 160°F (71°C). This lowered setting at reduced heating loads results in more continuous pump operation and more uniform space temperature than with operation at 250°F (121°C). Service water temperature of 140°F (60°C) is usually satisfactory for residences.

operative temperature. That temperature of an imaginary environment in which, with equal wall (enclosing areas) and ambient air temperatures and some standard rate of air motion, the human body would lose the same amount of heat by radiation and convection as it would in some actual environment at unequal wall and air temperatures and for some other rate of air motion.

optimal collector area. The solar collector area that corresponds to the lowest life-cycle cost system or to the highest net annual savings.

optimum air supply. Volume of air delivered to a burner that will produce the maximum thermal efficiency under specific operating conditions.

orange heat. A division of the color scale, generally given as about 1652°F (900°C).

order of magnitude. A factor of 10; used as a convenience in comparing large numbers.

organic. Being, containing, or relating to carbon compounds, especially in which hydrogen is attached to carbon whether derived from living organisms or not. Usually distinguished from inorganic or mineral.

organism. Any living human, plant, or animal.

orientation. Alignment of a building along a given axis to face a specific direction, such as along an east-west axis to face south. Number of degrees to the east or west of south that a collection surface faces.

orifice. Opening; commonly used to apply to discs placed in pipelines or radiator valves to reduce the fluid flow to desired amount; the opening in an orifice cap, orifice spud, or other device whereby the flow of gas is limited and through which the gas is discharged.

orifice plate. Device inserted in a pipe or duct which causes a pressure drop across it. Depending on orifice size, it can be used to restrict flow or form part of a measuring device.

orsat. A device for measuring the percentage of carbon dioxide, oxygen, and carbon monoxide in flue gases.

osmophyllic. Organisms which prosper in solutions with high osmotic pressure.

osmotic energy conversion. Osmotic energy conversion is an unusual concept for obtaining energy from salt water based on the principle of simple osmosis. If salt water and freshwater are placed in a tank, separated by a semipermeable membrane resembling an extremely fine filter, the freshwater will flow through the membrane into the salt water. The membrane has pores big enough to let water molecules through, but too small to allow the passage of salt molecules. A "battery" could be built by stacking these osmotic membranes, alternating the two types, and filling the spaces in between with either freshwater or salt water in an alternating pattern. Again, less salty water could be used instead of absolutely fresh water. By placing electrodes at each end of the battery, the accumulated charges could be collected and converted into electric current.

osmotic pressure. Pressure produced by or associated with diffusion of a solvent through a semipermeable membrane.

Otto cycle engine. The early form of internal combustion engine was named for Nikolaus Otto. The standard Otto piston burns a uniform fuel/air charge. The conventional American car has used the Otto cycle engine.

outage. (1) Production time lost because of breakdowns or other unforeseen causes; (2) the difference between the full-rated capacity and the actual contents of a barrel or tank; (3) the loss of a volatile liquid such as gasoline ascribed to evaporation or pilferage. Also called shrinkage.

outer continental shelf. The submerged lands extending from the outer limit of the historic territorial sea [typically 3 miles (4.8 km)] to some undefined outer limit, usually a depth of 656 ft (200 m). In the United States, this is the portion of the shelf under federal jurisdiction.

outfall. The mouth of a sewer, drain, or conduit where an effluent is discharged into the receiving waters.

outgassing. The process by which materials and components expel gases.

outlet. Where the air or water leaves a solar collector or other device.

out of plane. Denotes wind-energy conversion system blade motion perpendicular to the rotor plane.

outplanting. The establishment of a forest stand by planting tree seedlings or tree cuttings.

outside air. Air taken from outdoors and therefore not previously circulated through the system.

outside coil. Portion of the heat pump, located outside the building, that functions as a heat-transfer point for collecting heat from or dispelling heat to the outside air.

outside screw and yoke. A type of bonnet so constructed that the operating threads of the stem are outside the valve housing, where they may be easily lubricated and do not come into contact with the fluid flowing through the valve.

overcast day. A day when solar radiation is blocked by clouds or haze. A cloudy day may have only a fifth of the solar radiation of a clear day.

overfire air. Air forced into the top of an incinerator to fan the flame.

overhead. The relatively low-boiling-point liquids removed from the top of a distillation unit.

overheat. The tendency for some passive solar heating systems to get too hot in the late afternoon.

overload. A load or weight in excess of the designed capacity, as applied to mechanical and electrical engineering plants, to loads on buildings and structures, and to excess loads on haulage ropes and engines.

overspeed control. The action of a control system, which prevents overspeeding of the wind-energy conversion system rotor.

ovonic. A device for converting heat or sunlight directly to electricity. Invented by Standford F. Ovshinsky, it has a unique special glass composition which changes from an electrically nonconducting state to a semiconducting state.

oxidant. Any oxygen-containing substance that reacts chemically in the air to produce new substances. Oxidants are the primary contributors to photochemical smog.

oxidation. A chemical reaction in which oxygen unites or combines with other elements. Organic matter is oxidized by the action of aeròbic bacteria; thus, oxidation is used in wastewater treatment to break down organic wastes.

oxidize. The addition of oxygen to a molecule. In the case of organic matter the molecule is often broken down to a smaller one. In aerobic organisms the end product of oxidation of organic matter is carbon dioxide and water. When methane is burnt, it is completely oxidized to CO_2 and H_2O. The removal of hydrogen from a molecule is also a form of oxidation and thus the formation of methane is the methane-bacteria's way of obtaining oxidizing power.

oxidation pond. A man-made lake or pond in which organic wastes are reduced by bacterial action. Often oxygen is bubbled through the pond to speed the process.

oxygen (O). A gas occurring in the atmosphere, of which it forms about 21 percent volume. It is chemically very active and is necessary for combustion; the combination of the oxygen with other substances generally produces heat.

oxygen gasification. Produces a medium-energy gas composed mainly of carbon monoxide and hydrogen, without the large fraction of nitrogen found in low-energy gas produced by air gasification. Methanol, ammonia, gasoline, and other useful products can be produced by catalytic conversion of the carbon monoxide and hydrogen.

ozone (O_3). A molecule containing three oxygen atoms. It occurs in minute quantities in the air near the earth's surface and in larger quantities in the stratosphere as a product of the action of ultraviolet light of short wavelengths on oxygen (O_2). At the top of the earth's atmosphere, it acts as a protective layer by absorbing ultraviolet radiation. It is also a major component of photochemical smog; it has a sharp, unpleasant odor and is an eye irritant.

ozone layer. A region in the earth's upper atmosphere containing ozone which helps shield living organisms from the sun's ultraviolet radiation.

packed distillation column. A column or tube constructed with a packing of ceramics, steel, copper, or fiberglass-type material to increase surface area through which vapors or liquid may pass.

packed tower. An air-pollution control device in which polluted air is forced upward through a tower packed with crushed rock or wood chips while a liquid is sprayed downward on the packing material. The pollutants in the airstream either dissolve or chemically react with the liquid.

packerhead. A brass fitting which is installed over the top of the drop pipe in a water pumping windmill to prevent overflow. It also is used to avoid well contamination.

packing. A general term describing any yielding material used to effect a tight joint. Valve packing is generally a "jam packing"; it is pushed into a stuffing box and adjusted from time to time by tightening down on a packing gland or packing nut.

packing gland. A device which holds and compresses the packing, and provides for additional compression by manual adjustment of the gland, as wear of the packing occurs. A packing gland may be screwed or bolted in place.

paddle wheel fan. A centrifugal fan with radial blades.

panel. A group of modules fastened together, preassembled and wired, designed to serve as an installable unit in an array.

panel heating. See **radiant heating.**

panel radiator. A radiating surface (see **radiant heating**) primarily flat to increase its ability to transmit radiant energy.

panemone. A name for drag-type vertical-axis wind machines; coming from pan (all directions) and anemone (wind), it could describe Darrieus-type machines also, but is not generally used except for drag machines. System reacts to wind from any direction. (See diagram p. 216.)

parabola. Locus of a point moving in a plane so that its distances from a fixed point (focus) and a fixed straight line (directrix) are equal.

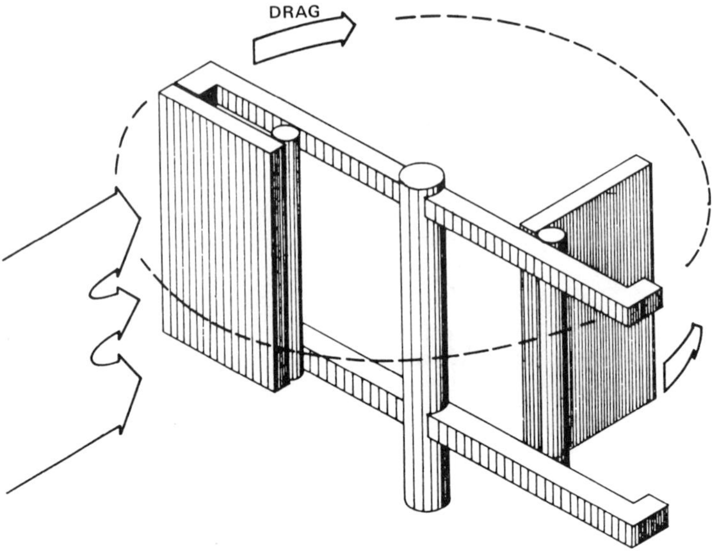

Courtesy DOE

Simple panemone type wind-energy turbine.

parabolic collector. A device for collecting the sun's energy that uses a bowl or disk-shaped reflector. The reflector, which is often mirrored, concentrates the radiation, producing temperatures as high as 3000°F (1649°C).

parabolic concentrator. A parabolic-shaped structure having a reflective surface that moves to track the sun and focuses the solar radiation onto the receiver.

parabolic dish. A parabolic-shaped device covered with a highly reflective or mirrored surface with focus at a single point; used in solar concentrators with concentration ratios typically from 1000 to 10,000,000. (See photograph p. 217.)

parabolic dish technology. The purpose of the parabolic dish technology program is to support and implement the commercial/industrial development of point focusing receiver technology. Parabolic dishes are point focusing concentrators and are typically 18-40 feet (5.5-12.2 meters) in diameter. (See diagram p. 217.)

parabolic focusing collector. A concentrating collector which focuses beam radiation with a parabolic reflector.

parabolic reflector. Reflector designed in the shape of a parabola to focus extra sunlight onto the absorber of a concentrating collector. (See diagram p. 218.)

parabolic trough. A trough-shaped, mirroring device focusing along a line, and used for concentration ratios less than about 100. (See photograph p. 218.)

parabolic trough technology. Parabolic troughs are linear concentrating collectors which focus the sun's radiant energy along a line at the focus of the trough. The system works by reflecting and concentrating sunlight onto an absorber tube placed at the focus of the parabolic trough, which then heats a working fluid flowing through the absorber tube to the desired temperatures [150°-750°F (66°-399°C)].

Courtesy DOE

The Raytheon parabolic dish solar concentrator system.

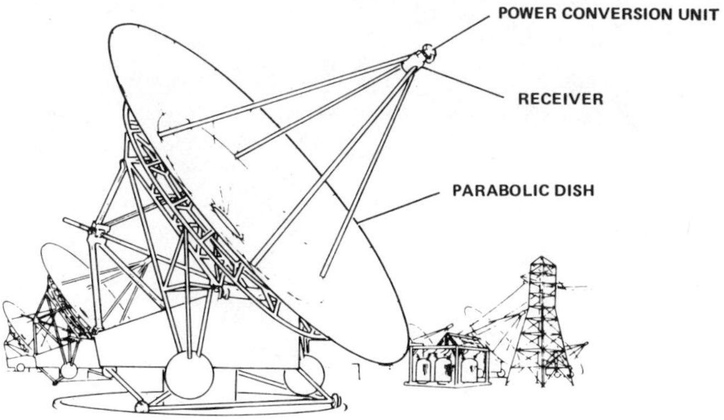

Courtesy DOE

The parabolic dish system's basic components.

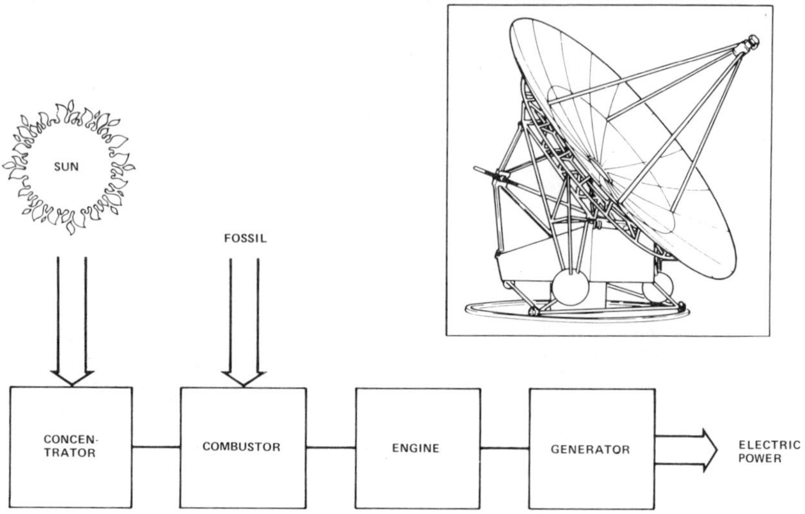

Block diagram of solar thermal electric power parabolic dish.

Courtesy DOE

The highly reflective curved metal plates on this solar collector cause the sun's rays to converge on the center tube.

This system can provide heat for midtemperature industrial applications (the utilization of hot water and steam for direct use in industrial processes and thermal energy as a substitute for fossil fuels); total energy systems (systems that provide both electricity or mechanical energy and process heat or space heating and cooling); and for other applications such as irrigation. In order to achieve maximum collection efficiency, some collectors are designed to track the sun across the sky. For optimum protection against degradation by the environment, the surface of the trough mirrors is coated with transparent protective materials which also contain optical properties that increase collector efficiency.

Courtesy DOE

The Suntec collector arrays (right foreground) have mirrored slats that track the sun and concentrate solar radiation onto a fixed receiver assembly. General Atomic's collectors (right background) have a stationary precision-cast concrete trough with silvered glass facets that focus the sun's rays on a sun-tracking receiver tube. The sun-tracking collectors on the left were designed and fabricated by Sandia.

parallel auxiliary. A scheme by which auxiliary energy is added to a solar system and the load is met entirely by auxiliary energy if the solar storage temperature is too low and entirely by solar energy if this temperature is adequate.

parallel configuration. Where a solar heater and auxiliary heater combine to heat a house; in contrast to a series or preheater arrangement. (See diagram p. 220.)

parallel connection. A method of interconnecting two or more electricity-producing devices, or power-using devices, such that the voltage produced, or required, is not increased, but the current is additive. Opposite of series connection.

partial pressure. That part of the total pressure of a mixture of gases contributed by one of the constituents.

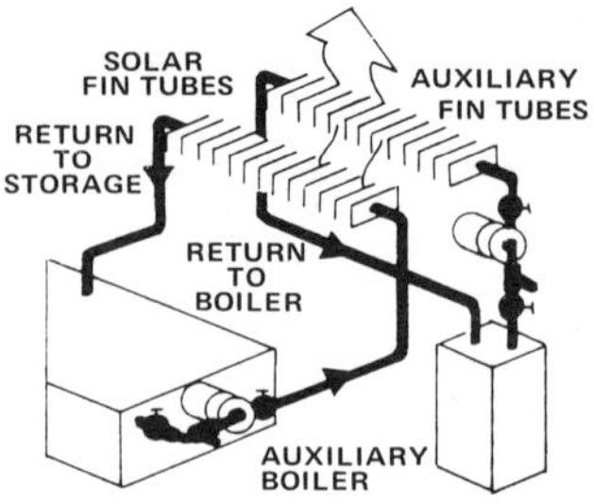

Courtesy National Copper Association
Illustration of parallel distribution configuration of heat exchanger.

particulate loading. The introduction of particulates into the ambient air.

particulates. Microscopic pieces of solids which emanate from a range of sources and are the most widespread of all substances that are usually considered air pollutants. Those between 1 and 10 microns are most numerous in the atmosphere, stemming from mechanical processes and including industrial dusts and ash.

pascal (Pa). Unit of pressure equal to 1 newton per square meter (N/m^2).

passive. A solar heating system that does not use fans or pumps to transfer solar heat to storage.

passive heating. A passive solar energy system uses the structural elements of the building to collect and store the sun's heat. The heat is then distributed to the living space from these elements by means of natural convection, conduction, and radiation. For example, solar heat can be collected through south-facing windows, absorbed and stored in a masonry floor, and later reradiated from the floor to the room. Fans and ducts are sometimes used to assist in heat distribution. Design of a passive system requires careful consideration of many factors. These include the local climate, use of the building site, construction materials, and the floor plan.

passive solar design. A structural design that makes use of the structural elements of a building, using no moving parts, to heat or cool spaces in the building.

passive solar energy. Passive solar heating applications generally involve energy collection through south-facing glazed areas; energy storage in the building mass or in special storage elements; energy distribution by natural means such as convection, conduction, or radiation with only minimal use of low-power fans or pumps; and a method controlling both high and low temperatures and energy flows. Passive cooling applications usually include methods of shading collector areas from exposure to the summer sun and provisions to induce ventilation to reduce internal temperatures and humidity. It is usually built as an essential component of the building rather than as an addition. (See diagrams pp. 221, 222, and 223.)

passive solar heating. Solar heating of a building accomplished by architectural design without the aid of mechanical equipment.

passive solar system. A system that uses gravity, heat flows, or evaporation to operate without mechanical devices to collect and transfer energy, such as south-facing windows.

Courtesy DOE

This home in Baltimore County, Maryland, uses passive solar concepts to provide "natural" heating and cooling.

passive water heating system. With reference to water heating, a thermosiphoning system. See **thermosiphoning.**

pathogen. Bacteria which cause disease both in animals and humans, e.g., *Salmonellae* cause typhoid.

pattern factor. An index which describes the status of units over a wide area, for example, a utility load pattern.

payback. A traditional measure of economic viability of investment projects. A payback period is defined in several ways—one of which is the number of years required to accumulate fuel savings which exactly equal the initial capital cost of the system. Payback often does not give an accurate representation of total life-cycle value.

peak day. The 24-hour day period of greatest total energy sendout.

peaking capacity. The load-carrying ability of the wind-energy conversion system during peak load periods (usually significant only if storage is an integral part of the facility). The generating capacity of facilities or equipment designed for use under extreme demand conditions.

peaking turbines. Utility generating units assigned solely to respond to the periods of highest load demand.

peak load. Maximum demand for electrical power experienced during a defined period by a utility company. Since peak load determines the total generating capacity required, some utilities have begun to encourage conservation by charging highest prices during peak demand periods. The design heating and cooling load used in mechanical system sizing. Usually set to meet human comfort requirements 93-97 percent of the time.

peak-load (or peaking) power plants. In an electric utility system, peak-load power

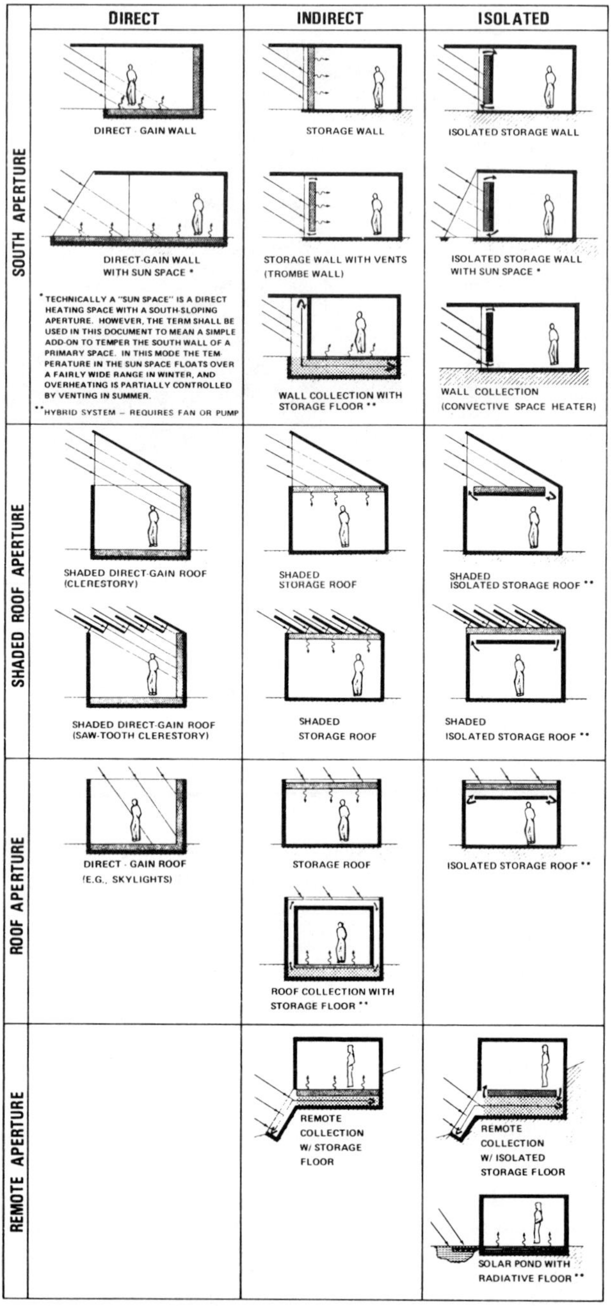

Courtesy SERI

Examples of passive solar heating system approaches.

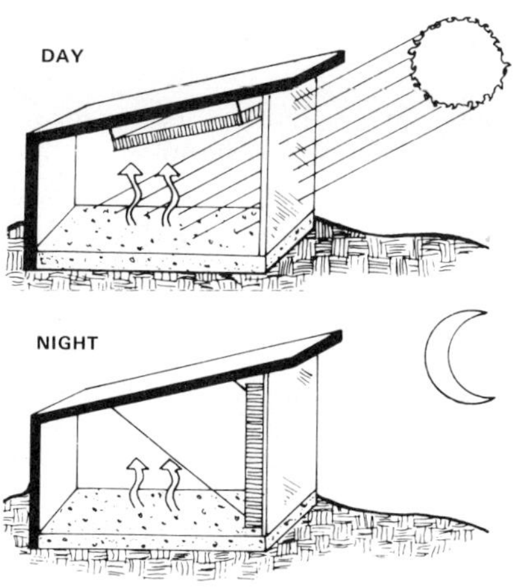

Courtesy DOE

During the day solar insolation passes through the glass and heat is stored in masonry floor; in the evening heat is released from the floor into the room.

plants operate only during periods of peak or near-peak demand. Because they operate only for relatively short fractions of the day, peak-load power plants produce relatively little energy compared to the total capacity and tend toward minimum capital cost. Low utilization of capacity and the use of expensive fuels cause the electric power produced by peaking plants to be more expensive than power from baseload power plants. See **baseload power plants.**

peak power (solar). Output of a solar-energy system on a clear day with the peak solar radiation at normal incidence. Peak solar radiation is usually assumed to be 1000 watts per square meter (W/m^2), or about 320 British thermal units per hour per square foot ($Btu/h/ft^2$).

peak-sharing. The process of supplying power from an extraneous source to help meet the peak demand on a system.

peak watt. A measurement unit used for performance rating of solar panels and arrays. A panel rated at 1 peak watt (Wp) will deliver 1 watt (W) under specified standard operating conditions with a solar insolation of 10^3 W/m^2.

peak wind speed. Maximum instantaneous wind speed occurring over the time under consideration.

peat. Partially decomposed organic material. Name given to layers of dead vegetation that is sometimes regarded as the youngest member of the ranks of coal.

pebble bed or rockbed. Large bin of rocks of uniform size used for storing solar heat in air-type heating systems.

pelletizing. The densification of finely sized wood and bark into a uniform cylindrical shape.

Pelton wheel. An impulse water turbine with buckets which are bolted to its periphery and struck by a high velocity jet of water.

penstock. A pipe which transports water to a turbine for the production of hydroelectric energy.

pentane. Any of three isomeric hydrocarbons of the methane group occurring in petroleum.

percentage humidity. Ratio of weight of vapor in a mixture of air and water vapor to the weight of vapor when the vapor is saturated at the same temperature. Contrast with relative humidity.

percent possible sunshine. The amount of radiation available compared to the amount which would be present if there were no cloud cover; usually measured on a monthly basis. The percentage of daytime hours during which there is enough direct solar radiation to cause a shadow. See **direct radiation.**

percolating filter. A form of aerobic waste treatment, involving the distribution of the water and dissolved organic matter over "beds" of stones, clinker, plastic media, etc., upon which the aerobes grow. As the organic waste trickles past them, they absorb it and break it down. As they grow, they gradually fall off the media and flow out with the effluent as a sludge. This is then settled out and the treated water can be discharged to the river, etc.

percolation. Movement, under hydrostatic pressure, of water through the pores or spaces of rock or soil.

percolation rate. The rate, usually expressed as a velocity, at which water moves through saturated granular material.

perfect combustion. Fuel burning when all combustibles are consumed with no excess air so that the theoretical amount of oxygen only is used.

performance ratings. Several organizations have developed thermal performance ratings based on collector efficiencies. These efficiencies are obtained by independent test laboratories using the collector test procedure (ASHRAE 93-77) developed by the American Society of Heating, Refrigeration and Air Conditioning Engineers (ASHRAE).

Ratings are usually expressed in British thermal units per hour (Btu/h), or British thermal units per day (Btu/day), for the collector panel under clear sky conditions at various fluid inlet and ambient temperatures.

A standard ratings procedure has recently been developed jointly by the Solar Energy Industries Association (SEIA) and Air-Conditioning and Refrigeration Institute (ARI).

perimeter heating. A term originally coined by McClatchy & Co., Philadelphia builders, to describe Warren Webster & Co.'s forced hot-water heating system employing series connected baseboard convectors placed the full length of the exposed walls of a building, and currently used in that sense by that manufacturer. The term is also applied by the warm-air heating industry to describe a variety of warm-air systems all of which introduce the warm air into the rooms by registers or warm-air baseboards located along the outer walls, preferably under the windows, close to, or along the floor.

Perlite®. An insulation material used for home insulation. It is usually installed by pouring the material into unfinished attic floor spaces. Perlite has the same insulation value as vermiculite.

Permafoam®. An insulating material product of Mid-America Industries Inc. which is a polystyrene foam.

permeance. The rate of water vapor transmission on a basis of unit time, unit area, and unit vapor pressure difference through a material. Permeability is the rate through a unit thickness of the material. The term Perm is the unit for 1 grain per (square foot) X (hour) X (inches of mercury).

pesticide. An agent used to control pests. This includes insecticides for use against harmful insects; herbicides for weed control; fungicides for control of plant diseases; rodenticides for killing rats, mice, etc.; and germicides used in disinfectant products, algaecides, slimicides, etc. Some pesticides can contaminate water, air, or soil and accumulate in humans, animals, and the environment, particularly if they are misused. Certain of these chemicals have been shown to interfere with the reproductive process of predatory birds and possibly other animals.

pesticide tolerance. A scientifically and legally established limit for the amount of chemical residue that can be permitted to remain in or on a harvested food or feed crop as a result of the application of a chemical for pest-control purposes. Such tolerances or safety levels, established federally by the Environmental Protection Agency (EPA), are set well below the point at which residues might be harmful to consumers.

petroleum. Material occurring naturally in the earth and composed predominantly of mixtures of chemical compounds of carbon and hydrogen with or without

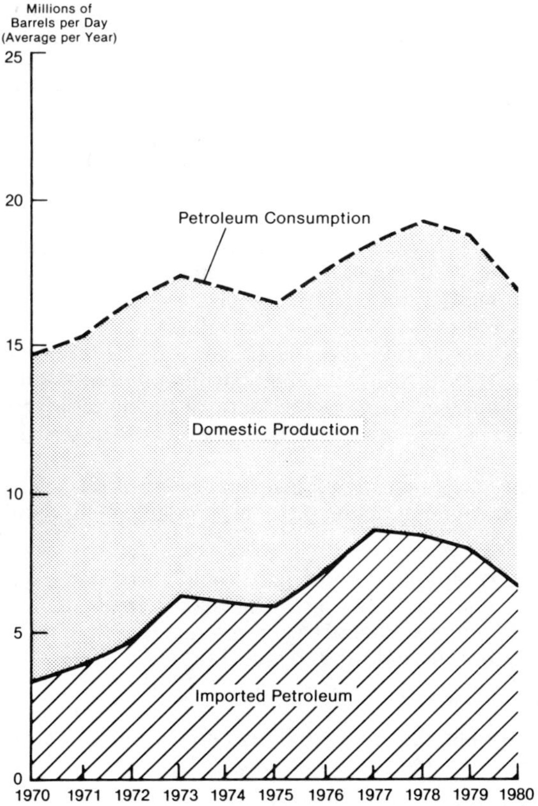

Courtesy U.S. National Alcohol Fuels Commission

U.S. petroleum consumption, production, and imports (1970–1980).

other nonmetallic elements, such as sulfur, oxygen, and nitrogen. Petroleum may contain or be composed of such compounds in the gaseous, liquid, and/or solid state, depending on the nature of these compounds and the existent conditions of temperature and pressure.

pF. The numerical measure of energy with which water is held in the soil.

pH. This is the symbol for the logarithm of the reciprocal of the hydrogen-ion concentration of a given solution. It represents the acidity or alkalinity of the solution, on a scale of 0 to 14. Since pure water, regarded as neutral, has a pH value of 7, 0 to 7 indicates acidity, while 7 to 14 indicates alkalinity. pH control is of critical importance in enzyme and yeast functions. The pH value can be roughly detected by litmus paper, with pH meters providing more exact readings.

phase change. The change from one solid, liquid, or gaseous state to another. Associated with such a change of state is a large energy input (or release), known as the heat of fusion or vaporization. See **latent heat.**

phase-change material. A material which is used to store thermal energy as latent heat.

phase-change storage. In order for a solar energy system to provide heat at night, or on cloudy days, some method is needed of storing excess solar energy collected during the day. Although the most common storage materials are water and rock, other storage materials are also being tested, including substances that store heat by melting. Such substances are known as "phase-change materials."

Heat must be added to a substance for it to melt; that is, for it to change from a solid to a liquid. Conversely, heat is released as the liquid cools to a solid. As a substance changes from a liquid to a solid or from a solid to a liquid it is said to change "phase." The substance that melts and solidifies is known as the "phase-change material."

Phase-change materials can be used to store solar energy. Excess heat collected by a solar system is used to warm the phase-change material, which melts. Later in the day, when the temperature drops, the phase change material cools, releasing heat which warms the room.

Phase-change materials have the advantage of being more compact than air or liquid storage and can easily be added to existing buildings. But the disadvantages of phase-change storage include the tendency of the materials to break down after many phase-change cycles, and the difficulty and cost of designing and building a container.

Scientists are currently investigating several types of phase-change materials, including inorganic salts, eutectic mixtures, and organic compounds such as paraffin waxes.

phenols. A group of aromatic compounds having the hydroxyl group directly attached to the benzene ring. They give the reactions of alcohols, forming esters, ethers, and thiocompounds. Phenols are more reactive than the benzene hydrocarbons. They are present in coal tar and wood tar.

phosphate. The salt or ester of a phosphoric acid.

phosphorus (P). An element that while essential to life, contributes to the eutrophication of lakes and other bodies of water.

photobiology. The branch of biology concerned with the study of the interactions of light with biological systems. An example is the study of photosynthesis.

photochemical oxidants. Secondary pollutants formed by the action of sunlight on the oxides of nitrogen and hydrocarbons in the air; they are the primary contributors to photochemical smog.

photochemical smog. Air pollution associated with oxidants rather than with sulfur oxides, particulates, etc. Produces necrosis, chlorosis, and growth alterations in plants and is an eye and respiratory irritant in humans.

photochemistry. The branch of chemistry concerned with the study of chemical reactions initiated by interactions of light with matter.

photoconversion. Conversion of light into other forms of energy by chemical, biological, or physical processes.

photoelectric cell. A device which registers changes in the amount of light falling on it by converting it to electricity.

photelectric effect. The complete absorption of a photon by an atom with the emission of an orbital electron (photoelectron).

photoelectrochemical cell. A special kind of photovoltaic cell in which the electricity produced is used immediately within the cell to produce a useful chemical product, such as hydrogen. The product material is continuously withdrawn from the cell for direct use as a fuel, or as an ingredient in making other chemicals, or it may be stored and used subsequently.

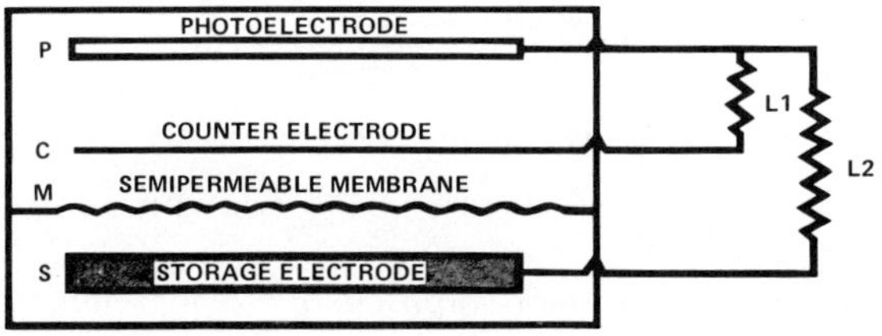

Courtesy SERI

Photoelectrochemical cell used for power generation and storage.

photoelectrochemistry. The branch of chemistry concerned with the study of oxidation and reduction reactions initiated by visible, ultraviolet, or near-infrared radiation at the surface of a photovoltaic device immersed in a conducting solution.

photoelectrolysis. The use of a photoelectrochemical cell to produce hydrogen and oxygen with water. (See **photoelectrochemistry** and diagram p. 228.)

photogalvanic processes. Light-induced production of electrical current.

photolysis. A chemical process initiated by light, in which molecular bonds are made or broken. (See photograph p. 229.)

photometer. An instrument for measuring the intensity of light, or for comparing intensities from two sources. The more accurate types are built around photoelectric cells.

photon. In the quantum theory of light, light consists of tiny bundles of energy. One of these bundles is called a photon. A particle of light, which acts as an in-

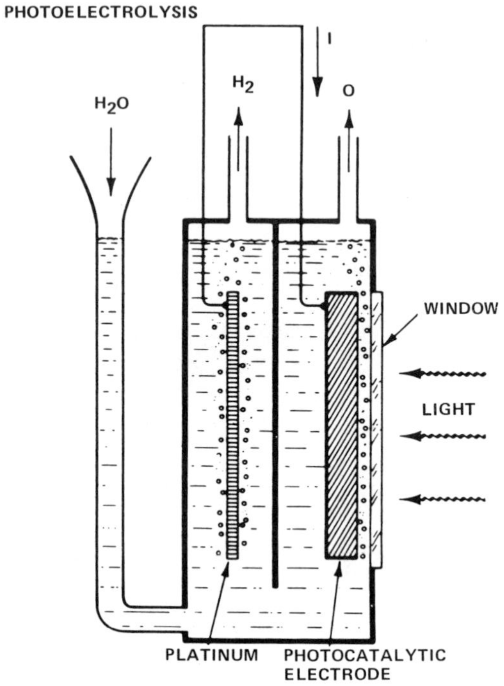

Source J.B. Goodenough, "The Options for Using the Sun," Technology Review (MIT), **79** *(1), 62–71, Oct./Nov. (1976)*

Process diagram of photochemical cell energy conversion unit.

divisible unit of energy, a quantum or corpuscle of radiant energy moving with the speed of light.

photophysical processes. Physical processes following the absorption of light after which there is no change in chemical bonds. Such events usually take place in the femtosecond (10^{-15}s) to nanosecond (10^{-9}s) time region.

photosensitize. To make sensitive to light.

photosynthesis. The process by which chlorophyll-containing cells in green plants convert incident light to chemical energy and synthesize organic components from inorganic components. (See diagram p. 229.)

photothermal conversion. The conversion of light to thermal energy.

photovoltaic. A solid-state electrical device whose current-voltage characteristic is a function of the incident radiation; capable of producing electric power when exposed to radiant energy, especially sunlight.

photovoltaic array. An interconnected system of photovoltaic modules that functions as a single electricity-producing unit. The modules are assembled as a discrete structure, with common support or mounting.

photovoltaic cell. A photovoltaic cell is one that generates electrical energy when light falls on it. This term distinguishes it from a photoconductive cell (photore-

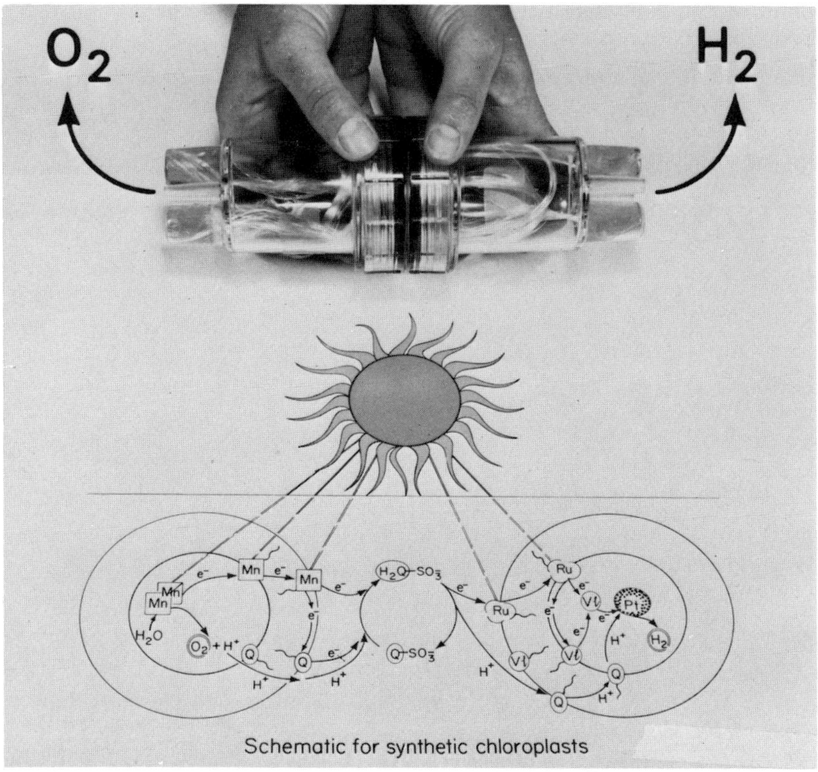

Courtesy DOE

Conceptual model for photolysis of water with synthetic chloroplasts.

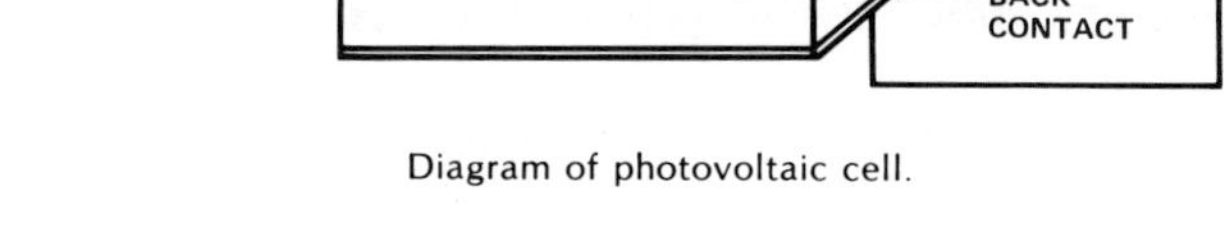

Courtesy SERI

Diagram of photovoltaic cell.

sistor), which changes its electrical resistance when light falls on it. A type of semiconductor in which the absorption of light energy creates a separation of electrical charges. The separation creates an electrical potential. The net effect is direct conversion of light, especially solar energy, into electricity. Typical materials used in the construction of photovoltaic cells are silicon, cadmium sulfide, and gallium arsenide.

photovoltaic collector. A photovoltaic module or array which receives sunlight and converts it into electricity.

photovoltaic concentrator arrays. Systems that use lenses or mirrors to concentrate the sun's rays onto solar cells which convert sunlight directly to electricity. The concentrated sunlight increases the cells power output; thus, fewer cells are required to produce a given amount of electricity.

Courtesy DOE

Photograph of Frensel lens on photovoltaic concentrator arrays.

photovoltaic conversion. Photovoltaic cells, or "solar cells," are devices that convert sunlight directly into electricity. Direct current (DC) electrical power is produced within the solar cell material and does not require any fuel other than sunlight. Photovoltaic cells do not require moving parts and produce no waste materials. Both private industry and the government are working to develop mass production techniques that will bring down the cost of photovoltaic cells, which are currently too expensive for most commercial and residential applications.

photovoltaic effect. The generation of an electromotive force when radiant energy falls on the boundary between certain dissimilar substances in close contact.

photovoltaic energy. Energy obtained from devices that convert sunlight directly into electricity.

photovoltaic module. A number of photovoltaic cells electrically interconnected and mounted together, usually in a common sealed unit or panel of convenient size for shipping, handling, and assembling into arrays.

photovoltaic process. Process of converting light rays directly into electricity without going through the intermediate steps involving turbines and generators. In photovoltaics, solar cells convert sunlight into a stream of electrons easily converted into household current.

photovoltaic system. A complete set of components for converting sunlight into electricity by the photovoltaic process, including array and balance-of-system components.

Courtesy DOE

Mead, Nebraska, irrigation project uses 120,000 individual solar cells to produce 25 kilowatts of electricity.

photovoltaic-thermal (PV/T) system. A photovoltaic system which, in addition to converting sunlight into electricity, collects the residual heat energy and delivers both heat and electricity in usable form. Also called total energy system. See **combined collector.**

physical compatibility. Ability of materials and components in contact with each other to resist degradation due to physical causes such as differences in thermal expansion.

phytoplankton. The plant portion of plankton.

piezoelectric. Having the ability to develop surface electric charges when subjected to elastic deformation, and, conversely, deforming in response to surface electric charges.

piggyback operation. Arrangement of chilled-water generation equipment whereby exhaust steam from a centrifugal chiller driven by a steam turbine is used as the heat source for an absorption chiller.

pinch valve. A valve used to pinch off, or stop, the flow of liquid, usually water, at a desired location.

pipe coil. Pipe used for heating, usually in (1) parallel lines connected to heaters at the end, forming a grid; (2) in parallel lines with ends alternately connected to adjacent lines by bends, forming a so-called continuous coil.

pipe, connecting socket. A holder of metal or plastic, which is usually threaded to mate with one or more pieces of pipe. A section of pipe may have a male thread at one end, and this is attached to a connecting socket having a female thread of the same pitch.

pipe friction. Wall roughness of pipes or ducts which causes a loss of pressure in the direction of flow and a resultant evolution of heat.

pipeline refrigeration. Piping of refrigeration from one central plant to adjacent buildings.

piston. The working part of a pump, hydraulic cylinder, or engine that moves back and forth in the cylinder, and is generally equipped with one or several rings or cups to control the passage of fluid. It ejects the fluid from the cylinder, as in a pump, or receives force from the fluid, which causes a reciprocating motion, as in an engine.

pitch angle. Angle between the chord and plane of the cone of rotation of a wind-energy conversion system blade at the $0.75R$ station, where R is the maximum blade radius.

pitch control. The mechanisms by which the change of the orientation of the wind turbine blade relative to the wind is achieved, thereby altering the aerodynamic efficiency of the blade.

pitching. This is a brewer's term, meaning to add the yeast to the wort. Pitching occurs when the wort (unfermented mash) has been converted to fermentable sugars and the temperature is brought down to approximately 80°F (27°C).

pitting. Localized loss of material by erosion or chemical decomposition.

plankton. The floating or weakly swimming plant and animal life in a body of water; they are often microscopic in size.

planned unit development. A development planned as a whole, where conventional subdivision regulations (such as type of housing, height limitations, setbacks, densities, and minimum lot sizes) are waived to allow more design flexibility and amenities. This kind of development has greater potential for solar access planning than does conventional development.

plant availability. The percentage of time a plant is able to provide power when required.

plant factor. The ratio of the average power load of an electric power plant to its rated capacity. Sometimes called capacity factor. See **capacity factor.**

plasmid. A small portion of one organism's genetic material which, when inserted into a second organism's cell nucleus, performs as it did in the original organism.

plate distillation column. A distillation column constructed with perforated plates or screens. (See diagram p. 233.)

plate-type heat exchanger. An advanced heat-exchanger design being developed for the Ocean Thermal Energy Conversion program which would have ammonia in-

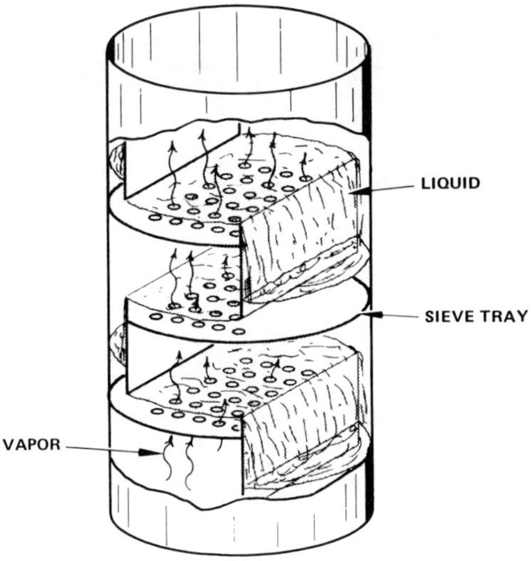

Courtesy DOE

Illustration of vapor–liquid flow through plate distillation column.

side the structure and water outside of it, eliminating the need for a large heavy shell. The state-of-the-art heat exchangers consist of configurations of tubes inside large heavy shells.

plenum. A cavity of air space through which air is moved. In some passive solar designs a plenum may be used to evenly distribute heat which otherwise would collect at a single point.

Plexiglas®. A glazing product of Rohm & Haas which is a transparent sheet of methyl methacrylate.

plug valve. One with a short section of a cone or tapered plug through which a hole is cut so that fluid can flow through when the hole lines up with the inlet and outlet, but when the plug is rotated 90°, flow is blocked.

plumbing. The installing of or the installation consisting of piping, fixtures, valves, and traps to carry water to the fixtures and carry liquid and other wastes from these and other fixtures.

plumbing system. Arrangements in a building of pipes, fixtures, fittings, valves, and traps to supply water and remove liquid wastes.

plume. The visible emission from a flue or chimney.

p-n junction. Junction of dissimilar semiconductor materials, where electrons move from one type to another, under specific conditions.

point source. A single point from which light emanates, such as the sun, the filament of an electric lamp, or other superheated metal; in air pollution, a stationary source of a large individual emission, generally of an industrial nature.

pole miles. Miles measured along the line of poles, structures, or towers carrying electric conductors regardless of the number of conductors or circuits carried.

polished rod. Brass rod connecting the red rod, in a water-pumping windmill, to the sucker rod. The polished rod is inserted through the packerhead to provide the mechanical movement for pumping.

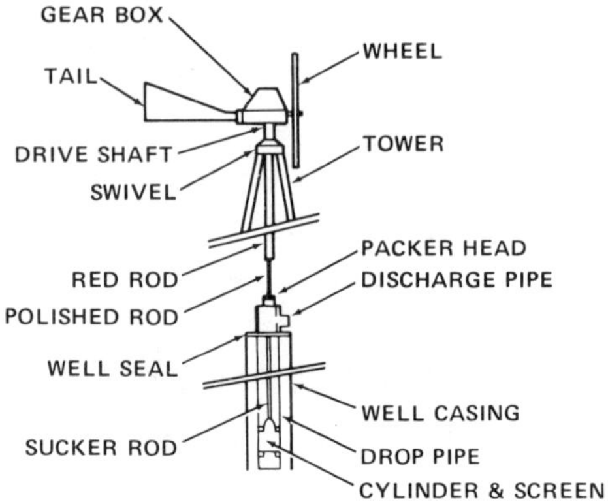

Courtesy DOE

Pictoral diagram of water-pumping windmill with polished rod.

pollen. A fine dust produced by plants; a natural or background air pollutant.

pollution. The presence of matter or of energy whose nature, location, or quantity produces undesired environmental effects.

polycrystalline silicon; polysilicon. Silicon which has solidified at such a rate that many small crystals (crystallites) were formed. The atoms within a single crystal are symmetrically arrayed, whereas in crystallites they are jumbled together.

polyelectrolytes. Synthetic chemicals used to speed flocculation of solids in urban waste.

Polyethylene®. A thin, transparent sheet of plastic used for moisture-proofing floors and walls, as a covering for greenhouses, and a cover material for solar collectors.

polymer. A macromolecule which is formed by the chemical union of five or more combining units called monomers. Cellulose is a polymer. In the case of cellulose, the combining units are glucose molecules. As many as 6000 to 8000 glucose units link together to form a cellulose molecule.

Poly-pane®. A glazing material of Worp Brothers which is a transparent plastic sheet 0.001 inch (0.0254 mm) thick.

polysaccharide. A combination of 11 or more monosaccharides, linked together by glycosidic bonds; for example, starch and cellulose. For alcohol fuel production, these can be broken down into fermentable sugars by various techniques.

polystyrene. Panels of a highly flammable foamlike plastic used for house insulation.

polyurethane. Plastic that can be applied as a foam and used as an insulating barrier.

pond. A body of water of limited size; a small lake.

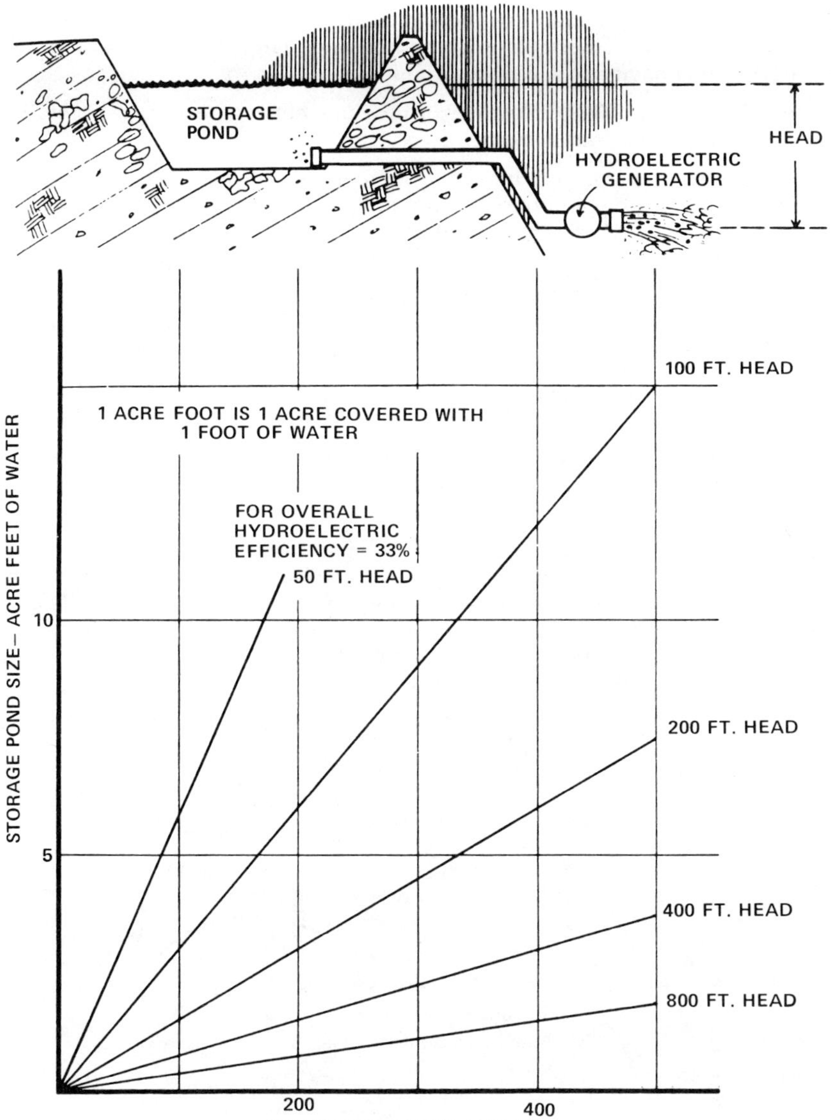

Courtesy DOE

Nomograph for energy storage pond sizing.

ponding. Retention of water due to deflection of (previously) horizontal surfaces.

positive drive. A driving connection in two or more wheels or shafts that will turn them at approximately the same relative speeds under any conditions.

potable water. Water that is suitable for drinking or cooking purposes, meeting the requirements of appropriate health officials.

potential. Electrical pressure, such as the electrical pressure normally existing between the conductors of a circuit on the terminals of a machine or apparatus; also a measure of the capacity of a well to produce oil or gas.

potential difference. The difference in pressure between any two points in an electrical circuit, measured in volts.

potential energy. The form of mechanical energy a body possesses by virtue of its position. If a body is being dropped from a higher to a lower position, the body is losing potential energy; if a body is being raised, it is gaining potential energy. Energy stored in matter because of its position or because of the arrangement of its parts; e.g., water stored behind a dam.

potential gradient. An ascending or descending value of voltage related to a linear measurement, as a distance along the earth surface or ground.

pound of steam. One pound (mass) of water in the vapor phase, not to be confused with the steam pressure which is expressed in pounds per square inch (psi).

pounds per square inch absolute (psia). The measurement of pressure, whether hydraulic (liquids) or pneumatic (gases), made without including the effect of atmospheric pressure, which is 15 pounds per square inch (psi) at sea level.

power. Any form of energy available for doing any kind of work; specifically, mechanical energy as distinguished from work done by hand; the rate at which work is done or energy is expended; measured in units of energy per unit of time such as calories per second (cal/sec), and in units such as watts (W) and horsepower (hp). The rate work is performed—mechanical power is force times velocity (see **horsepower);** electric power is volts times amps (V^4A).

power coefficient. The ratio of power extracted by a wind turbine to the power in the reference area of the free windstream.

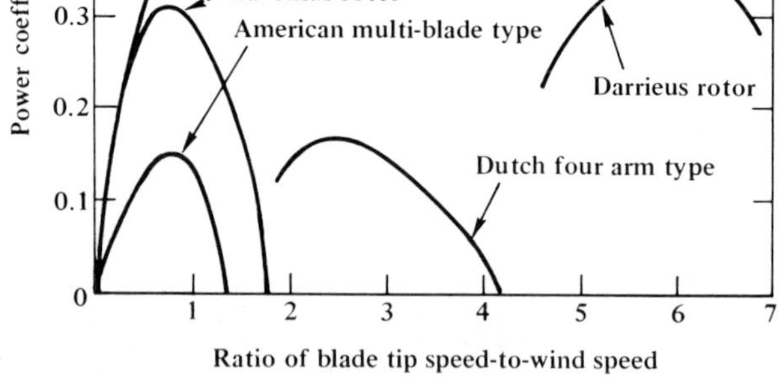

Courtesy DOE

Power coefficient of various types of wind-energy conversion systems.

power conditioning. Changing or modifying the characteristics of electrical power (i.e., DC to AC, or 12 volts to 32 volts).

power curve. A plot of wind-energy conversion system power output versus mean wind speed.

power density. The amount of power per unit area of the free windstream (the energy flux in the free windstream).

power duration curve. The cumulative probability curve of power output, expressed in probability units or in hours per year.

power factor. Relationship between volt-amperes and wattage. When the power factor is unity, volt-amperes equals wattage.

Powermizer®. A trade name of the Solarex Corporation for a battery charge controller that can be used with solar cells to prevent overcharging of a battery.

power output. Useful power delivered from a wind-energy conversion system at a stated mean wind speed.

power tower. The term used in the popular press to describe solar thermal central receiver power systems. A tower placed so that the reflected direct radiation from heliostat mirrors can be focused onto a receiver at its top. Heat exchange takes place at the top of the tower and the heated fluid is used in a conventional power system. See **Solar One.**

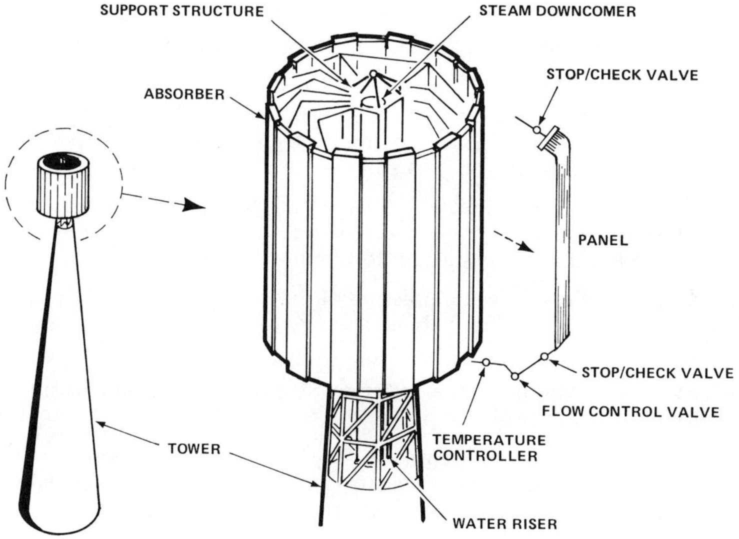

Courtesy DOE

Functional elements of receiver on power tower.

ppm. Parts per million. The unit commonly used to represent the degree of pollutant concentration where the concentrations are small. Larger concentrations are given in percentages. Thus, BOD is represented in ppm while suspended solids in water are expressed in percentages. In air, ppm is usually a volume/volume ratio; in water, a weight/volume ratio.

practical yield. The amount of product that can actually be derived under normal operating conditions; i.e., the amount of sugar that normally can be obtained from a given amount of starch or the amount of alcohol that normally can be obtained is usually less than theoretical yield.

precipitate. A solid that separates from a solution because of some chemical or physical change or the formation of such a solid.

precipitator. A device to remove fine ash, tars, dusts, or smoke particles from flue gases or other gaseous streams; the device may employ mechanical, electrostatic, or chemical means, or a combination of these; in pollution control work, any of a number of air-pollution control devices usually using mechanical/electrical means to collect particulates from an emission.

preheater. A way to use a solar heating system where solar heat warms the air or water and then an auxiliary heater or furnace finishes the heating by adding additional heat to the air or water.

premalting. A technique to increase the efficiency of the alcohol conversion process. Premalting is the addition of a small amount (10 percent) of malt or enzyme to the mash prior to cooking. Particularly useful for corn mash processing.

premature failure. Failure before the design lifetime.

present value. The value of a future cash flow discounted to the present; based on the premise that a dollar today is worth more than a dollar received in the future by virtue of the amount of interest (or return) it earns.

present worth. Amount of money that must be invested today in order to have a specified worth at a future time.

pressure. The load divided by the area over which it acts; the force per unit area applied to the outside of a body, normally expressed as force per unit area in pounds per square inch (psi) or grams per square centimeter (g/cm^2).

pressure differential. The pressure differential across the trap is determined at design conditions. If a vacuum exists on the discharge side of the trap, the vacuum is added to the inlet side pressure to determine the differential.

pressure drop. The loss in static pressure of the fluid (air, gas, or water) due to frictional obstruction in the pipe, valves, fittings, regulators, burners, appliances, or breeching.

pressurized. Any structure, area, or zone fitted with an arrangement that maintains nearly normal atmospheric pressure.

pretreatment. In wastewater treatment, any process used to reduce pollution load before the wastewater is introduced into a main sewer system or delivered to a treatment plant for substantial reduction of the pollution load.

primary air. Air that is mixed with fuel before the mix reaches the ignition zone to enhance combustion.

primary cell. A cell which generates or makes its own electrical energy from the chemical action of its constituents. Examples of primary cells are the voltaic cell, Daniell cell, LeClanche cell, and dry cell.

primary energy. Fuels as they are extracted from their original sources; i.e., fuels not derived from other fuels (coal, oil, and natural gas, for example).

primary fuel. Fuel consumed in original production (solar, coal, oil, natural gas, for example) of energy as contrasted to a conversion of energy from one form to another.

primary heating system. A conventional oil, gas, or electric heating system as opposed to a solar heating system.

primary treatment. The first stage in wastewater treatment in which substantially all floating or settleable solids are mechanically removed by screening and sedimentation.

primary voltage. The voltage of the circuit supplying power to a transformer, as opposed to the output voltage or load-supply voltage which is called secondary voltage.

priming. The act of adding water to displace air, thereby promoting suction, as in a suction line of a pump; in a boiler, the excessive carry-over of fine water particles with the steam due to insufficient steam space, faulty boiler design, or faulty operating conditions.

private placements. A security offering which is limited to a small group of investors and hence is not in the public markets or under the full scrutiny of the Security Exchange Commission; such investments may be in Sub-Chapter S corporations, standard corporations, limited partnerships, etc.

probable reserves. A realistic assessment of the reserves that will be recovered from known oil or gas fields based on the estimated ultimate size and reservoir characteristics of such fields. Probable reserves include those reserves shown in the proved category.

process development unit. An experimental facility with an input capacity up to 25 oven dry tons of biomass per day; establishes proof of concept, preliminary process economics, and engineering feasibility for a pilot or demonstration plant.

process guarantee. Refers to the financial ability of an engineering company to successfully make its process perform, within given tolerance levels, in the event that the process does not meet production levels originally agreed to in the contract.

process heat. Thermal energy which is used in agricultural, chemical, and industrial operations. (See diagram p. 240.)

process weight. The total weight of all materials, including fuels, introduced into a manufacturing process. The process weight is used to calculate the allowable rate of emission of pollutant matter from the process.

productivity. The efficiency with which economic resources (men, material, and machines) are employed to produce goods and services; the value added to goods divided by the man-hours input.

products of combustion. Constituents resulting from the combustion of a fuel with the oxygen of the air including the inerts but excluding excess air.

profile. A graphic representation of the variation of one property, such as wind speed or temperature, with respect to another property such as distance or position.

projected area. As seen from the direction of wind velocity, the area covered at

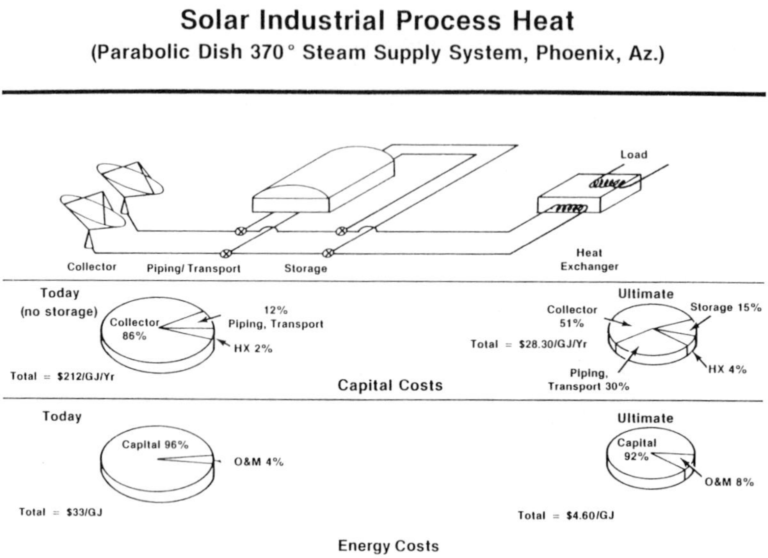

Courtesy Mitre Corporation

Capital and energy costs for solar industrial process heat.

any instant by the wind-energy conversion system rotor blades. (Area solidly covered by the blades as opposed to the swept area.)

projection. An estimation of probable future events.

proof. Ethyl alcohol content is measured by proof strength. Proof is defined as twice the percent of ethyl alcohol by volume when the liquid is at 60°F (15.56°C). By this measurement, 1 gallon of 95 percent (190 proof) alcohol is equivalent to 1.9 proof gallons of alcohol.

proof gallon. A U.S. gallon of liquid which is 50 percent ethyl alcohol by volume; also 1 tax gallon.

proof spirit. Designates the alcoholic content of a liquid.

proof testing. A proof hydrometer is used to measure liquids that are lighter than water. To test alcohol proof, sample the alcohol with the proof hydrometer. A reading on the hydrometer of 160 or higher indicates fuel grade alcohol. (See diagram p. 241.)

propeller anemomter. A horizontal axis wind turbine whose output is used exclusively to determine wind speed.

propeller fan. A fan with airfoil blades and which moves the air in the general direction of the axis of the fan.

protein. A class of plant and animal molecules with both structural and functional (muscle, enzymes) roles. They are long chains of up to 25 different amino acids whose sequence and number uniquely determine the protein's character and function.

proton. An elementary particle with a single positive electrical charge and a mass approximately 1847 times that of the electron; the nucleus of an ordinary or light

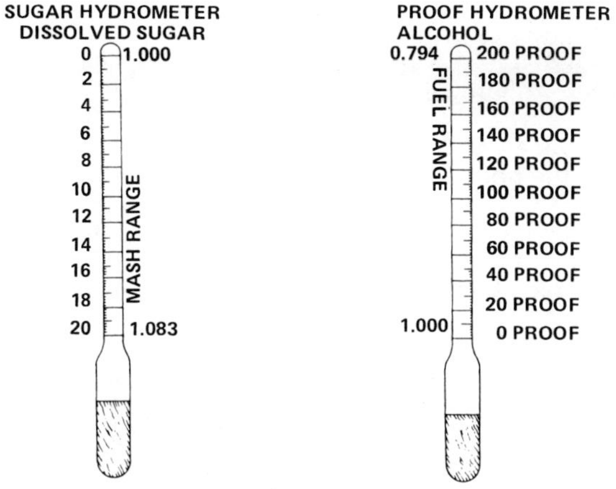

Courtesy DOE

Proof testing can be easily performed by using a proof hydrometer.

hydrogen atom. Protons are constituents of all nuclei. The atomic number of an atom is equal to the number of protons in its nucleus.

***p*-silicon.** Silicon containing a minute quantity of impurity, or dopant, such as boron, which provides insufficient electrons to exactly complete the crystal structure. There is no electrical imbalance, however.

psychrometer. A device for measuring the vapor pressure and the relative humidity of the air or the quantity of moisture in the air, employing a wet-bulb and a dry-bulb thermometer.

psychrometric chart. A graph showing the properties of airstream mixtures and used in air conditioning.

pulverization. The crushing or grinding of material into small pieces.

pump. A device which forces water through an active solar heating system.

pump capacity. The volume of fluid, at a specified pressure, that a pump can transfer or lift when powered by an engine or motor of any given horsepower.

pumped storage. Method of generating hydroelectricity by using low cost (usually off-peak) power to pump water to an elevated storage area and releasing it at peak demand periods into a river or other body of water.

pumped-hydro storage. A utility storage system in which off-peak power is used to pump water to an elevated reservoir for later reconversion to electricity by passage through turbines at peak-demand periods. Storage of energy potential in the form of a water head.

pumping station. A station at which sewage is pumped to a higher level. In most sewer systems pumping is unnecessary; wastewater flows by gravity to the treatment plant.

pure ethyl alcohol. Ethyl alcohol that has not been denatured and is usually sold at 190 proof and 200 proof.

putrescible. Capable of being decomposed by microorganisms with sufficient rapidity to cause nuisances from odors, gases, etc.

PV. Abbreviation for photovoltaic(s).

pyramidal optical condenser. A solar heating system that has reflectors mounted in the roof peak of the house.

pyranometer. A radiometer that is used to measure the total incident solar energy per unit time per unit area upon a surface which includes the beam radiation from the sun, the diffuse radiation from the sky, and the shortwave radiation reflected from the foreground.

Courtesy SERI

Solar flux experiment being conducted by Solar Energy Research Institute with a pyranometer.

pyrheliometer. Instrument that measures the intensity of the direct beam radiation from the sun; the diffuse component is excluded.

pyrolysis. Transformation of a material into another compound(s) via application of heat; also called destructive distillation or the heating of organic matter, e.g., coal, wood, petroleum, solid wastes, in the absence of oxygen. Initial products are water vapor and various volatile compounds. At higher temperatures, organic matter recombines into complex hydrocarbons and water. Principal products are gases, oils, and char.

pyrolytic gasification. Pyrolysis (the breakdown of matter by heat in the absence of oxygen) is used to convert solid fuels into gas, oil, and char (charcoal from biomass and coke from coal). If the oil and char are not considered principal products, extensive additional processing is needed to convert them to gas. Pyrolytic gasification produces a medium-energy gas (MEG) that contains carbon monoxide, hydrogen, methane, and other hydrocarbons.

Q

quad. One quadrillion (10^{15}) Btu. Commonly used as measure of annual energy consumption, usually expressed as primary fuel equivalent. Present U.S. Consumption is about 78 quads.

quality. The ratio by weight of vapor to liquid plus vapor in a mixture, as in steam.

quantum. Unit quantity of energy according to the quantum theory. It is equal to the product of the frequency of radiation of the energy and 6.6256×10^{-27} erg-seconds. The photon carries a quantum of electromagnetic energy.

quantum yield. A technical parameter from quantum physics. For practical purposes, the quantum yield of a solar cell is the spectral response of the cell; that is, the relative output at different wavelengths of light.

quarter chord. The point on a wind-energy conversion system airfoil chord line one quarter chord back from the leading edge.

quarter chord line. A line passing through the locus of quarter chord points of a wind-energy conversion system rotor blade.

quarter chord moment. Aerodynamic moment that causes a torque about the quarter chord point of a wind-energy conversion system airfoil in a wind flow.

quasisteady. Describes the state of the solar collector test when the flow rate and temperature of the fluid entering the collector are constant, but the exit fluid temperature changes "gradually" due to the normal change in insolation that occurs with time for clear sky conditions.

quenching. Cooling by immersion in oil or water bath or spray; also applies to cooling in salt and molten metal baths, or by means of an air blast; rapid cooling; the act of terminating an ionizing current by the use, for example, of an appropriate gas or vapor filling or by momentary reduction of the applied potential difference.

quench tank. A water-filled tank used to cool incinerator residues.

quilt. Several flexible sheets of cloth or plastic (or plastic foam) attached to one another in series to form a thick, soft, flexible assembly. Quilts constitute the heart of some window insulating shades and curtains.

R

race. An aqueduct or channel for conducting water to or from the place where it performs work.

raceway. Conduits, moldings, and other hollow material, often concealed, through which wires are run from one outlet to another.

rad. A unit of measurement of any kind of radiation absorbed by man.

radial blade fans. Radial blade fans are generally narrower than other types of centrifugal fans. Consequently, they require a larger diameter wheel for a given capacity. This increases the cost and is the main reason why they are not used for air conditioning duty. The radial blade fan is well suited for handling low volumes at relatively high static pressures and for material handling. Absence of surge and a nearly straight horsepower curve with linear relationship with cubic feet per minute are its other advantages. This proportional relationship allows capacity control to be actuated from motor power input. Disadvantages of this type of fan are higher cost and lower efficiency.

radial station. An airfoil section of a wind-energy conversion system blade a certain distance along the quarter chord line.

radiant energy. Energy that radiates or travels outward in all directions from its source.

radiant heating. Usually a system of heating by surfaces at higher than body temperatures whereby the rate of heat loss from human beings by radiation is controlled; the heat gained in a house or room by heat emitted from large, sun-warmed surfaces, such as walls or floors.

radiant power (watts). The time rate of flow of radiant energy.

radiant sky cooling system. A space conditioning system that circulates a fluid through the collectors during those hours which have the appropriate conditions for cooling (usually night and early morning). The cool liquid or gas is stored and cool air is transferred to the building interior during the cooling period.

radiant systems. A solar space conditioning system that transfers solar heated or climatically cooled fluid to a building component (floor or wall), thereby providing space conditioning to the building interior by radiation.

radiation. The emission and propagation of energy through space (or through material medium) in the form of waves (or photons). Flow of energy from one body to another across a space. Radiation may proceed via visible light or via invisible infrared rays, depending on the temperature of the radiating body. In wood heating, heat transfer from a stove directly to another surface such as a wall, furniture, or a person without affecting the temperature in between. See **extraterrestrial radiation; solar radiation.**

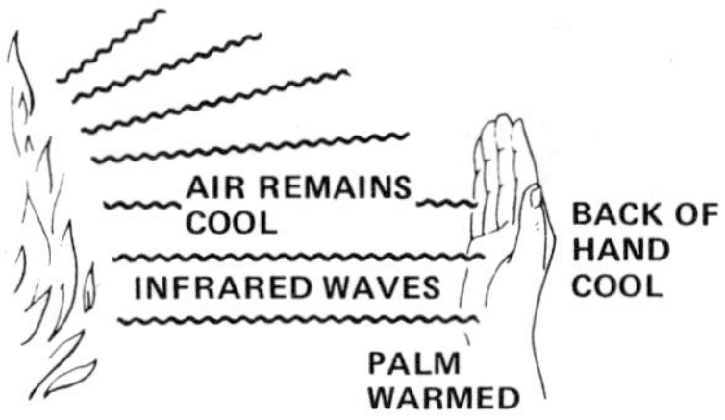

Courtesy N. Soderstrom, Heating Your Home with Wood, *Popular Science Skill Book.*
Illustration of effects of heat radiation.

radiative cooling. Cooling provided by warm surfaces radiating excess heat to cool surfaces. Water bodies (roof ponds) and massive construction materials (concrete and stone) absorb heat from interior spaces during daytime hours and radiate it away at night.

radiator. A device in a home to let heat flow easily from heated water into the air. Radiators usually transfer more heat by convection than by radiation heat flow.

radiobiology. The study of the principles, mechanisms, and effects of radiation on living matter.

radioecology. The study of the effects of radiation on species of plants and animals in natural communities.

radiometer. Instrument measuring the intensity of any kind of radiation.

radiosonde. A balloon-borne instrument for the simultaneous measurement and transmission of atmospheric pressure, temperature, and humidity.

rainwater collector. A tray used to catch rainwater; analogous to a solar collector.

rainwater equivalent. A part of or function on a rainwater collector system that is analogous to a solar collector system.

Rankine cycle. The complete expansion cycle of a gas which is brought into a structure, such as a turbine engine, where the gas is compressed and then expands rapidly to give off energy.

Rankine-cycle solar air conditioner. A small solar-powered turbine is used to run a conventional air-conditioning unit. A working fluid is heated until it expands and turns a turbine. The turbine can either drive a conventional air conditioner or turn a generator and make electricity. Solar-heated hot water at 216°F (102°C) enters the system from the solar panels. It flows through coils in three sealed tanks (vapor separator, boiler, and preheater), where the liquid refrigerant picks up heat and vaporizes. The Freon vapor at 199°F (93°C) and 54 pounds per square inch (psi) flows to the turbine. The Freon is cooled in a heat exchanger and condenser, and returns through the feed pump and gearbox back to the boiler where the cycle repeats. When there is no solar heat to run the turbine, the air conditioner can run on line power.

Rankine engine. Heat engine using various components, including a working fluid pumped under pressure to a boiler where heat is added; an expander (turbine) where work is generated; and a condenser used to reject low-grade heat to the environment. A steam engine is a Rankine engine.

rasp. A device used to grate solid waste into a more manageable material, ridding it of much of its odor.

rate base. The value established by a regulator authority, upon which a utility is permitted to earn a specified rate of return.

rated horsepower. Theoretical horsepower of an engine based on dimensions and speed.

rated load. The kilowatt power output which can be delivered continuously at the rated output voltage; may also be designated as the 100 percent load or full-load rating of the unit.

rated output capacity. The output power of a wind machine operating at the constant speed and output power corresponding to the rated wind speed.

rated power. The power output obtained from a wind-energy conversion system operating at its rated wind speed.

rated wind speed. Wind speed at which the rated power is specified (usually the lowest wind speed which gives maximum wind-energy conversion system output).

rate of flow. The volume or units of a material passing a given point in a system per unit of time.

rate schedule Rules determining the cost of electricity or gas used during a billing period. Typically, the first group or block of kilowatt-hours (kWh) or of therms costs more than the second block; the second more than the third, and so on.

rating. Limits placed on the operating conditions of a machine, apparatus, or device based on its design characteristics.

ratio of specific heats. For gases, the ratio of the specific heat at constant pressure to the specific heat at constant volume.

rawinsonde. A radiosonde system with the additional capability of evaluating wind speed and direction aloft by means of tracking by radar or radio direction finder.

raw sewage. Untreated domestic or commercial waste water.

raw source energy. The quantity of energy input at a generating station required to produce electrical energy, including all thermal and power conversion losses.

Rayleigh frequency distribution. A mathematical idealization giving a ratio of time the wind blows within a given wind speed band to the total time under consideration. This distribution is dependent only on mean wind speed

$$p(x) = \frac{\pi x}{2\bar{x}^2} \exp[-\pi x^2/4\bar{x}^2]$$

where $\bar{x}$ is the average value of the variate x.

reactive power. The quantity in alternating current circuits that is obtained by taking the square root of the difference between the square of the kilovolt-amperes and the square of the kilowatts. It is expressed as reactive volt-amperes.

reagent. Any material that causes a chemical reaction when added to a second substance.

real disposable income. The figure, in dollars of constant value, obtained by subtracting corporate earnings not paid out as dividends, depreciation, and all taxes from gross national product and then adding welfare-type transfer payments and government interest payments. This figure is intended to represent the income which the public has available for making purchases.

real gross national product. Gross national product in constant dollars.

real growth rate. The growth rate net of the general economy inflation rate.

real income. Current dollar income corrected for price level changes relative to some base period as measured by a suitable price index, or the implicit Gross National Product deflator. Also termed "constant dollar" income.

real power. The rate of supply of energy, measured in kilowatts.

receiver. That element of a solar central receiver system to which solar radiation is directed by the heliostats and where it is absorbed and converted to thermal energy.

receiver fluid. The cooling fluid that is circulated through the receiver to absorb the solar radiation as thermal energy, also called heat transport fluid.

receiving waters. Rivers, lakes, oceans, or other bodies that receive treated or untreated wastewaters.

recessed radiator. One set into the wall but not enclosed.

reciprocating engine. An apparatus which converts the energy in a fluid to mechanical energy by means of the expansion of the fluid (gas) against a piston. It normally includes a cylinder, closed by a piston connected by means of a connecting rod to a crankshaft; a valve mechanism admits and discharges fluid at appropriate times in the cycle.

recirculated air. Air returned from a space to be heated, conditioned or cleaned, then redistributed to the space.

recombination. A free electron being reabsorbed into a hole.

recombination loss. Energy loss which occurs when electron-hole pairs recombine without contributing photovoltaically.

recoverable resource. That portion of a resource expected to be recovered by present-day techniques and under present economic conditions.

recovered energy. Reusing heat or energy that would otherwise be lost.

rectenna. A device that converts microwave energy into direct current power;

consists of a number of small dipoles, each having its own diode rectifier network, which are connected to direct current buses.

rectification. With regard to distillation, the selective increase of the concentration of a component in a mixture by successive evaporation and condensation.

rectifier. A device for converting alternating current to direct current; one use of this current is for external corrosion control of pipe and other metals; also the container or chamber in a cooling system where, typically, ammonia vapor is separated from water vapor.

rectifying column. The portion of a distillation column above the feed tray in which rising vapor is enriched by interaction with a countercurrent falling stream of condensed vapor.

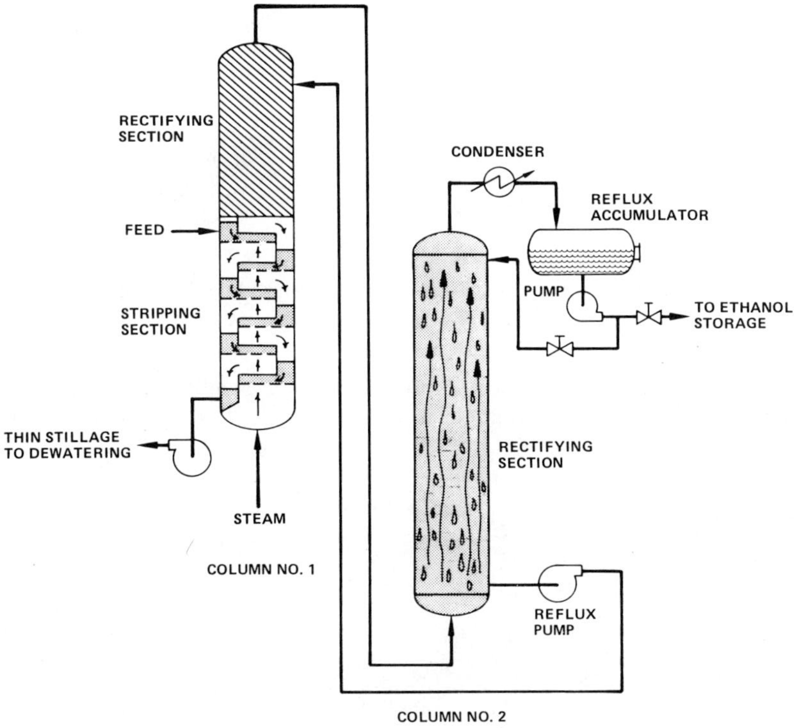

Courtesy DOE

Schematic of small scale fermentation ethanol distillation system with rectifying column.

recuperator. A continuous heat exchanger in which heat is conducted from the products of combustion to incoming air through flue walls.

recycling. The repetition of a particular process; the return of a stream or part of a stream to a previous process or location for additional recovery of the desired components.

redox. Abbreviation for reduction-oxidation.

redox battery. In this battery, the electrolytes in the anode and/or cathode departments contain ions of metals which can exist in more than one valence state. At the anode, such a redox couple is oxidized, while another couple is reduced at the cathode.

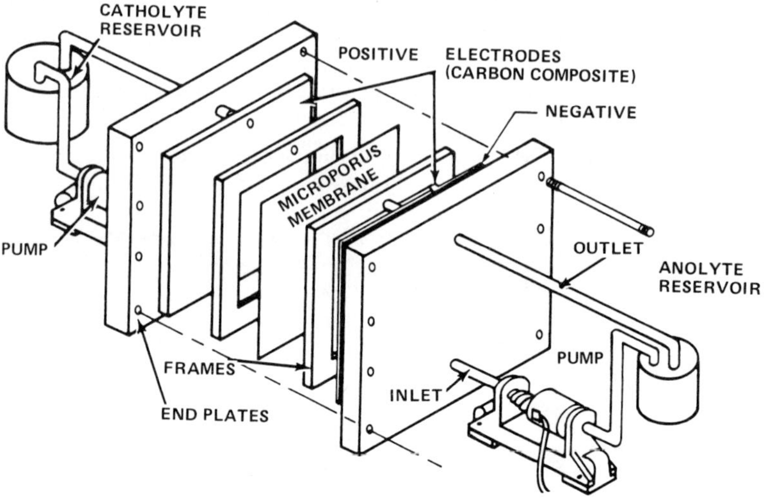

Source Gel, Inc., Durham, NC

Iron-redox battery structure and plumbing connections for energy storage.

red rod. The uppermost part of the water-pumping assembly. It is designed to be the weakest link in the pump train and is usually constructed of wood to permit it to break in case of system malfunction.

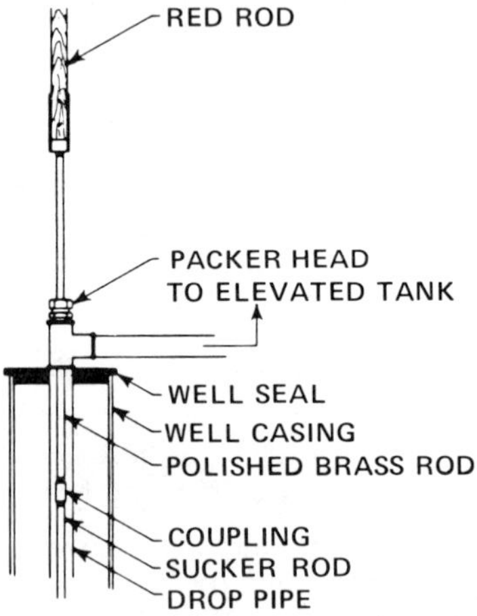

Courtesy New Mexico Energy Institute

Illustration of water pumping windmill red rod connection.

reducer. A pipe fitting with a larger size at one end than at the other; the larger size is designated first. Reducers are threaded inside, unless specified flanged, welded, or for some special joint.

reference area. For a wind-energy conversion system it is πR^2, where R is the maximum blade radius, not including effects of coning, flapping, or tethering.

refinery. A device (usually a tower) or process which heats crude oil so that it separates into chemical components, which are then distilled off as more usable substances. Simple structure components vaporize first. Typical crude fractions are: methane and ethane (the gasolines); propane and butane; kerosene, fuel oil, and lubricants; jelly paraffin, asphalt, and tar. The term is also applied to the plant in which the refinery process is carried out.

reflectance. The ratio or percentage of the amount of light reflected by a surface to the amount incident on the surface. The remainder that is not reflected is either absorbed by the material or transmitted through it. Good light reflectors are not necessarily good heat reflectors.

reflected light. When light reaches an object some of it may be absorbed, and some may be bounced off the surface. The light which is bounced off the surface is called reflected light.

reflected radiation. Solar radiation received after reflection from the ground or adjacent structures.

reflection. The process by which solar radiation, on striking a light-colored surface, is not absorbed—it reflects away; also refers to radiation heat flow on striking a shiny surface.

reflective loss. The energy which strikes a surface and is not absorbed but reflected from it.

reflectivity. The ratio of the specular component of reflected solar energy to the incident solar energy.

reflectometer. A device for measuring the reflectances of light or other radiant energy.

reflector. A shiny surface that reflects sunlight into a collector to increase the solar radiation it absorbs.

reflector control. Reflector control by adjustment of the properties, position, or size of the reflector in such a way as to change the reflectivity.

Reflecto-shade®. A glazing product of Modico Inc. which is a slightly transparent sheet including an extremely thin coating of aluminum.

reflux. A distillate fraction having a certain boiling range which is introduced into fractionating equipment to bring about a cooling effect, resulting in more intimate contact between the vapors and thus in more efficient fractionation.

reforming. The cracking of petroleum naphtha or of straight-run gasoline of low octane number, usually to form gasoline containing lighter constituents and having a higher octane number.

reforming processes. A group of proprietary processes in which low-grade or low-molecular-weight hydrocarbons are catalytically reformed to higher grade or higher-molecular-weight materials; also applies to the endothermic reforming of methane for the production of hydrogen by the reaction of methane and steam in the presence of nickel catalysts.

refraction. A change of direction of a ray of light when it passes from one medium to another of different optical density.

refractive index. See **index of refraction**.

refractory. A ground-up substance mixed with concrete and applied to the bottom of a firebox to provide added protection against burn-out.

refrigerant. Fluid, such as Freon® , that is used in heating or cooling devices, such as heat pumps, air conditioners, or solar collectors.

refrigeration capacity. The rate of heat removal by a refrigerating system, usually expressed in British thermal units per hour (Btu/h) or in tons.

refrigeration cycle. The full sequence of condensation and evaporation. The heat of evaporation is obtained from the material to be cooled.

refrigeration ton. The heat required to melt one short ton (2000 pounds) of ice in 24 hours. The latent heat of ice being 144 British thermal units per pound (Btu/lb):

$$1 \text{ refrigeration ton} = \frac{144 \times 2000}{24} = 12{,}000 \text{ Btu}$$

refuse reclamation. The process of converting solid waste to saleable products. For example, the composting of organic solid wastes yields a saleable soil conditioner.

regain. Moisture regain is the ratio of normal moisture weight to the weight of material holding the moisture, expressed in percent.

regenerative cooling system. Cooling system operated by using a charging and discharging cycle with a thermal or latent heat storage device.

regenerative heating. The process of utilizing heat which must be rejected in one part of the cycle to perform a useful function in another part of the cycle.

regulated hot water. Hot water whose temperature is held more or less fixed.

regulator. An opening in a wall or door in the return airway of a district to increase its resistance and reduce the volume of air flowing. It consists of a sliding shutter which can be adjusted to any proportion of the maximum aperture.

rehabilitation. Returning the land to a form in conformity with a prior land-use plan, including a stable ecological state that does not contribute substantially to environmental deterioration.

relative density. See **specific gravity.**

relative humidity. The amount of moisture contained by an atmosphere compared with the maximum amount that it could contain at the same temperature, expressed as a percentage.

relative wind. Velocity of air with reference to a body in it.

reliable. A solar heating system that is rugged and simple and expected to operate a long time without need of repair.

relief valve. A valve designed to open automatically to relieve excess pressure.

remaining reserves. Those quantities of crude oil, natural gas, natural gas liquids, and sulfur as estimated under proved or probable reserves after deducting those quantities produced up to the respective date of the estimate.

remote site. Not connected to a utility grid. See **distributed power**; **stand-alone**.

renewable energy source. Solar energy and certain forms derived from it, such as wind, biomass, and hydro.

renewable resources. Sources of energy that are regenerative or virtually inexhaustible, such as solar energy.

repowering. Repowering involves the installation of a solar thermal generating system in parallel with existing fossil fuel powered boilers at an intermediate or peaking electric utility plant or at an industrial site. The repowering concept conserves fossil fuels by utilizing fossil boilers as an auxiliary system only during periods of inadequate solar insolation. The fact that solar thermal central receiver systems may be used in conjunction with the reliable, existing fossil boiler subsystems should invoke favorable interest for investment on the part of the electric utilities and industry. In addition, repowering offers the use of existing power conversion systems, minimal site development, and plant installation requirements.

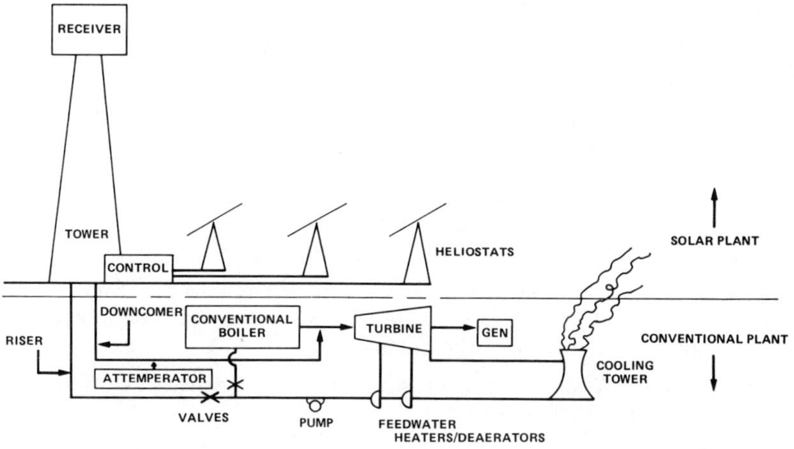

Courtesy DOE

Solar thermal repowering plant configuration diagram.

required radiation. The area of radiator surface required, based on the heat-loss computation for the space to be heated.

reradiation. The transfer or loss of energy by radiation of thermal energy from a prior absorption of energy. Often, the reradiation energy is at a longer wavelength than the absorbed energy, which allows for control of this transfer.

reserve capacity. The difference between the total system dependable capacity and the actual or anticipated total system peak load for a specified period.

reserve generating capacity. Extra capacity maintained to generate power in the event of unusually high demand or a loss or scheduled outage of regular generating capacity.

reserves. Resources which are known in location, quantity, and quality and which are economically recoverable using currently available technologies.

reservoir. A space such as a pond, lake, tank, or basin which is used for storage and control of water; a natural underground container of liquids, such as oil or water, and gases.

residential demonstration program. From 1974 to 1979 the U.S. Department of Housing and Urban Development sponsored a five-year program to demonstrate the solar heating and cooling of residential buildings, as part of the National Program for Solar Heating and Cooling of Buildings.

Under this program, proposals were solicited from builders, developers, architects, engineers, universities, and local governmental agencies for the construction of solar-equipped housing units, or the retrofitting of solar systems to existing structures. Demonstration grants were generally made to cover the additional cost of the solar system, including design and installation.

residual fuel oil. The heavier oils that remain after the distillate fuel oils and lighter hydrocarbons are boiled off in refinery operations. Included are products known as ASTM grades Nos. 5 and 6 oil, heavy diesel oil, Navy Special Oil, Bunker C oil, and acid sludge and pitch used as refiner fuels. Residual fuel oil is used for the production of electric power, for heating, and for various industrial purposes.

residual heat. The total heat source remaining in a shutdown reactor, including after-heat.

residue. That which remains after a part has been separated or otherwise treated.

resistance. The tendency of a material, especially insulation, to retard the flow of heat. Often called R-value. The ability of all conductors of electricity to resist the flow of current, turning some of it into heat. Resistance depends on the cross-section of the conductor (the smaller the cross-section, the greater the resistance), and its temperature (the hotter the cross-section, the greater its resistance).

resistance heating. A method of converting electricity into heat for the purpose of home heating. Heat generated by an electric current as it is conducted through a substance opposing the flow of the current.

resistance to flow. The same as pressure drop. It is the resistance offered by a conduit or device to the flow of fluid.

resistance weld pipe. Pipe made by bending a plate into circular form and passing electric current through the material to obtain a welding heat.

resistivity. The reciprocal of conductivity.

resistor. An electric device which "resists" electric current flow, used to control current (e.g., field-current in a generator).

resonance. Potentially destructive condition where the frequency of the applied load equals the natural vibration frequency.

resource recovery. The process of obtaining materials or energy, particularly from solid wastes.

resources. The total estimated amount of a mineral, fuel, or energy source, whether or not discovered or currently technologically or economically extractable; quantities of an energy commodity that may be reasonably expected to exist in favorable geologic settings, but that have not yet been identified by drilling;

also reserves and/or materials that have been identified, but cannot now be extracted because of economic or technological limitations, as well as economic or subeconomic materials that have not as yet been discovered.

restrictive covenants. The most common form of private agreement that can be used to protect solar access; a restrictive covenant is a contract between two or more people which involves mutual promises of reciprocal benefits and burdens among the contracting landowners.

retrofit. The subsequent installation of equipment after initial construction is completed, such as the installation of solar collectors and other hardware onto an existing house to convert it for solar heating.

return air. Air returning to a heater or conditioner from the heated or conditioned space.

return hose. A hose in a thermic cycle that lets cool storage water return to the collector.

return on investment. Market discount rate that will result in zero life-cycle savings.

return time. Time before the wind returns to a higher, specified value, such as the cut-in speed of a windmill.

reverse osmosis. An advanced method of waste treatment relying on a semipermeable membrane to separate waters from pollutants.

reverse thermosyphon. A way in which a passive thermosyphon system can lose heat at night by the air or water circulating in the reverse direction.

Reynolds' number. A dimensionless ratio proportional to pipe or duct diameter, velocity, and density of fluid and inversely proportional to the viscosity of the fluid. With English units, a Reynolds' number over 2500 indicates turbulent flow; below that, streamlined flow.

R-factor. A unit of thermal resistance used for comparing insulating values of different materials; the reciprocal of the conductivity; the higher the R-factor of a material, the greater its insulating properties. See **resistance.**

ribbon. A thin sheet of crystalline or polycrystalline material, such as silicon, produced in a continuous process by withdrawal from a molten bath of the parent material.

ribbon cells. Photovoltaic cells produced by a continuous ribbon process which provides greater panel packing potential efficiency.

rigid PVC. Polyvinyl chloride or a polyvinyl chloride/acetate copolymer characterized by a relatively high degree of hardness, formulated with or without a small percentage of plasticizer.

riparian rights. Rights of landowners to the water on or bordering their property, including the right to prevent diversion or misuse of upstream water.

riser. A flow channel or pipe that distributes the heat-transfer liquid from the header across the face of an absorber plate. See **header.**

rising stem. A threaded stem which is unscrewed or screwed through the bonnet to open or close the valve. The handwheel may rise with the stem, or the stem may rise through the handwheel.

river basin. The total area drained by a river and its tributaries.

road octane. A numerical value based upon the relative antiknock performance of an automobile with a test gasoline as compared with specified reference fuels. Road octanes are determined by operating a car over a stretch of level road or on a chassis dynamometer under conditions simulating those encountered on the highway.

rock-bin. The storage medium for an active air-heating collector system. Rocks are crushed and held in an airtight container with appropriate ducts.

rock wool. A form of natural mineral home-insulation material, which may be installed in batts or blankets. Rock wool is fire and moisture resistant.

Rocky Flats Test Facility. DOE test facility for performance evaluation of small wind energy conversion systems. The Rocky Flats facility is located in Golden, CO.

roll bond. Bonding sheets of metal together by simultaneously rolling and using a heat treatment.

roof credit. Savings from roof/structures rendered unnecessary by photovoltaic arrays.

roof pitch. The measure of steepness of a roof expressed as the number of inches the roof rises for each foot of horizontal run.

roof pond. A passive solar collection system in which plastic bags of water on the roof are appropriately exposed or insulated during the day and night depending on the thermal requirements of the building for summer and winter operation. Heat transfer (into or out of the bags) to the structure is directly through the ceiling.

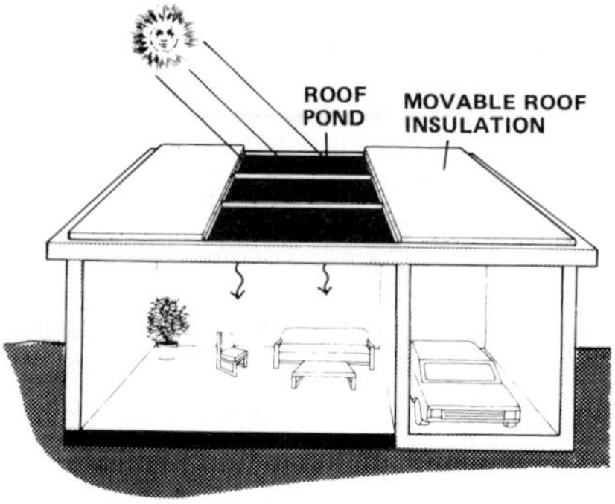

Courtesy DOE

Movable roof insulation permits sun's direct radiation to heat bag of water contained on roof pond.

roof spray. A system that reduces heat gain through a roof by cooling the outside surface with a water spray.

room heater. A self-contained, freestanding, nonrecessed, air-heating appliance intended for installation in the space being heated and not intended for duct connection.

rotation. The planned number of years between the formation or regeneration of a forest stand and its final cutting at a specified stage of maturity.

rotor. The power-producing blades and rotating members of a wind turbine (e.g., the blades). A system of rotating aerodynamic elements attached to a single shaft that converts the kinetic energy in the wind into mechanical shaft energy.

rotor axis. Axis of the rotor shaft through which power is extracted from a wind-energy conversion system rotor.

rotor diameter. Twice the distance from the rotor axis to the outermost point on the blade.

rotor efficiency. The efficiency of the rotor only; does not include transmission pumps, generators, or line or head loss.

rotor plane. A plane perpendicular to the wind-energy conversion system rotor axis passing through a point where the blade quarter chord lines intersect.

rotor power coefficient. Power delivered to the rotor main shaft, rotor axis, divided by the power available in the flow field at a specified wind speed. See **rotor efficiency.**

rotor tilt. In a horizontal axis wind-energy conversion system, the angle between horizontal and the rotor axis.

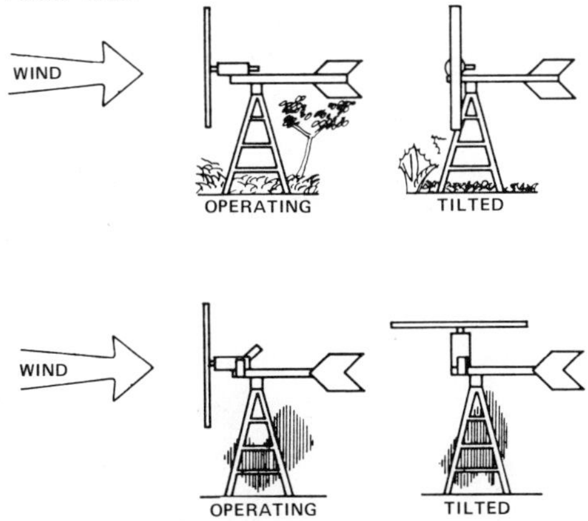

Courtesy DOE

Two methods of tilting the rotor for wind-energy conversion system control.

rotor torque. The moment (torque) produced by a wind-energy conversion system rotor about its axis.

rubbish. A general term for solid waste—excluding food waste and ashes—taken from residences, commercial establishments, and institutions.

rumen. Herbivores, such as cows and sheep, have a multiple-chambered stomach, the rumen, which has evolved so that they can digest the grass and cellulose that they eat. Inside the rumen live anaerobic bacteria (among them methanogens) which break down the cellulose to smaller carbohydrate molecules which the ruminant can absorb.

run. A length of pipe made of more than one piece of pipe; a portion of a fitting having its ends in line or nearly so, in contradistinction to the branch or side opening, as of a tee.

run duration. Length of time after wind speed or power goes below (or above) a certain level (called the run level), until it returns above (or below) that level.

runoff. That portion of the rainfall not absorbed by the deep strata. It is used by vegetation or lost by evaporation, or it may find its way into streams as surface flow.

run of the wind. The distance the wind travels during a specific time period; this usually refers to the dial reading from a wind anemometer.

rural windmills. Wind power was once used widely on American farms but was replaced during the 1930s by inexpensive electricity and natural gas. Now, with rising fuel costs, wind power is being reexamined as a power resource for farms, rural homes, and remote rural regions. Old-fashioned water-pumping windmills with multiple blades are available from several manufacturers.

R-value. An insulation material resistant to temperature changes or heat flowing through it. The R-value required to adequately insulate a home will depend on such factors as the climate and the type of heating system. The larger the R-value is, the better the insulation. Adequate insulation in the attic and walls saves money on both heating and cooling bills. By the installation of adequate insulation, it has been estimated that 20 to 30 percent of the energy used to heat homes in winter, and 10 percent required to cool them in summer can be conserved. See **thermal resistance.**

S

saccharide. A simple sugar or a compound that can be hydrolyzed to simple sugar units.

saccharify. To hydrolyze a complex carbohydrate into simpler soluble fermentable sugars, such as glucose.

Saccharomyces. A class of single-cell yeasts which selectively consume simple sugars.

saddle fitting. A clamp which is attached over a piece of pipe like a saddle on a horse.

saddle flange. A flange curved to fit a boiler or tank and to be attached to a threaded pipe. The flange is riveted or welded to the boiler or tank.

safety factor. Good design practice dictates the use of safety factors in steam-trap selection. Safety factors from 2 to 1 to as high as 8 to 1 may be required, and for the following reasons. The steam pressure at the trap inlet or the back pressure at the trap discharge may vary. This changes the steam-trap capacity. If the trap is sized for normal operating load, condensate may back up into the steam lines or apparatus during start-up or warm-up operation. If the steam trap is selected to discharge a full and continuous stream of water, the air could not be vented from the system.

safety glass. Glazing materials predominately inorganic in character which meet the appropriate requirements of the ANSI Standard and include laminated glass, tempered glass, and wired glass. Glazing materials so constructed, treated or combined with other materials so as to minimize the likelihood of cutting and piercing injuries resulting from contact.

saline soil. A soil containing excessive quantities of the neutral or nonalkaline soluble salts.

salinity. The total amount of solid material in grams contained in 1 kilogram (kg) of seawater when all the carbonate has been converted to oxide, the bromine and iodine replaced by chlorine, and all organic matter completely oxidized. Expressed as grams per kilogram (g/kg) of seawater or parts per thousand (ppt).

salt. A chemical compound formed when the hydrogen ion of an acid is replaced by a metal, or, together with water, when an acid reacts with a base. Salts are named according to the metal and the acid from which they are derived.

salt hydrate. A crystalline solid compound of one or more salts combined with water in definite proportions. Upon melting, it forms a salt/water solution.

saltwater intrusion. The invasion of salt water into a body of freshwater, occurring in either surface or groundwater bodies. When this invasion is caused by oceanic waters, it is called seawater intrusion.

salvage. The utilization of waste materials.

sand interceptor. A trap designed to retain sand.

Sandia Central Receiver Test Facility. The Sandia Laboratories Central Receiver Test Facility has a field of 222 heliostat arrays, each containing 25 4-foot-square (1.2-meter-square) mirror facets which concentrate the sun's rays on solar receivers (boilers) mounted on the 200-foot-high (61-m) tower. Water circulates through a boiler and is converted to steam by the high-intensity solar radiation. The steam can be used to drive a turbogenerator for the production of electricity.

Sandia Vertical Axis Wind Turbine. Sandia Laboratories have developed and tested the eggbeater shaped vertical axis wind turbine (VAWT). The new blades on this 55-foot (17-meter) diameter turbine are shaped like the cross-section of an airplane wing. The aerodynamic lift from the blades provides driving torque regardless of wind direction. In a 30 mph wind, these new 24-inch (0.6-meter) blades produce 60 kW of electricity as opposed to 50 kW of electricity that the old 21-inch (0.5-meter) blades produced. Differing from the conventional horizontal axis machine, the VAWT has its generating equipment at ground level which reduces tower construction costs. Another advantage of the VAWT is that it does not need mechanisms to turn it into the wind or to control blade rotation in high winds.

Courtesy DOE

Darrieus vertical axis wind-energy conversion system.

sanitary landfill. An engineered method of solid waste disposal on land in a manner that protects the environment; waste is spread in thin layers, compacted to the smallest practical volume, and covered with soil at the end of each working day.

sanitary sewer. One carrying sewage but not surface or groundwater.

satellite power system. Concept for providing large amounts of electricity for terrestrial use from one or more satellites in geosynchronous earth orbit. A very large array of solar cells on each satellite would provide electricity which would be converted to microwave energy and beamed to a receiving antenna on the ground. There it would be reconverted into electricity and distributed as any other centrally generated power, through a grid.

Courtesy Marshall Space Flight Center

Rendering of a satellite power system beaming energy to earth.

saturated air. Air containing saturated water vapor with both air and vapor at the same dry bulb temperature.

saturated steam. Steam at a temperature and pressure such that any lowering of temperature or increase in pressure will cause condensation.

saturation. Completely filled; a condition reached by a material, whether it be in solid, gaseous, or liquid state, which holds another material within itself in a given state in an amount such that no more of the material can be held within it in the same state.

saturation pressure. That pressure for a given temperature at which the vapor and the liquid can exist in stable equilibrium.

scenario. A set of projections used as an assumption to conduct future planning.

schedule number. In piping, an American Standards Association designation of the strength of pipe. It is 1000 × (internal service pressure ÷ maximum tensile stress), where the pressure and stress are in pounds per square inch (psi).

Schottky barrier. A cell barrier established at the interface between a semiconductor, such as silicon, and a sheet of metal.

screen. A thin object, usually of wire mesh, that excludes flies, mosquitoes, etc., but permits passage of light and air.

screening. The removal of relatively coarse floating and suspended solids by straining through racks or screens.

screwed bonnet. A type of bonnet so constructed that it attaches to the body by means of a screwed joint. It may screw over the body, or inside the body, or may be attached to the body by means of a union-type screwed connection.

screwed flange. A flange screwed on the pipe which it is connecting to an adjoining pipe.

screwed joint. A pipe joint consisting of threaded male and female parts screwed together.

scrubber. An air-pollution control device that uses a liquid spray to remove pollutants from a gas stream by absorption or chemical reaction. Scrubbers also reduce the temperature of the emission.

scrubbing equipment. Equipment for countercurrent liquid-vapor contact of flue gases to remove chemical contaminants and particulates.

Sealed Air Solar Pool Blanket®. A swimming pool insulation cover material made by the Sealed Air Corporation which is a 0.3-inch (7.62-mm) thick assembly consisting of plastic bubbles 0.5 (12.7 mm) inch apart on centers.

seamless pipe. Pipe or tube formed by piercing a billet of steel and then rolling.

seasonal control. Controlling solar heat through the various seasons; for example, awnings block the summer sun but allow solar heat in winter.

seasonal efficiency. The ratio, over an entire heating season, of solar energy collected and used to the solar energy striking the collector.

secondary air. The air for combustion externally supplied to the flame at the point of combustion.

secondary market. Market of institutional investors who buy government guaranteed debt for their portfolio.

secondary treatment. Wastewater treatment, beyond the primary stage, in which bacteria consume the organic parts of the wastes. This biochemical action is accomplished by use of trickling filters or the activated sludge process. Effective secondary treatment removes virtually all floating and settleable solids and approximately 90 percent of both BOD and suspended solids. Customarily, disinfection by chlorination is the final stage of the secondary treatment process.

second law efficiency. The ratio of the minimum amount of work or energy necessary to accomplish a task to the actual amount used.

second law of thermodynamics. Heat cannot be changed directly into work at constant temperature by any cyclic process; one of the two "limit laws" which govern the conversion of energy, which describes the inevitable passage of some energy from a useful to a less useful form in any energy conversion.

sediment. Solid material, both mineral and organic, that is in suspension, is being transported, or has been moved from its site of origin by air, water, gravity, or ice, and has come to rest on the earth's surface either above or below sea level.

sedimentation. In wastewater treatment, the settling out of solids by gravity.

sedimentation tanks. In wastewater treatment, tanks where the solids are allowed to settle or to float as scum. Scum is skimmed off; settled solids are pumped to incinerators, digesters, filters, or other means of disposal.

sediment basin. A reservoir for the retention of rock or other debris from a sediment-producing area.

seepage. Water that flows through the soil.

segment. A portion of a river basin, the surface waters of which have common hydrologic characteristics (or flow regulation patterns); common natural, physical, chemical, and biological processes and common reactions to external stresses such as discharging of pollutants.

seismic. Characteristics of, or produced by, earthquakes or earth vibration.

seismicity. Tendency to result in earthquakes, or earth vibration.

selective absorber. A surface which has high absorptance at wavelengths corresponding to the solar spectrum and low emittance in the infrared range.

selective black paint. A material more absorbent of the infrared long wavelengths of sunlight than nonselective black paint; used for coating the absorber plates in solar collectors.

selective surface coating. Specially adapted coating with high solar radiation absorptance and low thermal emittance, used on the surface of an absorber plate to increase collector efficiency.

self-contained cooling unit. A combination of apparatus for room cooling complete in one package; usually consists of compressor, evaporator, condenser, fan motor, and air filter. Requires connection to electric line.

self-reliance. The ability to meet basic needs locally; it implies maximal development of local resources, but neither isolation nor complete self-sufficiency.

sellback option. A residential photovoltaic system designed for sale to utility grid of excess photovoltaic power not required for direct single-family residence loads.

semiautomatic valve. A valve that is opened manually and closed automatically, or vice versa.

semiconductor. A crystalline material whose atomic structure is such that it has an electrical conductivity intermediate between an insulator and a metal. Its conductivity can be greatly influenced by the addition of doping agents. The doped semiconductor can have either an excess of energy-conducting electrons (*n*-type) or a deficiency of them (*p*-type).

semidiurnal. Having a period or cycle of approximately one-half lunar day (12.42 solar hours). The tides and tidal currents are said to be semidiurnal when two flood periods and two ebb periods occur each lunar day.

semipermeable membrane. A thin, soft, pliable sheet through which some molecules (usually smaller ones) can pass while others (usually larger) cannot.

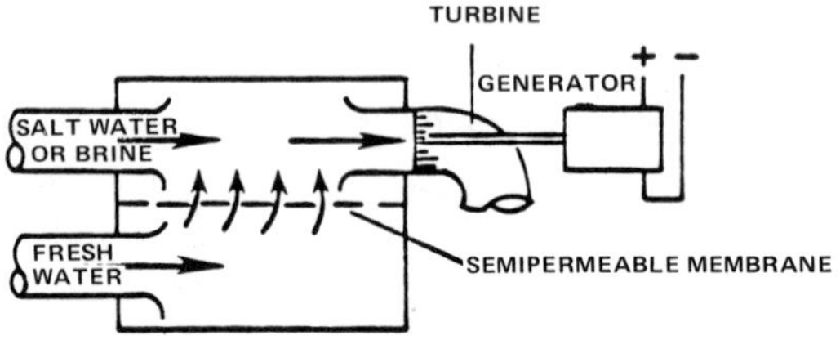

Courtesy SERI

Application of semipermeable membrane in salinity gradient energy conversion module.

senescence. The process of growing old. Sometimes used to refer to lakes nearing extinction.

sensible cooling effect. The difference between the total cooling effect and the dehumidifying effect.

sensible heat. The energy associated with changes in the temperature of a material. Thermal storage as sensible heat is the most mature concept, with tanks of hot water and beds of hot rocks used in both active and passive solar-thermal systems. The quantity of material required to store a given amount of thermal energy depends on a property of the material known as specific heat and a characteristic temperature change determined by the system. Typical values are on the order of 10 to 100 British thermal units of energy storage per pound of material. As a storage technique, sensible heat has a major disadvantage in that energy is returned at an ever-decreasing temperature. Its major advantages are low cost and relative simplicity.

SENSIBLE HEAT STORAGE MATERIALS	SPECIFIC HEAT BTU/LB. °F.	DENSITY LB./FT.^{3}	UNIT HEAT CAPACITY BTU/FT.^{3}/°F.	
			NO VOIDS	30% VOIDS
WATER	1.0	62.5	62.5	(62.5)
WATER (30%)– ETHYLENE GLYCOL (70%) MIXTURE (230°F.)	.8	64.1	51	36
SCRAP IRON	.112	489	55	38
MAGNETITE	.165	320	53	37
SCRAP ALUMINUM	.215	168	36	25
CONCRETE	.27	140	38	26
ROCK (BASALT)	.20	180	36	25
BRICK	.2	140	28	20
ROCK SALT (NaCL) (30-70% BY WEIGHT AT 230°F.)	.219	136	30	21

Courtesy National Copper Association

Table of sensible heat storage materials.

sensible heat storage. A heat-storage medium, typically water or gravel, in which the addition or removal of heat results in a temperature change only (as opposed to, say, a chemical reaction).

sensors. Devices such as pressure transducers, thermocouples, and flowmeters used to detect changes in a system. See **temperature sensors.**

separation energy. The energy per unit mass required to achieve a specified degree of separation of a given mixture of two or more isotopic species. It can be either the theoretical minimum thermodynamic value or the actual value for a given process.

septic tank. A single-story settling tank in which the organic portion of settled sludge is allowed to decompose anaerobically without removal or separation from the bulk of the water flowing through the tank. Only partial liquefaction and gasification of the organic matter is accomplished and eventually the undecomposed solids will accumulate to the extent that solids removal is necessary.

sequence control. An electrical method of control whereby once action has been initiated, a number of electrical circuits will automatically function in a prescribed order.

series auxiliary. A scheme by which an auxiliary energy source boosts the temperature of the solar-heated fluid whenever the fluid flowing from solar storage is too cool.

series configuration. A solar heater used as a preheater to an auxiliary heater.

series connection. A method of interconnecting devices that generate or use electricity so that the voltage, but not the current, is additive one to the other. Opposite of parallel connection.

service fitting. A street ell or street tee with male threads at one end and female threads at the other.

service pipe. A pipe connecting water to a building.

service water heating. If a flow-control valve is provided at the boiler outlet, boiler water temperature may be maintained continuously so that a coil inserted in the water may be used to heat service water. The coil may be small, in which case it is used with a storage tank, or the coil may have enough surface so that water is heated to the required service temperature by one passage through the coil. This latter type is advantageous where a storage tank may overheat the basement, and because the tank may corrode rapidly if the water is corrosive. Year-round operation of the system keeps the boiler warm continuously and keeps any soot deposits dry, thereby preventing corrosion of the boiler surfaces or chimney connection pipe.

In residence heating, the load imposed by the instantaneous water heater may be disregarded in determining boiler size because it is of such short duration. In some cases, however, the water-heating load of the instantaneous heater, 3-5 gpm heated 100 degrees, may be larger than the heating load and will therefore determine the boiler or burner capacity required.

set back. The liquid portion of the stillage that is recycled as a portion of the process liquid in the mash preparation.

set temperature. The temperature at which a thermostat is set.

settleable solids. Bits of debris and fine matter heavy enough to settle out of wastewater.

settling tank. In wastewater treatment, a tank or basin in which settleable solids are moved by gravity.

settling time. In a controlled system, the time required for entrained or colloidal material to separate from the liquid.

sewage. The total of organic waste and wastewater generated by residential and commercial establishments.

sewer. Any pipe or conduit used to collect and carry away sewage or storm-water runoff from the generating source to treatment plants or receiving streams. A sewer that conveys household and commercial sewage is called a sanitary sewer. If it transports runoff from rain or snow, it is called a storm sewer. Often storm-water runoff and sewage are transported in the same system or combined sewers.

sewerage. The entire system of sewage collection, treatment, and disposal. Also applies to all effluent carried by sewers whether it is sanitary sewage, industrial wastes, or storm-water runoff.

shade. A device that consists mainly of one or more flexible sheets or quilts and can be unrolled (or vertically raised or lowered) to cover a window and, when not needed, can be moved out of the way. The device may be used to provide privacy, reduce heat-loss, or prevent solar radiation from penetrating deep into the room. It may be attached at the top or bottom of the window and may be opaque, translucent, or transparent. The roll-up process may be accomplished by an automatic spring-return or by other means.

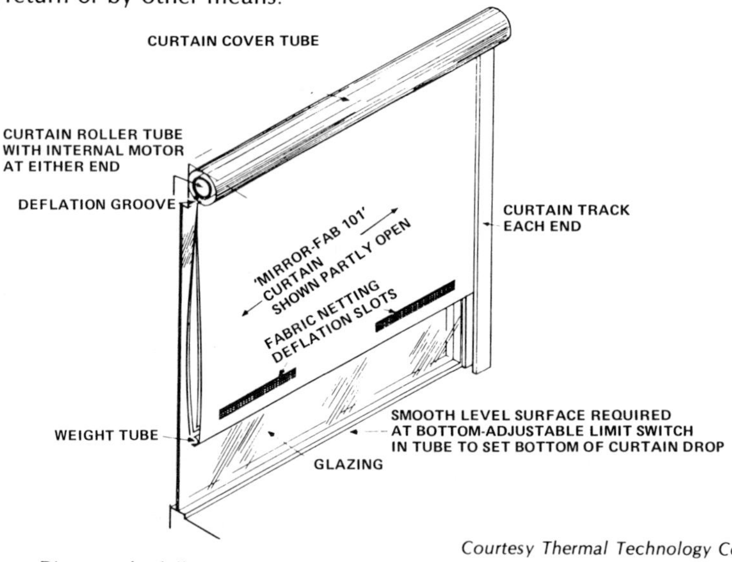

Courtesy Thermal Technology Corporation

Diagram of a fully automated self-inflating movable insulated window shade.

shading. An effective way to keep a house cool in the summer is to shade it from the outside by selective planting of trees, installation of awnings, or initial house design.

shading coefficient. The ratio of the solar heat gain through a specific glazing system under a given set of conditions, to the total solar heat gain through a single layer of clear, double-strength glass under the same conditions. The smaller the number, the greater the reduction in solar heat gain.

shading mask. A section of a circle that is characteristic of a particular shading device. This mask is superimposed on a circular path diagram to determine the time of day and the months of the year when a window will be shaded by the device.

shadowing. The shading of the reflective surface of one heliostat from the sun's rays by another heliostat.

shear. A relative motion parallel to the surface of contact.

sheathing. The first wooden cover, usually of a house made or framed of wood, before the roof and outside finish of the walls are applied.

shelter belt. A tree row planted in windy country to shelter a home, crops, and soil.

Shenandoah Total Energy Solar Project. The Solar Total Energy Large Scale Experiment in Shenandoah, Georgia, uses a heat-transfer fluid which circulates through the receiver tubes of the parabolic-dish solar collectors. Solar radiation is concentrated on the receivers by the collectors, and it heats the transfer fluid to 600°F (316°C). The transfer fluid is pumped into a heat exchanger or thermal storage unit for later use. In the exchanger, the transfer fluid boils and superheats the working-fluid steam. While the transfer fluid is returned to the collectors to repeat the cycle, the superheated working fluid drives a multistage steam turbine which in turn drives an electrical generator which produces electricity for the system's electrical requirements. Steam is extracted for the knitwear manufacturing processes. The working fluid exhausted from the prime mover is cooled as it passes through the water-cooled condenser. Condenser-cooling water is then used for heating, air conditioning, or hot water.

short circuit current. The current flowing freely from a photovoltaic cell through an external circuit which has no load or resistance; the maximum current possible.

short nipple. A nipple whose length is a little greater than that of two threaded lengths, or somewhat longer than a close nipple, so that it has some unthreaded portion between the two threads.

shoulder nipple. A nipple of any length which has a portion of pipe between two pipe threads. As generally used, however, it is a nipple halfway between the length of a close nipple and a short niple.

shroud. A structure used to concentrate or deflect a windstream.

shut down. A safe condition for a wind-energy conversion system which is not its power-producing mode but either protects the wind-energy conversion system from unusual circumstances, or protects those working on the machine.

shut-down wind speed. Wind speed at which the control system will shut down a wind-energy conversion system.

shutter. A rigid insulating plate, or set of closely connected rigid plates or rigid strips, used to cover a window for the purpose of reducing heat-loss. It may be applied to the outdoor side of the window or the indoor side, or between glazing sheets. It may be permanently attached, e.g., by hinges, or may remain unattached

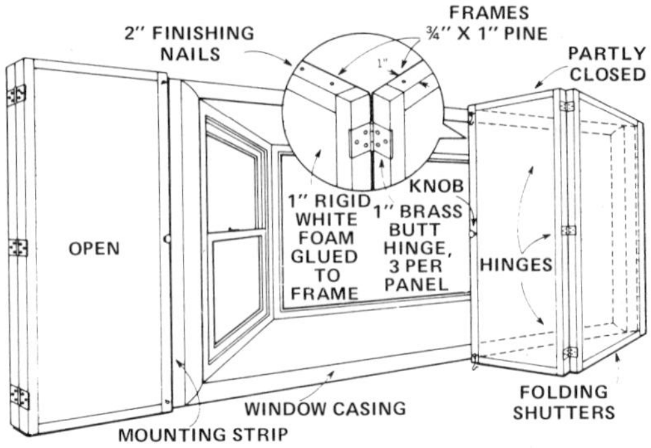

Courtesy A.J. Hand, Home Energy How-To, *Popular Science Book/Harper and Row*
Illustration of window insulating shutter.

so as to be instantly removable. It may be thick or thin and may be opaque, translucent, or transparent.

Siemens process. A commercial method of making purified silicon.

sight gauge. A clear calibrated cylinder through which liquid level can be observed and measured.

sigma function. The enthalpy of an air-steam mixture less the heat of the liquid.

sigma heat. The sum of sensible heat and latent heat in a substance or fluid above a base point, usually 32°F (0°C).

silica gel. A desiccant; a hydroscopic material that readily absorbs substantial quantities of moisture and is used to reduce the relative humidity of air or gas.

silicon (Si). A chemical element, atomic number 14; semimetallic in nature; dark gray; an excellent semiconductor material. A common constituent of sand and quartz (as the oxide). Crystallizes in face-centered cubic lattice like diamond. See **polycrystalline silicon.**

silicon cells. Photovoltaic cells made principally of silicon, a semiconductor. (See diagram p. 269.)

silt. Finely divided particles of soil or rock. Often carried in cloudy suspension in water and eventually deposited as sediment.

silvered surfaces. Shiny, mirrorlike surfaces which reflect both radiation heat flow and solar radiation.

silver oxide cell battery. A cell in which depolarization is accomplished by oxide of silver.

silvicultural farm. An energy farm composed of trees.

silviculture. The technology of raising trees or forest management.

simple distillation. The best example of simple distillation is the moonshiner's still. The vapors from the pot are condensed and withdrawn from the still without any rectification. The disadvantage of this type of distillation is that, at the beginning of the run, a high proof can be obtained (around 130, give or take some, depending upon the percentage of the alcohol in the mash), but this immediately falls off, and diminishes to nothing at the end of the run. As a result, the total from the

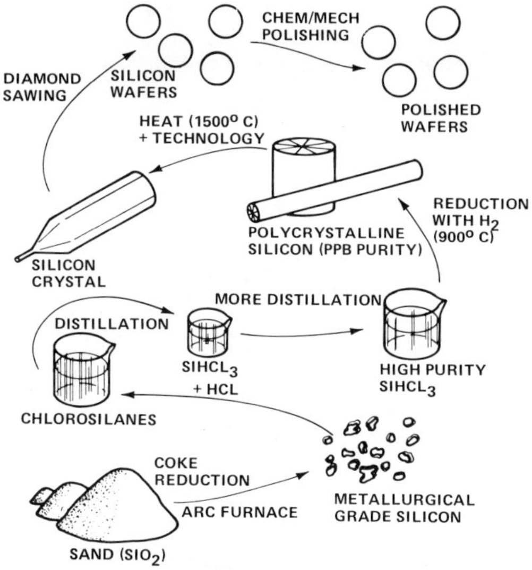

Courtesy SERI

Steps in the production of photovoltaic silicon cells.

entire run will be only around 40-50 proof. This will have to be run again several times to reach the 170 to 180 proof needed for alcohol fuel.

simple engine. A reciprocating engine from which steam or compressed air is exhausted to atmosphere after expansion in one cylinder only.

simple sugars. See **fermentable sugars.**

SIMSHAC. A program developed by Colorado State University for computer-aided design of solar systems.

simulation. The use of mathematical representations of real systems to determine what is likely to happen under various possible sets of conditions.

sine wave. The type of alternating current generated by utility companies, rotary inverters, sophisticated solid-state inverters, and alternating current (ac) generators.

single-glazed. Covered by one pane of glass or other transparent material.

site survey. The evaluation of a proposed or existing energy site with respect to those parameters important to its operation.

sizing. Determining the size or capacity of separate components in a solar water-heating system (especially collector area and storage tank capacity) that will provide the greatest benefit per dollar of investment.

skimming. The mechanical removal of scum from the surface of water.

skydome. The visible hemisphere of sky, above the horizon, in all directions.

skylid. Skylight or window in the roof that can be opened.

skylight. A clear or translucent panel set into a roof to admit sunlight into a building.

skyspace. That portion of the sky which must remain unobstructed for a collector to operate effectively. Protecting solar access simply means locating objects, such as buildings and trees, where they will not shade a collector's skyspace. Skyspace is specified by using latitude-dependent skyspace angles, which give the sun's position at critical times. Skyspace requirements vary with latitude and the use pattern of the collector.

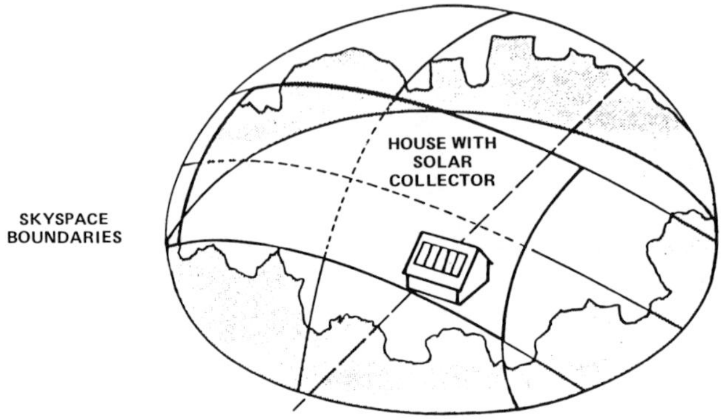

Courtesy HUD

Portion of unobstructed sky for proper collector operation.

sky temperature. The temperature of the sky. Usually it is cooler than the outdoor temperature, especially when there are no clouds and where it is very dry (as the desert).

Skytherm system. A form of movable insulation and a roof pond system developed by Harold Hay. A passive system with rooftop water bags storing heat and insulating slabs that slide over the water bags at night to prevent heat loss. The system involves motor-driven sliding insulation panels.

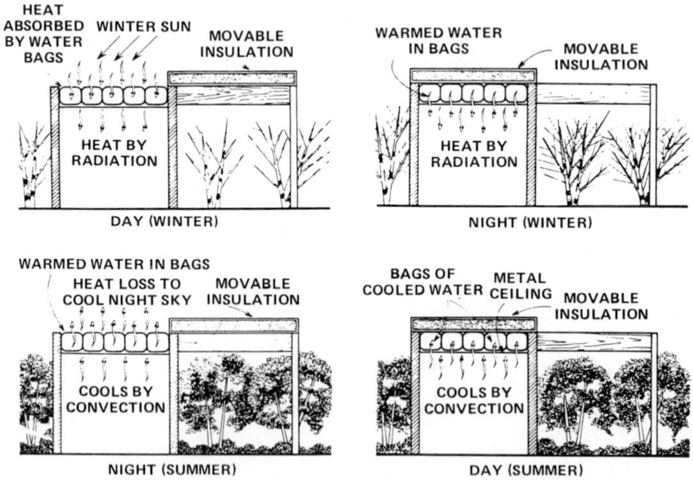

Courtesy A.J. Hand, Home Energy How-To, *Popular Science Book/Harper and Row*

Harold Hay's Skytherm process provides solar heating and cooling.

slash. The residue left on the ground after harvesting or thinning, or resulting from storms, fire, etc.; it includes unutilized logs, uprooted stumps, broken or uprooted stems, tops, and branchwood.

sleeve weld. A joint made by butting two pipes together and welding a sleeve over the outside.

slip-on flange. A flange slipped over the end of the pipe and then welded to the pipe.

slip rings. Assembly used to transfer electrical power or signals from a rotating shaft.

sludge. The collection of solids removed from sewage during wastewater treatment. Sludge disposal is then handled by incineration, dumping, or burial.

small-scale biomass energy project. Defined in the Energy Security Act (P.L. 96-294), 1980, as "a biomass energy project with an anticipated annual production capacity of not more than 1,000,000 gallons of ethanol per year, or its energy equivalent of other forms of biomass energy."

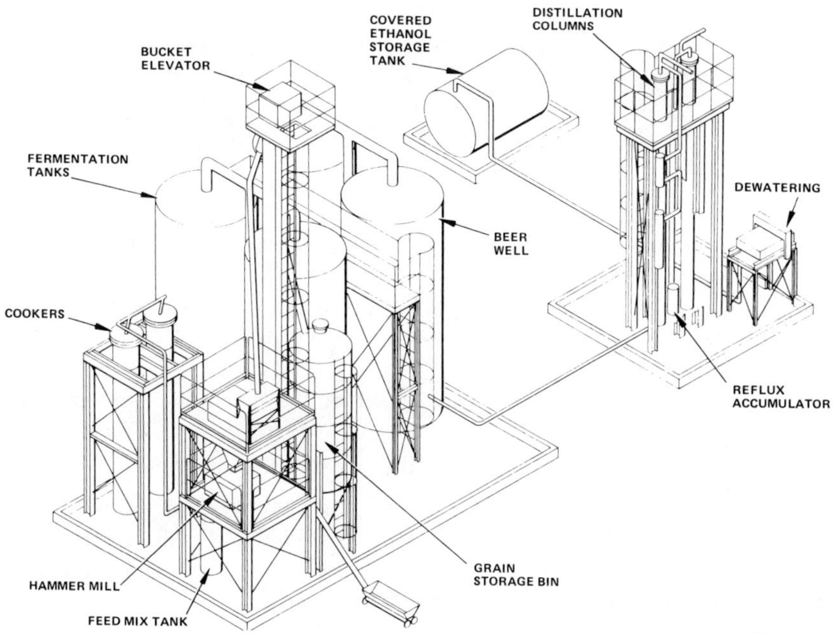

Courtesy DOE

Schematic layout of a small scale plant to convert biomass (corn) into fuel grade alcohol.

small wind-energy system. Wind-energy conversion system that outputs a maximum useful power output less than an equivalent of 100 kilowatts. This power, stated here in kilowatts (kW), may be in any useful form, e.g., horsepower (hp), British thermal units (Btu), etc. (See photograph p. 272.)

Smith-Putnam machine. The experimental wind turbine generator, named after its designer and builder built in 1941 near Rutland, VT, generated electricity and fed

Courtesy DOE

The Enertech 1-kW prototype small scale wind-energy conversion system.

power into the commercial electrical network for 3½ years until structural damage occurred.

smog. A mixture of smoke and fog; a fog made heavier and usually darker by smoke and chemical fumes; generally used as an equivalent of air pollution, particularly associated with oxidants.

smoke. Solid particles generated as a result of the incomplete combustion of materials containing carbon.

smoke pipe. A single wall pipe venting the stove to the chimney.

socket weld. A pipe joint made by use of a socket weld fitting which has a prepared female end or socket for insertion of the pipe to which it is welded.

sodium (Na). The sixth most abundant element in the earth's crust and the most common alkali metal (Group I*A*); a soft electropositive metal.

sodium-sulfur batteries. An advanced battery system, consisting of molten sodium in the negative electrode and a mixture of sulfur and sodium polysulfide in the positive electrode. It has a solid ceramic electrolyte. The sodium-sulfur battery operates at a temperature of about 572° to 662°F (300° to 350°C). Engineering cells are presently delivering about 50 to 60 watt-hours per pound (Wh/lb).

soft radiation. Ionizing radiation of long wavelength and low penetration.

soil conditioner. A biologically stable organic material such as humus or compost that makes soil more amenable to the passage of water and to the distribution of fertilizing material, providing a better medium for necessary soil bacteria growth.

soil conserving crops. Crops that prevent or retard erosion and maintain rather than deplete soil organic matter.

sol-air temperature. That temperature of the outdoor air which, in contact with a completely shaded building surface, would give the same rate of heat entry into the outdoor surface of that material as exists with the actual dry bulb temperature of the outdoor air, the actual intensity of solar and sky radiation incident upon that surface, and the actual wind velocity.

solar. Of and pertaining to the sun. See **solar energy.**

SOLAR. A computer program, developed by the A. D. Little Corporation, for detailed prediction of solar system performance.

solar absorptance. The ratio of the amount of solar radiation absorbed by a surface to the amount of radiation incident on it.

solar access. Allowing sunlight to strike a solar collector. This is accomplished by locating obstructions, such as buildings and trees, where their shadows will not fall on a collector during critical periods of operation. The concept of skyspace defines that portion of the sky which must remain unobstructed and defines specified critical angles for use in solar planning. See **skyspace.**

solar air-conditioning. Using solar heat to cool a building. Often evacuated tube collectors with low heat loss must be used.

solar air heater. Heaters that force across pipes containing sun-heated water and blow the air out through registers into rooms for heating. The faster the air is blown through a system, the lower the temperature of the air.

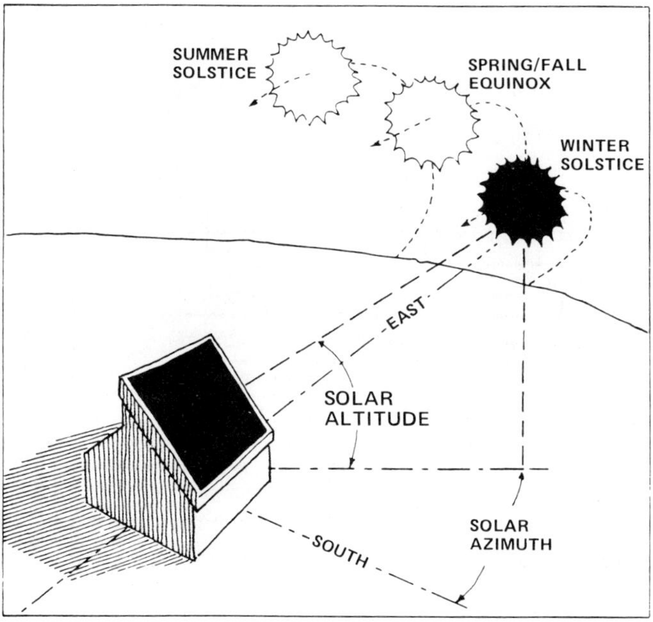

Courtesy National Copper Association

Diagram of solar-altitude angle. (See **solar-altitude angle.**)

solar-altitude angle. The angle between the line joining the center of the solar disc to the point of observation at any given instant and the horizontal plane through that point of observation. (See diagram p. 273.)

solar angles. Angles used to specify the sun's position at a given time.

solar-assisted absorption cooling system. An air-conditioning system that uses solar heat rather than conventional forms of energy as its primary power source for driving the refrigeration cycle.

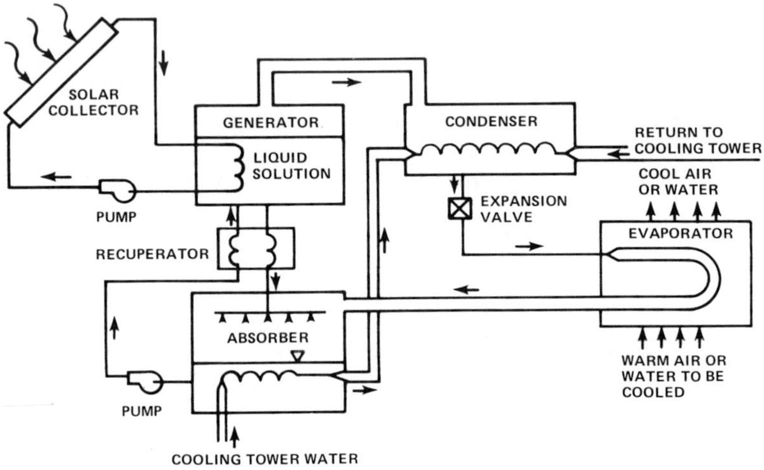

Courtesy F. Kreith and J.F. Kreider, Principles of Solar Engineering, *Hemisphere Publishing Corp.*
Flow diagram of solar-assisted absorption cooling system.

solar-assisted heat pump system. A heat pump that uses solar-heated fluid to increase its coefficient of performance (COP).

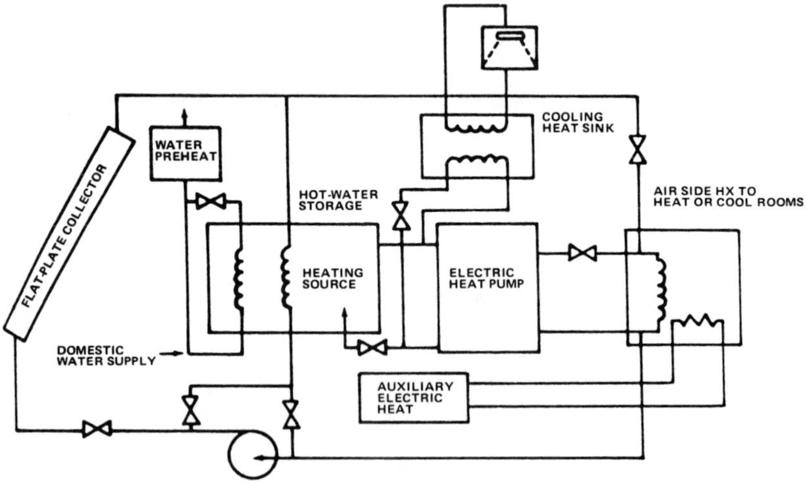

Courtesy F. Kreith and J. F. Kreider, Principles of Solar Engineering, *Hemisphere Publishing Corp.*
Block diagram of solar-assisted heat pump.

solar azimuth. The horizontal angle measured between the sun and due south.

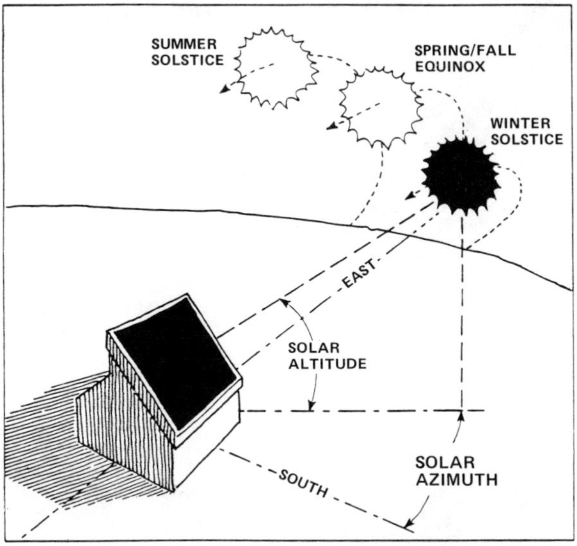

Courtesy National Copper Association

Pictorial diagram showing solar azimuth relationship.

solar balloon. Using the sun's energy to heat air within an envelope (a balloon) to develop buoyancy.

solar building. A building which uses solar energy for heating and/or cooling.

solar building load ratio. The ratio of the insolation on a tilted collector to the building thermal load, on a monthly or annual basis.

solar cell. A photovoltaic cell that is designed specifically to produce electrical energy from sunlight.

solar cell battery. An electric cell that converts energy from the sun into electrical energy.

solar collection area. An expanse of transparent or translucent material that is located on that side of the structure which faces south.

solar collection efficiency. The ratio of the amount of energy collected to the total solar irradiation during the period under consideration.

solar collector. A device used to gather and accumulate the sun's energy or solar radiation. Nearly all collectors have a layer of glass on top (some have plastic) to trap the heat once it passes into the collector. The medium used to transfer the heat to the rest of the system varies. Collectors using water to transport the heat have layers of honeycomb or corrugated or crimped-copper conductors and insulators. Air-transport collectors vary in their consistency. Another type of collector, the focusing collector, employs a concave mirror turned by a motor to follow the sun as the earth moves in order to maintain a focus on the sun. The basic function of the solar collector is to capture the sun's heat for household heating and cooling. (See diagram p. 276.)

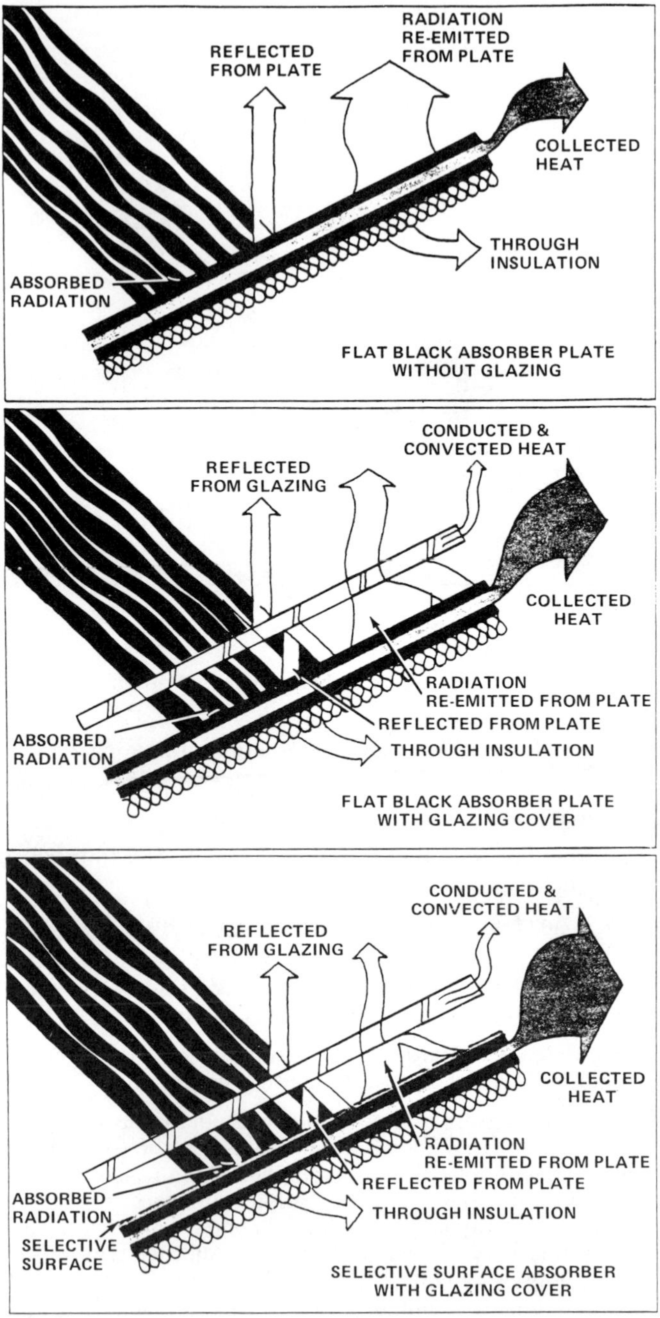

Courtesy National Copper Association

Three typical flat-plate solar collector system designs.

solar collector panel. A sandwichlike structure with a black surface to absorb solar heat. It may be used to heat either liquid, usually water or an antifreeze mixture, or air as the transfer medium.

solar compass. Synonym for sun compass (q.v.).

solar concentrator. Reflector or lens designed to focus a large amount of sunshine into a small area, thus increasing the temperature.

Courtesy DOE

Photograph of solid compound parabolic concentrator coupled to photovoltaic cells.

solar constant. The normal insolation in near-earth space. The intensity of solar radiation beyond the earth's atmosphere, at the average earth-sun distance, on a surface perpendicular to the sun's rays. The value for the solar constant is 1353 W/m^2, 1.940 $cal/cm^2/min$, or 429.2 $Btu/ft^2/hr$ ($\pm$ 1.6%).

solar cooling. Use of sunlight for cooling. Basically, solar cooling uses absorption equipment of the type employed in gas-burning refrigerators and air conditioners. Solar heat is applied to a solution, such as ammonia and salt water, to create a vapor that cools and condenses, and absorbs heat. As the vapor cools an area, the heat it absorbs causes it to vaporize again, thus repeating the process.

solar cooling systems. Solar-cooling techniques, like solar-heating techniques, may be classified as either passive or active.

Passive cooling is accomplished through the control of solar radiation and the use of other natural forces such as breezes and ground temperature. This first step in controlling solar radiation is the shading of windows by roof overhangs, awnings, or vegetation. Another heat rejection strategy involves the use of reflective films applied to window surfaces. These films block the heat contained in rays of sunlight but allow light to pass through.

The next important consideration is the ability of the house to vent unwanted heat build-up. Heat rises naturally through the house. Vents placed near the roofline will allow heat to escape. At the same time, air admitted to the building can be cooled by a variety of techniques. One method, most applicable in dry climates, is to draw incoming air through buried tubes which cool the air by contact with the lower, constant temperature [approximately 50° to 55°F (10° to 13°C)] of the earth. To achieve optimum air circulation through the house, careful planning and use of space is crucial to enable a strong, natural air flow.

Active solar systems also can be used for cooling. Three types of these systems are currently under development: absorption cooling, Rankine-cycle cooling, and dessicant cooling. Of these, the only system currently being marketed in the United States is an absorption cooler which operates on the same principle as a gas-fired refrigerator. The simple flat-plate solar collector usually cannot produce a high enough temperature to run an air conditioner, so high-efficiency solar collectors must be used. Until the cost of solar-cooling hardware is reduced, solar-actuated cooling techniques are not likely to be cost-effective for residential use. Cooling for commercial buildings in regions where cooling seasons are extremely long or power costs are abnormally high seems to be more promising.

solar declination. The angle of the sun north or south of the equatorial plane. It is plus if north of the plane and minus if south. The earth's tilted axis results in a day-by-day variation of this angle.

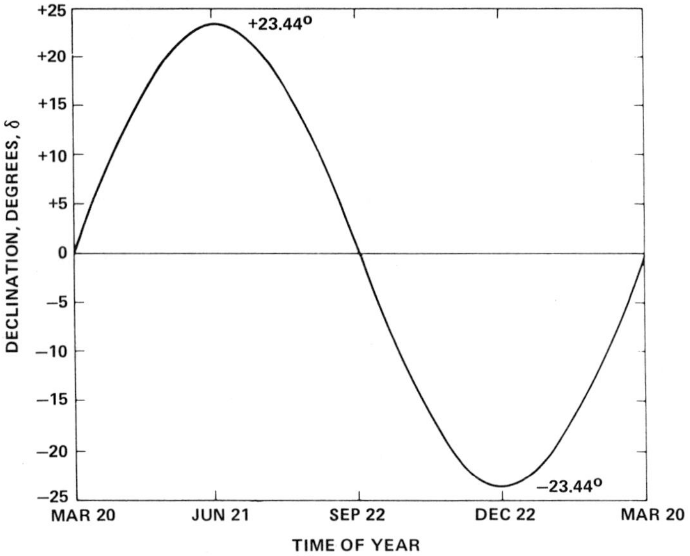

Courtesy W. C. Dickinson and P. N. Cheremisinoff, Solar Energy Technology Handbook, *Marcel Dekker*

Yearly variation of the solar declination.

solar degradation. Deterioration of materials and components by exposure to sunlight.

solar distillation. Solar distillation of alcohol is slow and inefficient. The solar distillation process uses an air tight box with a large surface area to collect solar radiation. The beer evaporates to the collector panel and trickles down the glass for collection.

solar electric cell. A device which converts sunlight directly into electricity.

solar electric conversion. Technique in which energy from the sun is transformed into electricity by way of solar thermal, solar photovoltaic, ocean thermal, and wind conversion.

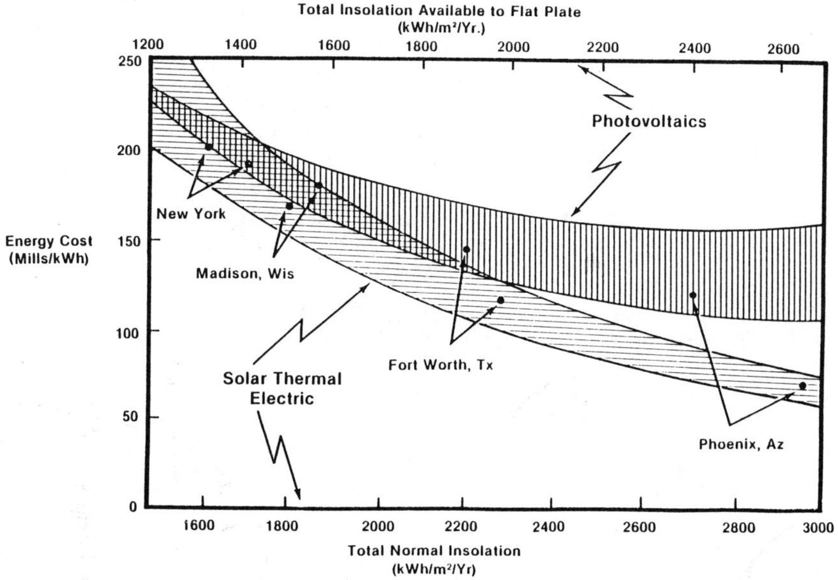

Courtesy Mitre Corporation

Solar electric generation delivered energy cost versus total normal insolation.

Solar Energizer®. A trade name of the Solarex Corporation for a solar panel.

solar energy. The energy transmitted from the sun in the form of electromagnetic radiation. Although the earth receives about one-half of one billionth of the total solar energy output, this amounts to about 420 trillion (10^{12}) kilowatt-hours (kWh) annually. (See diagram p. 280.)

solar energy equipment. Those system components attached to premises in such a manner as to deliver heating, cooling, or electricity by the system deriving its energy from the sun.

Solar Energy Industries Association (SEIA). Organization of manufacturers of solar devices, primarily for heating and cooling of buildings, headquartered in Washington, DC.

Solar Energy Information Data Bank (SEIDB). A comprehensive solar energy information data bank providing a solar energy library and a variety of solar energy information files. It is located at the Solar Energy Research Institute in Golden, CO.

Solar Energy Research Institute (SERI). The Solar Energy Research, Development and Demonstration Act of 1974 called for the establishment of a Solar Energy

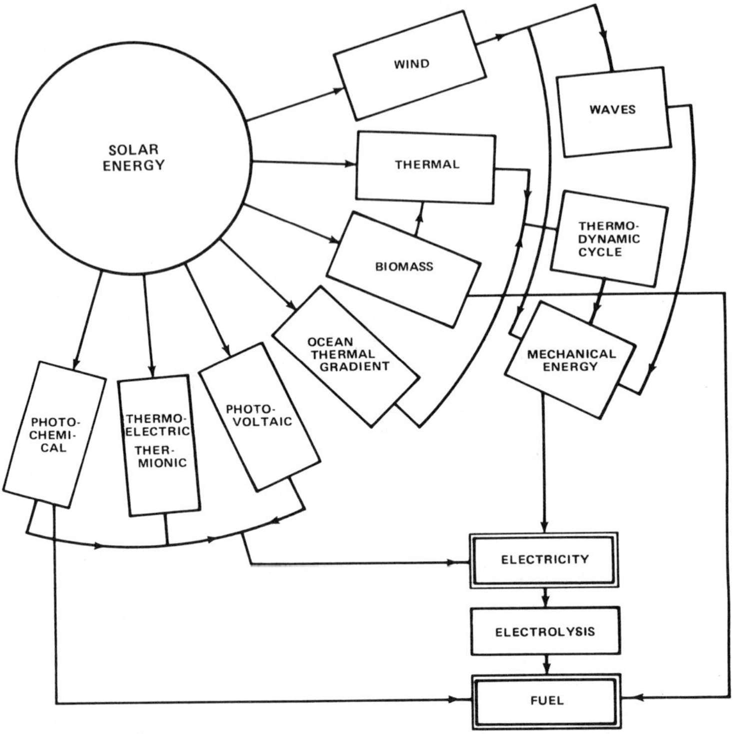

Courtesy DOE

Diagram of solar energy technologies and applications.

Research Institute whose general mission would be to support DOE's solar energy program and foster the widespread use of all aspects of solar technology, including direct solar conversion (photovoltaics), solar heating and cooling, solar thermal power generation, wind, ocean thermal conversion, and biomass conversion.

solar energy utilization. Solar radiation may be utilized by photovoltaics, solar thermal electric, ocean thermal energy conversions, and wind-energy conversion systems. These alternative energy sources utilize heliothermal, helioelectric, and heliochemical processes as additional means of producing energy.

solar enhanced oil recovery. One of the most cost-effective methods of enhancing oil recovery from underground reservoirs is the process of injecting steam into the reservoir, which then heats and pressurizes the formation allowing oil to flow to recovery wells where it can be pumped out by normal means. Solar energy can be employed to produce steam for this method. The environmental consideration, especially in California where most enhanced oil recovery is practical, makes solar thermal an attractive source of steam production. (See diagram p. 281.)

solar eyeball. Type of solar concentrating collector developed in England, which uses Fresnel lens to focus on solar cell. As the sun's rays move off focus, one of four gas reservoirs is heated. Gas expands, moving the magnetic slug in curved pipe below. Since the slug is held stationary by external magnet, it stays put while the

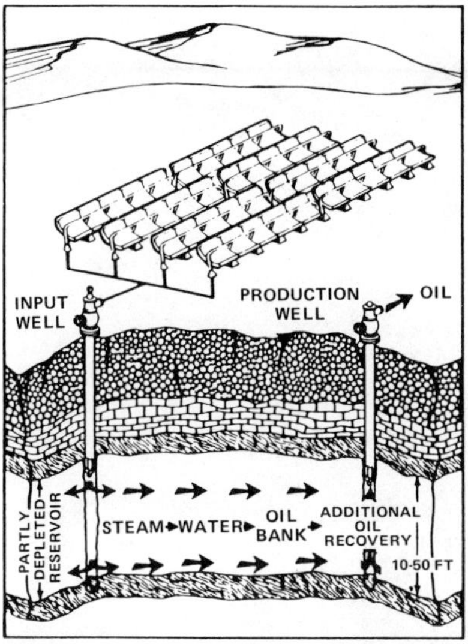

Courtesy DOE

Pictorial diagram of solar-enhanced oil recovery process.

pipe and whole assembly move, bringing the sun back into focus. "Eyeball" floats in a shallow tray of water, so movement is almost friction-free.

solar farm. A suggested power utility, based in a desert and covering a considerable area, where large amounts of electrical energy would be generated from solar energy.

solar feasibility. Solar-heating systems are now being used in every state of the union. However, the type and size of these systems vary with the climate and application.

Solar energy systems in most parts of the country can be designed to economically provide 40 to 70 percent of a building's heating needs. Above this range, each additional dollar spent yields a progressively smaller increase in the system's heating capacity. Other factors affecting the amount of heat provided by the system include the building site, the type and size of the building, and the specific characteristics of the solar energy system selected.

solar fraction. The percentage of a building's seasonal heating requirement provided by a solar system.

solar furnace. A solar device used to obtain high temperatures by focusing the sun's rays onto a small receiver. The largest solar furnace now in existence is located at Odeillo in the French Pyrenees and is capable of obtaining temperatures of nearly 6000°F (3316°C). (See photograph p. 282.)

solar gain. The part of a building's heating load, or an additional cooling load, which is provided by solar radiation striking the building or passing into the building through windows.

solar heating and cooling. The development, design, construction, and operation

Courtesy DOE

Solar furnace located near Odeillo in southern France was built to test extremely high temperature materials.

of systems that use and/or store the radiant energy of sunlight to provide comfort control and heated water for household, industrial, or agricultural use.

SYSTEMS & YEAR	NUMBER OF SYSTEMS INSTALLED (MILLIONS)	PRIMARY ENERGY DISPLACED (QUADS/YEAR)
Solar Water Heating Systems		
1985	1-1.5	0.05
2000	20-26	0.65
Solar Space & Water Heating Systems (Includes Solar Heat Pumps)		
1985	1.5-2.0	0.13
2000	8-9	0.75
Solar-Driven Heating and Cooling Systems (Includes Water Heating)		
1985	0.3-0.5	0.02
2000	2-3	0.60

Courtesy DOE

The table shows solar systems installed and the energy produced.

Solar Heating and Cooling Demonstration Act. Federal legislation which became law in September 1974. This legislation establishes the necessary industrial capacity for the rapid expansion of the use of solar energy for heating and cooling. It calls for demonstration of the practical use of solar energy on a large scale as follows:

solar heating technology within 3 years, and combined solar heating and cooling technology within 5 years.

Solar Heating and Cooling of Buildings (SHACOB). Both the process and the name of one of the budget categories in the Department of Energy's solar program.

solar heating and hot-water systems. The solar heating and hot-water system provides two basic functions: capturing the sun's radiant energy, converting it into heat energy, and storing this heat in an insulated energy storage tank; and delivering the stored energy as needed to either the domestic hot-water or heating system. The parts of the system which provide these two functions are referred to as the collection and delivery subsystems. The key components in the collection subsystem is the collector, whose basic function is to trap the sun's energy. The delivery subsystem is divided into two parts; one for providing heat to the hot-water tank and another for providing heat to the building heating system. When either the domestic hot-water or heating system requires heat, hot water from the energy storage tank is pumped to a heat exchanger in the domestic hot-water tank or in the building ducts.

solar heating system. Assembly of subsystems and components necessary to convert solar energy into thermal energy for heating purposes, in combination with auxiliary energy when required.

solar house. A dwelling that obtains a large part, though not necessarily all, of its heat from the sun.

solar irradiation. The amount of radiation, both direct and diffuse, which can be received at any given location.

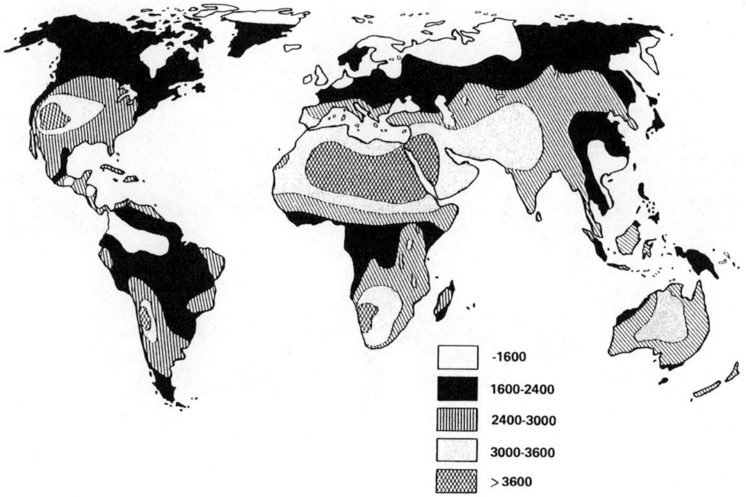

Courtesy DOE

Illustration of worldwide annual number of sunshine hours.

solarium. A living space enclosed by glazing; a greenhouse (q.v.).

solarization. An effect of strong sunlight (or artificial ultraviolet radiation) on some glasses, causing a change in their transparency. Glasses free from arsenic, and of low soda and potash content, are less prone to this effect.

solar loan. In order to obtain a loan for incorporating solar energy in your home, it is best to approach several financial institutions in your area. An increasing number of banks and savings and loan associations have financed or are willing to finance solar projects in both residential and commercial structures. When approaching a lending institution for financing a solar installation, it is advisable to have already selected a reputable contractor, and to have complete plans for both the house and the proposed solar system.

solar mass wall. See **trombe wall**.

solar multiple. The ratio of thermal power down the tower at the design point to the peak thermal power required by the turbine generator.

solar noon. Time of day when the sun reaches its maximum altitude for the day and is due south; this occurs exactly midway between sunrise and sunset.

Solar One. The nation's first solar electric central receiver powerplant, Solar One, has been completed in Barstow, CA. The 1818 computer-directed heliostats (mirrors) have been installed to direct sun rays to a steam boiler for electric power generation.

Major funding for the solar portions of the $140 million "solar power tower" was provided by the U.S. Deaprtment of Energy, The California Energy Commission, Southern California Edison Company, and the Los Angeles Department of Water and Power.

On its completion in 1981, Solar One was the world's largest solar generating

Courtesy DOE

Solar One is the nation's first solar electric central receiver power plant.

plant. It will produce 10,000 kilowatts of electric power during peak daylight hours, enough electricity to serve about 6000 typical customers.

The field of mirrors, covering some 100 acres (40 hectares) in the Mohave Desert, will reflect heat to a 300-ft (91-m) central tower. Steam as hot at 950°F (510°C) will be transferred to the ground to drive a conventional steam-turbine electric generator.

solar only. The operation of a hybrid power plant on the solar energy subsystem output alone.

solar panel (photovoltaic). A collection of solar cells connected in series, in parallel, or in series-parallel combination to provide greater voltage, current, or power than can be furnished by a single solar cell. Solar panels can be provided to furnish any desired voltage, current, or power. They are made up as a complete assembly. Larger collections of solar panels are usually called solar arrays.

solar photovoltaic conversion. Conversion technique which involves sunlight passing through a flat solar cell to generate electricity directly.

solar pond. A pond with a blackened base containing (stratified) brackish water and used to collect and store heat.

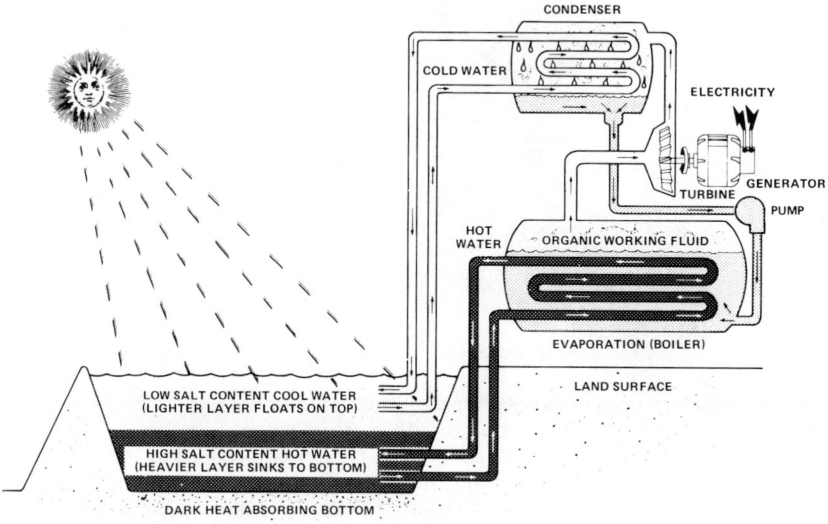

Courtesy DOE

Pictorial block diagram of solar pond generating concept.

solar power. Useful power derived from solar energy. Both steam and hot-air engines have been operated from solar energy. The energy transmitted from the sun in the form of electromagnetic radiation. Although the earth receives about one-half of one billionth of the total solar energy output, this amounts to about 420 trillion kilowatt-hours (kWh) annually.

solar power direct conversion systems. The production of electricity from solar radiant energy by direct conversion devices such as an array of silicon cells. One direct conversion scheme involves the launching of a satellite-mounted array of such cells in synchronous orbit, permanently placing the cells over a preselected position on the earth's surface. Radiant energy would be converted to direct current, which in turn would be converted electronically into microwave energy. Microwave energy would be beamed to hugh antennas located on the earth's surface beneath the satellite. The energy could then be converted to alternating current.

solar-powered irrigation. Distributed collectors or small central-receiver systems are adaptable for pumping irrigation water in the Southwest, either by direct-coupled thermally driven heat engines or by generating electricity. A 25-horsepower (hp) solar turbine (nonelectric) pumping system has been constructed on a farm near Willard, NM, for shallow-well irrigation.

Solar Power Satellite. A proposed satellite in geosynchronous orbit (22,000 miles) to convert solar energy to electricity and then to microwave energy, which is beamed to earth for later conversion to direct current electricity.

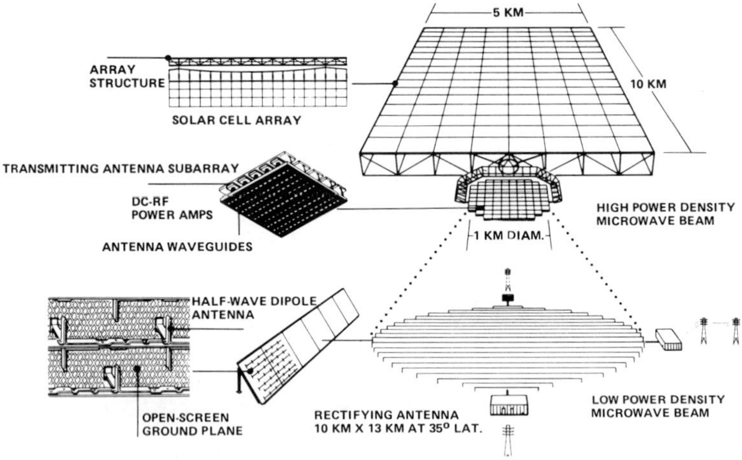

Courtesy NASA

Systems diagram of solar power satellite.

solar power thermal conversion systems. Thermal conversion systems involving an extensive array of steel pipes coated with materials heated by the sun's rays. In one concept, nitrogen flowing through the pipes would gather the heat and transport it to molten salt. The molten salt can be heated to a temperature of about 1000°F (538°C) for production of steam, which would power conventional turbines at a projected efficiency of about 30 percent. The area required to supply energy to a 1000-megawatt (MW) power plant would be about 10 square miles of collection surface, plus a 300,000-gallon (1.14×10^7 liters) reservoir of molten salt. Energy storage would be necessary for nights and cloudy days.

solar radiation. Energy from the sun which comes to earth in the form of direct, diffuse, or reflected rays. (See diagram p. 287.)

solar radiation intensity. The amount of free energy outside the earth's atmosphere averaging 442 British thermal units per square foot per hour.

solar Rankine system. A cooling system in which a solar-heated fluid, operating in a Rankine cycle, drives the compressor of a vapor-compression refrigeration system. (See diagram p. 288.)

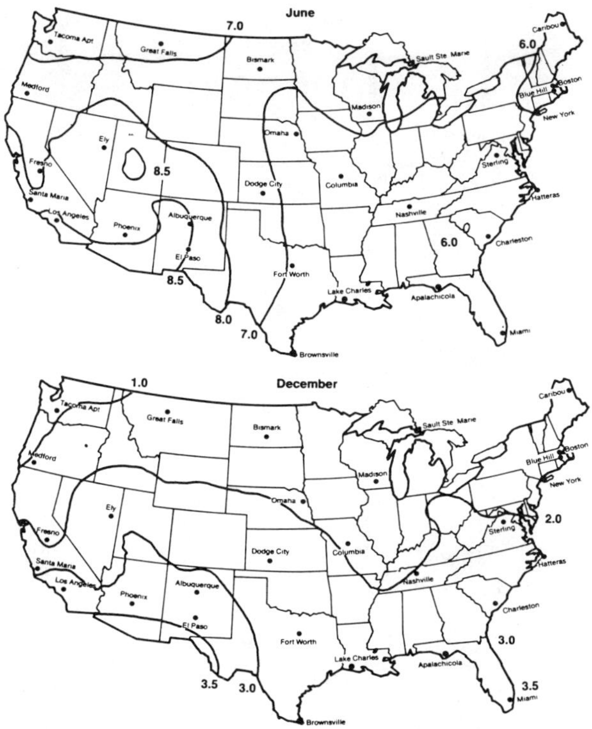

Source: E. Boes et al., "Distribution of Direct and Total Solar Radiation Availabilities for the U.S.A.," Sandia Laboratories (Sand76-0411), pp 19 and 25

Mean daily total horizontal solar radiation in kilowatt-hours per square meter.

solar repowering concept. An existing electric-generation unit is equipped with a solar-fired boiler so that it can operate on solar energy whenever the solar-derived electricity is available.

solar rights. Refers to the legal question regarding the right of a person who uses a solar energy device not to have his sunlight blocked by another structure.

solar salt. Salt obtained by solar evaporation of seawater.

solar salt pond generating project. A solar pond is a still body of water that collects solar energy and stores it as thermal energy. The introduction of salt into a pond allows the density of the pond water to be controlled. A salinity gradient within the pond allows thermal storage in the dense high-salinity water near the pond bottom. Natural convection, the process by which ponds normally cool, is thus overcome. The stored thermal energy is used to drive a Rankine-cycle heat engine. The hot water from the pond bottom is used to vaporize an organic fluid. The resulting high-pressure vapor turns a turbine and an electric generator. Cool water from the pond surface is used to condense the vapor back into liquid. (See diagram p. 289.)

solar sea power. The concept involving the use of temperature differences between the sun-heated surfaces of the ocean and the colder water deep under the

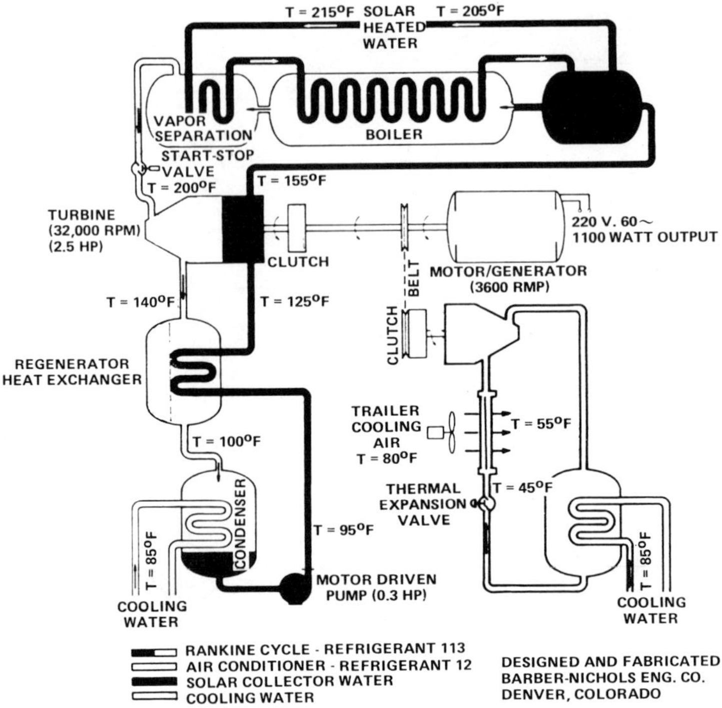

Courtesy Barber-Nichols Eng. Co., Denver, Colorado

Flow diagram of the solar-heated Rankine cycle electric power air-conditioning system.

surface to power heat engines. Since the water retains the heat of the sun, such plants, unlike other types of solar plants, could operate at night and during cloudy periods. The technology for such plants is envisioned to be large, extending a half mile or more under the water to reach the deep cold water. Since the temperature difference between this cold water and the surface waters is only in the range of 35.6°F (2°C), such a power plant would have a thermal efficiency less than 1/10th of the efficiency of a conventional, modern fossil-fueled plant. See **ocean thermal energy conversion**.

solar simulators. Equipment to simulate the solar flux for test purposes.

solar space heaters. A solar-energy heating system which uses a black metal surface under glass on the roof of a building to absorb sunlight and create a "greenhouse" effect to heat the water or air which is circulated during the day and stored at night.

solar spectrum. The total distribution of electromagnetic radiation emitted from the sun, minus those wavelengths that are absorbed by the solar atmosphere.

solar still. A distillation apparatus consisting of an airtight space through which solar energy penetrates and is partially absorbed by brackish water in a basin at the bottom of the space.

solar system cost. The costs of solar energy systems vary widely according to the

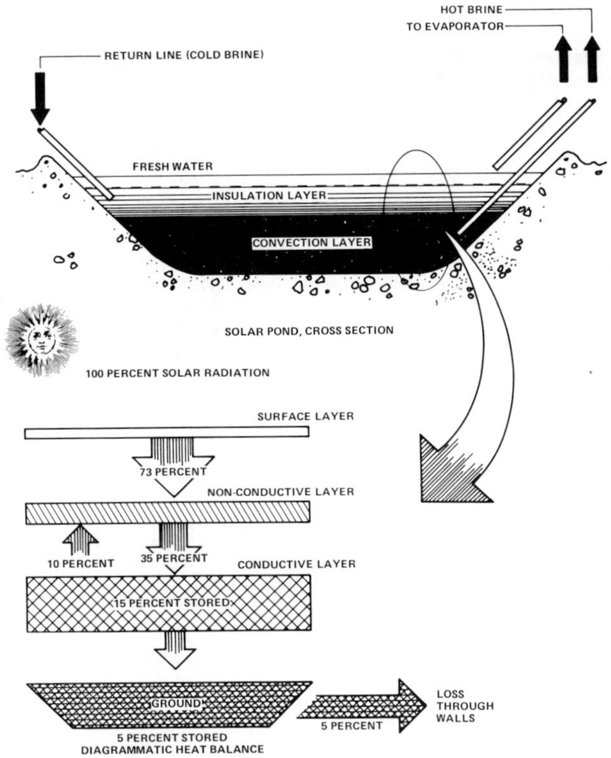

Courtesy DOE

Illustration of solar salt pond generating system.

type of system, the size of the building, the percentage of the building's heating needs that the system is designed to meet, and local labor and materials costs.

An active solar system requires hardware (such as flat-plate collectors and pumps) to collect and distribute solar energy. In general, the price of a professionally installed, active solar space heating and hot water system for an average-size, one-family home is between $5,000 and $20,000. For a solar hot water system alone, the range is between $1,500 and $4,000. Another way to estimate costs is to assume that the installed cost of the system will be between $30 and $45 per square foot (0.1 square meter) of collector area.

solar system efficiency. The amount of British thermal units lost from the time the sun's rays hit the solar collector to the moment they are used to heat the house or the water supply, compared with the original number coming in.

solar system size. Although rules of thumb for sizing solar energy systems should not replace specific calculations for an individual project, some general guidelines are possible. Keep in mind that these figures are only averages which do not apply to all situations.

(1) A home heating system needs 0.25 to 0.5 square feet (0.02 to 0.05 square meter) of collector area for every square foot (0.1 square meter) of heated floor space.

(2) A liquid system requires 1 to 2.5 gallons (3.8 to 9.5 liters) of water as heat storage for every square foot (0.1 square meter) of collector. The larger values are for lower temperature distribution systems, such as radiant slab heating.

(3) For every square foot (0.1 square meter) of collector an air system requires 0.5 to 1 cubic foot (0.014 to 0.028 cubic meters) of rock as heat storage.

solar thermal central receiver power system. A solar power system which concentrates the available solar energy, using an array of individually controlled heliostats to redirect the sun's rays to a tower-mounted receiver.

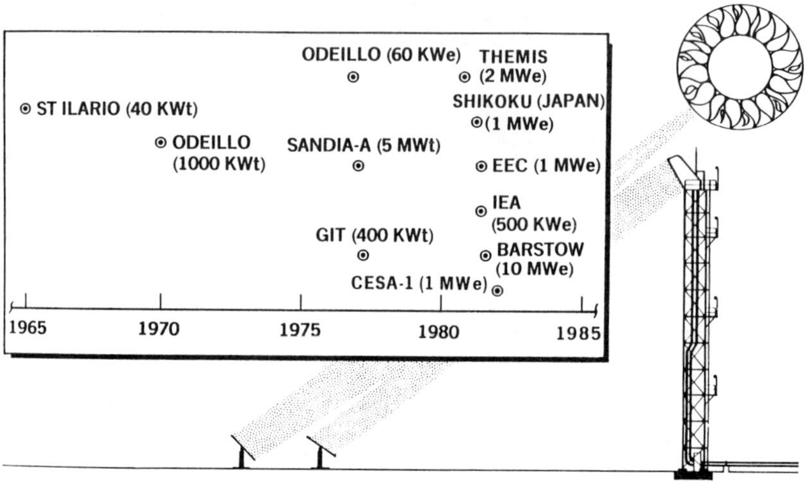

Courtesy DOE

Central reciever projects.

solar thermal conversion. The conversion of thermal energy (starting from solar energy) to electricity using various heat engines, e.g., the Solar One central receiver or small decentralized receivers with Brayton engine cycles, Stirling engine cycles, etc.

solar thermal electric. The basic concept underlying solar thermal electric power generation is the utilization of solar radiation to heat a working fluid to a temperature high enough to be used (either directly or indirectly) to power a turbine which will, in turn, drive an electric power generator and thus produce electricity which can be integrated into existing electric power networks. To attain the high temperatures required for this application, concentrated and direct solar radiation is usually necessary (diffuse radiation cannot be focused), thus signifying the need for some sort of focusing collector configuration. Two methods presently exist for attaining these requisite temperatures in working fluid quantities large enough for electric power generation; the central receiver system and the distributed collector system. (See diagram p. 291.)

solar thermal electric power. The indirect conversion of solar energy into electricity by solar collectors, a heat engine, and electrical generators.

solar time. Varying period, marked by successive crossings of the meridian by the sun; hour angle of the sun at the observer's point (apparent time) corrected to true time by use of the equation of time. Sundials show apparent solar time. Solar noon

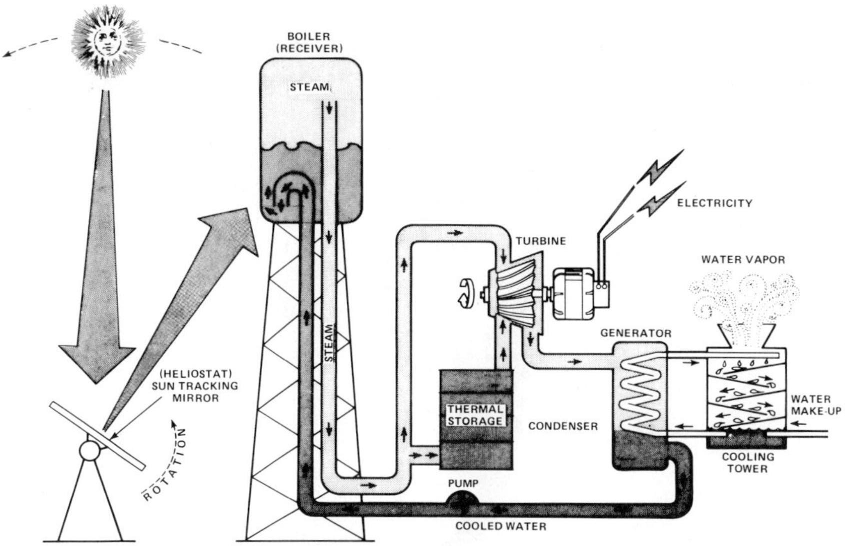

Courtesy DOE

Pictorial diagram of solar thermal electric power plant.

is that instant on any day at which time the sun reaches its maximum altitude for that day. Solar time is very rarely the same as local standard time in any locality.

solar total energy system. A system of solar (photovoltaic or solar thermal) production of electricity with the use of the waste heat for such thermal applications as heat, hot water, and absorption air conditioning.

solar window. Openings that are designed or placed primarily to admit solar energy into a space.

Solatex®. A glazing material, which is a low-absorptance glass, from ASG Industries Inc.

solcost. Method for estimating the size and cost of solar systems for heating domestic water, and making comparisons with conventional systems.

SOLDAY. A meteorological data base containing daily values for meteorological and solar radiation values. Available from the National Climatic Center.

solder joint. A method of joining tubes by use of solder.

solenoid. An electromechanical device so designed that when the electric current is applied to a coil of wire wound around a cylinder, the electromotive force causes a bar or plunger inside the cylinder to move; a series of coils or wire carrying a current of electricity; a cylindrical coil used to produce a magnetic field.

solenoid valve. An automatic valve that is opened or closed by an electromagnet.

solid fuels. Any fuel that is a solid such as wood, peat, lignite, bituminous and anthracite coals of the natural variety, and the prepared variety such as pulverized coal, briquettes, charcoal, and coke. These are divided into two broad classes: naturally occurring solid fuels and manufactured solid fuels.

solidifying. Changing phase from a liquid to a solid. Most phase-change materials (like wax) give up extra or latent heat on solidifying.

solidity. Ratio of rotor blade surface area to frontal (or swept) reference area of the entire windwheel.

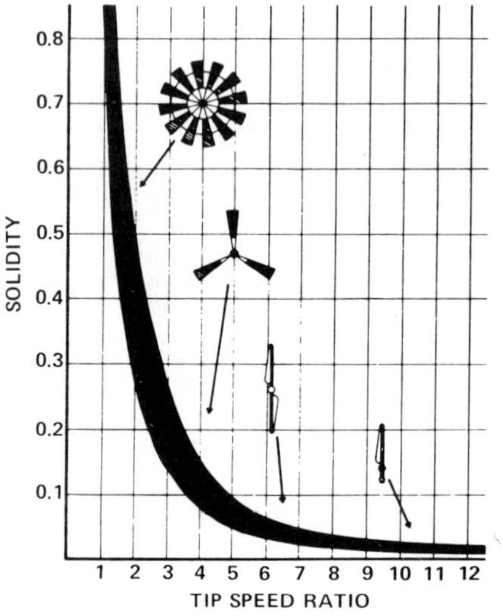

Courtesy DOE

Solidity versus tip speed ratio curve.

solid waste disposal. The ultimate disposition of refuse that cannot be salvaged or recycled.

solid waste management. The purposeful, systematic control of the generation, storage, collection, transport, separation, processing, recycling, recovery, and disposal of solid wastes.

SOLMET. A meteorological data base containing hourly values for meteorological and solar radiation values. Available from the National Climatic Center.

SOLSIM and SOLCOST. A pair of computer programs, developed by the Martin Marietta Corporation, that simulate the performance of, and most economical solar collector area for, solar systems.

solstice. Those times of the year when the sun is furthest north or south of the equator. In the northern hemisphere, it occurs about June 21 to begin summer and on about December 21 to begin winter.

SOLSYS. A computer program, developed by Sandia Laboratories, Albuquerque, NM, that simulates the transient performance of solar systems.

solution. The change of matter from a solid or gaseous state into a liquid. All solutions are composed of a solvent (water or other fluid) and the substance dissolved is called the "solute." A true solution is homogeneous, as salt in water.

solvent. A substance used to dissolve another substance; that component of a solution which is present in excess, or whose physical state is the same as that of the solution.

solvent extraction. A technique for drying or dehydrating alcohol fuel. Benzene is used to remove the water; however, this process is too toxic and expensive for small-scale alcohol producers.

soot. Agglomerations of tar-impregnated carbon particles that form when carbonaceous material does not undergo complete combustion.

sorbent. A material which extracts one or more substances present in an atmosphere or mixture of gases or liquids with which it is in contact, due to an affinity for such substances.

sorption. A term including both adsorbtion and absorption. Sorption is basic to many processes used to remove gaseous and particulate pollutants from an emission.

south-facing. Toward the equator; the best direction to point a solar collector in the northern hemisphere. In the southern hemisphere, for example, Australia, the best collector orientation is north-facing.

space charge. Same as cell barrier (q.v.), depletion zone.

space heating. Heating of the air inside a building. ("Space cooling" is the converse.)

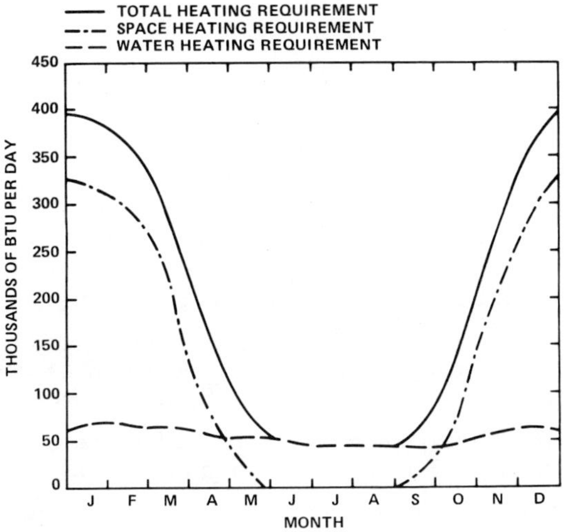

Courtesy HUD

Diagram of seasonal impact of space heating requirements.

special fuel. Defined in the Crude Oil Windfall Profit Tax Act of 1980 (26 USC 44E, P.L. 96-223) as "any liquid fuel (other than gasoline) which is suitable for use in an internal combustion engine."

specially denatured alcohol. Ethyl alcohol to which sufficient quantities of various denaturants have been added, pursuant to formulas prescribed by federal regulations, to render it unfit for beverage purposes without impairing its usefulness for other purposes; specially denatured alcohol may be distributed only to persons holding BATF permits.

specific energy. In cutting or grinding, the energy expended or work done in removing a unit volume of work material, usually expressed as inch-pound per cubic inch (in.-lb/in.^{3}) or horsepower per minute per cubic inch (hp/min/in.^{3}).

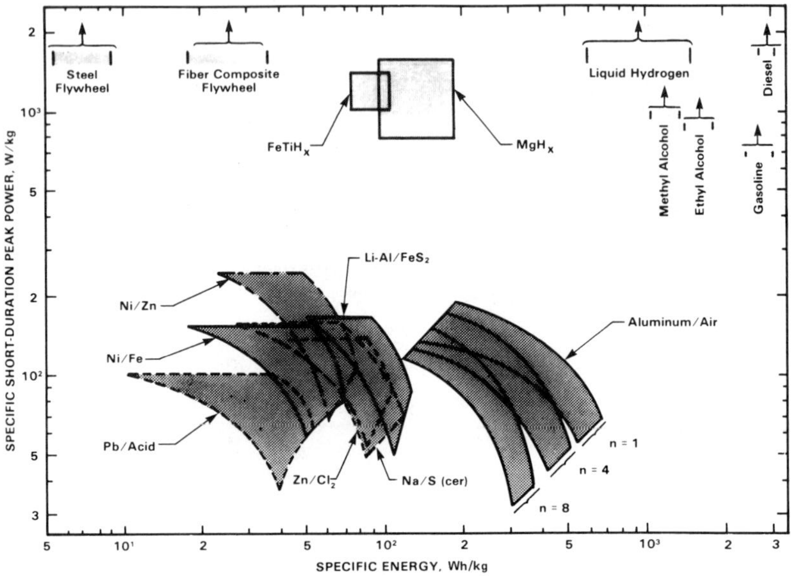

Notes: 1) Secondary Battery Energies are for 3-Hour Discharge Rate
2) Power and Energy Estimated at Output of Conversion Device (Engine, Motor/Controller, CVT)
3) Ranges Reflect Conservative Through Optimistic Estimates
4) n = Ratio of On-Board Aluminum to Other Reactants

Courtesy DOE

Illustration of specific energy of various battery materials.

specific gravity. The ratio of the mass of a solid or liquid to the mass of an equal volume of distilled water at 39.2°F (4°C).

specific heat. Ratio of the amount of heat required to raise a unit mass of material 1 degree to that required to raise a unit mass of water 1 degree.

specific heat capacity. The quantity of heat needed to change the temperature of 1 pound of a material by 1°F (or 1 kilogram of a material by 1°C).

specific humidity. Weight of water vapor, in grains or pounds, per pound of dry air.

specific output. Total energy output in kilowatt-hours (kWh) over a specified time period, divided by the rated power of the wind-energy conversion system in kilowatts (kW).

specific performance bonds. The ability of a contractor, through a third-party insurer, to eliminate the risk of nonperformance on a construction job.

specific speed. The revolutions per minute at which a fan, pump, or turbine would run if it were of such diameter as to deliver a unit volume of fluid per minute against a unit pressure.

specific volume. The volume of a unit weight of a substance at specific temperature and pressure conditions; the volume occupied by a unit of air, measured in cubic feet per pound (ft^3/lb).

specific weight. The weight of a unit volume of a substance, usually expressed as pounds weight per cubic foot.

spectral energy distribution. A curve showing the variation of spectral irradiance with wavelength.

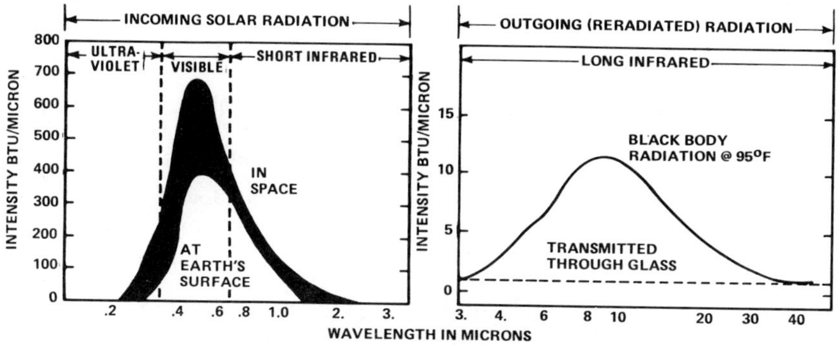

Courtesy DOE

Illustration of incoming and outgoing radiation.

spectral irradiance. The monochromatic irradiance of a surface per unit bandwidth at a particular wavelength. Units often W/m^2-nanometer bandwidth.

spectral reflectance. Ratio of the energy reflected from a plane surface in a given defined waveband to the energy incident in that waveband.

spectral reflection. Mirrorlike reflection; incident and reflected angles are equal.

spectrum. A plot of the distribution of the intensity of a given type of radiation as a function of its wavelength, energy, frequency, momentum, mass, or any related quantity.

spent grains. The nonfermentable solids remaining after fermentation of a grain mash.

spillage (radiation). Radiation reflected from the collector subsystem, but which misses the absorber surface of the receiver.

spillway. A passage for surplus water over or around a dam or other hydraulic structure.

spinning reserve. Utility generating capacity on line and running at low power in excess of actual load.

spiral pipe. Pipe made by coiling a plate into a helix and riveting or welding the overlapped edges.

split spectrum cell. A compound photovoltaic device in which sunlight is first divided into spectral regions by optical means. Each region is then directed to a different photovoltaic cell optimized for converting that portion of the spectrum into electricity. Such a device achieves significantly greater overall conversion of incident sunlight into electricity. See **vertical multiple function cell.**

split system. Historically, a combination of warm-air heating and radiator heating; the term is also used for other combinations such as hot water-steam, steam-warm air, etc.

split system heat pump. Heat pump with components located both inside and outside of a building; the most common type of heat pump installed in a residence.

split wedge. A wedge consisting of two pieces, into which the valve stem is screwed to expand the two pieces against the valve seating surfaces to assure a tight seal when the valve is closed.

spoil. Dirt or rock that has been removed from its original location, specifically materials that have been dredged from the bottoms of waterways.

square foot of radiator. The amount of heating surface in the form of radiators, convectors, unit heaters or other devices which will emit 240 British thermal units (Btu) per hour. This term has no exact relationship to the actual area of the surface.

square wave. Type of alternating current output from low-cost solid-state inverters; usable for many appliances, but may affect stereos and TV sets.

squirrel cage fan. A centrifugal blower with forward-curved blades.

stabilization. The process of converting active organic matter in sewage sludge or solid wastes into inert, harmless material.

stabilization ponds. See **lagoon; oxidation pond.**

stable air. An air mass that remains in the same position rather than moving in its normal horizontal and vertical directions. Stable air does not disperse pollutants and can lead to high buildups of air pollution.

stack. A smokestack; a vertical pipe or flue designed to exhaust gases and suspended particulate matter.

stack effect. The ability to set up a large enough temperature difference to effect the displacement of warm air by cooler air in a thermal chimney, such that the lighter warm air rises through a distribution space. See **stratification; thermal chimney.**

stack height. The height of a convector enclosure measured from the bottom of the enclosure to the top of the outlet.

stack loss. The flue gas loss; the sensible and latent heat lost up the chimney in the flue gas.

stack vent. An extension of a soil or waste stack above the highest horizontal branch connected to the stack.

stagnation depth. The depth a rainwater collector gets when no rainwater is being saved. The stagnation depth is higher when more rain is caught; analogous to stagnation temperature.

stagnation temperature. A high-temperature condition obtained in a solar collector when the sun is shining and no fluid is flowing through the collector; temperatures range from 250° to 400°F (121° to 204°C), depending on collector design. Any condition under which a collector is losing as much heat as it gains.

stainless-steel pipe. An alloy steel pipe with corrosion-resisting properties, usually imparted by nickel and chromium.

stall. Aerodynamic condition where the flow over an airfoil is separated. (Loss of lift and increased drag are indications of stall.)

stand. A contiguous group of trees sufficiently uniform in species composition, arrangement of age classes, and condition to be a homogeneous and distinguishable unit.

stand-alone. An isolated photovoltaic system not connected to a grid; may or may not have storage, but most stand-alone applications require battery or other form of storage. See **remote site.**

stand alone inverter. Inverter that operates without being connected to a utility or other grid for frequency and/or voltage signals.

stand-alone option. A residential photovoltaic system designed to cover all single-family residence loads without utility grid supplement.

standard air. Air weighing 1.2 kg/m^3 (0.075 lb/ft^3) and equivalent in density to dry air at a temperature of 21.1°C (70°F) and a barometric pressure of 1.01×10^5 N/m^2 (29.92 in. of Hg).

standard barometric pressure. Air pressure equal to 1.01×10^5 N/m^2 (29.92 in. of Hg).

standard conditions. In refrigeration, an evaporating temperature of 5°F (–15°C), condensing temperature of 86°F (30°C), liquid temperature before the expansion value of 77°F (25°C), and suction temperature of 12°F (−11°C).

standard corporation. A corporation in which the investment tax credits, depreciation, or corporate losses stay within the corporate shell and cannot be deducted from personal income taxes.

standard cubic foot. The column of a gas at standard conditions of temperature and pressure. The American Gas Association uses moisture-free gas at 60.8°F (16°C) and 30 inches of mercury (1.0037 atmospheres) as its standard conditions. The pressure standard is not universal in the gas industry; 14.7 pounds per square inch absolute (1.000 atmosphere) and 14.4 pounds per square inch absolute (0.980 atmosphere) are also used. The scientific community uses 32°F (0°C) and 1 atmosphere as standard conditions.

standard industrial classification manual. A book, prepared and issued by the Office of Statistical Standards, U.S. Bureau of the Budget, to enable classification of business establishments by the type of activity in which they are engaged.

standard liquid (or air) system. A reference solar system used in hourly solar heating simulations.

standard pressure. Formerly used to designate cast-iron flanges, fittings, valves, etc., suitable for a maximum working steam pressure of 125 lb. In practice, 14.696 psia is used in determining density at standard pressure and temperature.

standby heat loss. Heat lost through storage tank and piping walls (rate of heat loss differs from standing still and when moving).

starch. A carbohydrate polymer comprised of glucose monomers linked together by a glycosidic bond and organized in repeating units; starch is found in most plants and is a principal energy storage product of photosynthesis; starch hydrolyzes to several forms of dextrin and glucose.

starch granule. Large amounts of starch are laid down in the storage organs of plants (the tubers, seeds, roots, and stem pith, among others). Starch, itself, is a long chain of glucose molecules chemically bonded together by a water molecule. This chain is enclosed in a sac (very much like the covering of a baseball), known as the starch granule. The appearance of a starch granule has distinct characteristics, depending upon its plant origin. Potato granules, for example, resemble round or oblong spheres, while those of corn are crystalline in shape.

start up. The procedure whereby a wind turbine starts from parked standby status, is made operationally ready, achieves synchronization with the grid and begins to satisfy load. Start up requires several minutes.

stat. A suffix meaning apparatus to render stationary. For example, thermostat, a device to maintain a constant temperature; pressurestat, device to maintain constant pressure, etc.

static pressure. The force per unit area acting on the surface of a solid boundary parallel to the flow. The static pressure is constant in all directions and is the component normal to the direction of flow, the force exerted per unit area by a gas or liquid measured at right angles to the direction of flow, or the pressure when no liquid is flowing.

stationary source. A pollution emitter that is fixed rather than moving as an automobile.

steam. Water in the form of a vapor. It is often the medium by which heat energy liberated from fuel by combustion is converted into mechanical work.

steam accumulator. A vessel to smooth out violent steam loads on a boiler. While the steam demand is relatively low, the accumulator is charged and acts as a buffer when sudden steam demands occur.

steam engine. A reciprocating engine, worked by the force of steam on the piston; the steam expands from the initial pressure to the exhaust pressure in a single stage.

steam generator. The equipment which uses a heat source to change water into steam.

steam jet refrigeration. A method of cooling involving the use of steam nozzles to reduce the pressure in a water chamber so that the water boils at a low temperature; since heat is drawn from the water, it is thus cooled.

steam power. Energy or power generated through the use of a steam engine.

steam power plant. A plant in which the prime movers (turbines) connected to the generators are driven by steam.

steam pressurization. The principle of steam pressurization in which the expansion tank provides space for expansion of system water, means for pressurization by steam in the upper part of the tank, and storage space for heated water from the boiler.

steam trap. A device for allowing the passage of condensate, or air and condensate, while preventing the passage of steam.

steam turbine. A machine powered by high-pressure steam and used to drive mechanical apparatus. It has a rotary motion in contrast to a reciprocating motion.

Stefan-Boltzmann law. The energy radiated in unit time by a black body, given as

$E = K(T^4 - T_0^4)$, where T is the absolute temperature of the body, T_0 is the absolute temperature of the surroundings, and K is a constant.

stem. The usually threaded shaft to which is attached the handwheel at the top and the disc or wedge at the lower end. The stem may also be called the spindle.

still. An apparatus in which a substance is changed by heat into vapor, with or without chemical decomposition. The vapor is then liquefied in a condenser and collected in another part of the apparatus. Used generally for separating the more volatile parts of liquids and obtaining them in pure form.

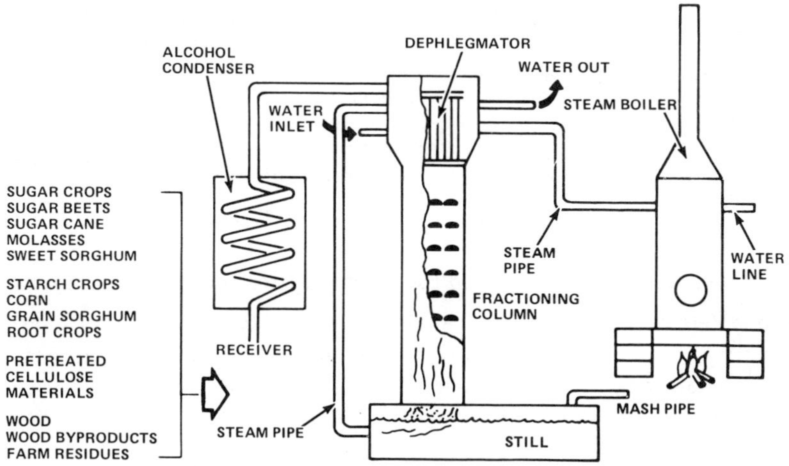

Courtesy DOE

Illustration of common still.

stillage. The nonfermentable residue from the fermentation of a mash to produce alcohol.

still air. Air which is not moving. Heat loss from a surface in still air is by natural convection and is much less than if the air is moving.

Stirling engine. An external-combustion engine, invented in the nineteenth century. The engine uses pistons driven by heated gas which travels in a sealed system from heat receiver to cylinders. It is potentially more efficient than a steam engine or gas turbine.

stock. An accumulation of petroleum, or products in storage, awaiting transfer to the point of utilization; also, a bar used to support a windmill sail.

stock rake. Tool shaped like a hoe used to manage ashes and embers in a wood stove.

stoichiometric ratio. The ratio of chemical substances necessary for a reaction to occur completely.

stop plug. An adjusting screw, extending through the body of a check valve, which adjusts and controls the extent of movement of the disc or clapper.

storage. Refers to storing heat collected during the day in excess of immediate requirements for use overnight or on cloudy days. Unless infinite storage is provided, a system cannot be fully utilized at all times except during periods of high demand.

storage battery. A combination of secondary cells or accumulators which, once charged, may be used for a considerable time as a source of electric current.

storage capacity. The amount of energy that can be stored by a solar heating system to be used at a later time for space or water heating.

storage-coupled. The utilization of an energy storage system to permit operation of the end-use system during nonsolar hours.

storage mass. A dense, heavy material that receives and holds heat from an absorber and later releases the heat to the interior of the structure; is of sufficient volume, depth, and thermal energy capacity to store and deliver adequate amounts of solar heat for the structure in which it is incorporated; is located so that it is capable of distributing the stored heat directly to the habitable areas of the structure through a heat distribution method; and has an area of directly irradiated material equal to or greater than the solar collection area.

storage material. Material that stores heat in a solar-heating system. Storage materials are rock-beds, water tanks, concrete walls and water-filled drums.

storage subsystem. Assembly used for storing thermal energy for later use.

storage system cost. A current estimated cost of a storage system or a projected future cost.

storage temperature. The average temperature of the heat stored (it might be a little hotter at the top and a little cooler at the bottom because of stratification); analogous to tank depth.

storage utilization factor. Fraction of the storage thermal capacity that can be delivered at rated conditions (nearly constant thermodynamic quality).

storage well. A cased bore hole extending from the surface into the storage reservoir, which is used primarily for oil or gas input and/or output purposes.

storm sewer. A conduit that collects and transports rain and snow runoff back to the groundwater. In a separate sewerage system, storm sewers are entirely separate from those carrying domestic and commercial wastewater.

storm windows. Glass or plastic applied over existing windows to save energy. Plastic can be self-installed for 50 cents per window each year. Single panes of glass can cover the window at a cost of $10 to $15 per window. Triple-track glass combination storm windows cost $30 each. These have screens and can be opened and closed easily. They are for installation on conventional double-hung windows.

stove coal. In anthracite only; two sizes of stove coal are made, large and small. Large stove coal, known as No. 3, passes through a 2¼- to 2-inch mesh and over a 1-7/8- to 1½-inch mesh; small stove coal, known as No. 4, passes through a 1-7/8- to 1-3/8-inch mesh and over a 1-1/8- to 1-inch mesh. Only one size of stove coal is usually made. It passes through a 2-inch square mesh and over a 1-3/8-inch square mesh.

stover. The dried stalks and leaves of a crop remaining after the grain has been harvested.

stow. The position or act of reaching the position of storage for the heliostats.

stratification. The temperature gradient distribution of a material or substance. A water-filled container used in solar heat storage usually will be warmer at the top than at the bottom. This difference is expressed as stratification or layering.

stray current. Electrical current, normally direct current, from either natural or man-caused source, which could cause corrosion of structures such as pipelines, if not reduced properly.

streamline flow. Also termed viscous or laminar flow. When the fluid particles are moving in a straight line parallel to the axis of the pipe or duct. With streamline flow, the friction varies directly as the velocity. See **critical velocity; turbulent flow.**

street elbow. An elbow with male thread on one end, and female thread on the other end. See **elbow.**

stress. The resultant force that resists change in the size or shape of a body acted on by external or internal forces.

stripper. A pressure vessel in which the carbon dioxide and other heavy hydrocarbons are stripped from the liquid methanol by passing a clean stream of methane up through the methanol.

stripping section. The section of a distillation column below the feed in which the condensate is progressively decreased in the fraction of more volatile components by stripping.

Styrofoam®. A home insulating sheet product by Dow-Corning Co.; it is a polystyrene foam material, whose color is often light blue.

Sub-chapter S Corporation. A corporation which is limited to 15 or fewer investors who can use the tax credits, depreciation, or losses which the corporation incurs in such a manner as to decrease personal income tax liability; the Sub-S election can be rescinded to convert the corporation at a later date to a standard 1244 corporation in which any loss investors incur can be deducted from personal income taxes.

subcooling. The cooling of a liquid to below its saturation temperature for the pressure under consideration.

sublimation. The vaporization of a solid (especially when followed by the reverse change) without the intermediate formation of a liquid.

sublime. To cause to pass from the solid state to the vapor state by the action of heat, and to condense again to solid form.

submarginal resource. That portion of a resource that cannot presently be extracted economically. As the economic picture and technology change, some submarginal resources may become recoverable resources.

submetering. The practice of remetering purchased energy beyond the customer's utility meter, generally for distribution to building tenants through privately owned or rented meters.

subsoil drain. One for collecting subsurface or seepage water and carrying it to a point of disposal.

substation. An electrical installation containing generating or power-conversion equipment and associated electric equipment and parts, such as switchboards, switches, wiring, fuses, circuit breakers, compensators, and transformers.

substitute natural gas (SNG). A high Btu gas composed primarily of methane with a heating value of about 1000 Btu per standard cubic foot (Btu/scf).

substrate. In alcohol fuel production, the substance upon which an enzyme acts. Starch is the substrate for amylase enzymes, while sugar is the substrate for yeast's zymase enzyme complex.

subsurface water. Water that exists below the surface of the solid earth; may be liquid, solid, or gaseous.

subsystem. A major, separable, functional assembly of a system, such as a complete collector, thermal storage device, etc.

sucker rod. In a water-pumping windmill, provides the lifting force connected to the pump cylinder.

sucrose (table sugar; saccharose; $C_{12}H_{22}O_{11}$). This is the most abundant disaccharide for human consumption. It is produced from sugarcane, beets and, to some extent, from sorghum cane. Yeast produces an enzyme (sucrase) which will hydrolyze sucrose to one molecule of glucose and one molecule of fructose, thus making it a fermentable sugar.

suction anemometer. A device that measures wind velocity by the degree of exhaustion caused by the flowing of the wind through or across a tube.

suction burner system. A system applying a vacuum to a combustion chamber to draw in the air and/or gas necessary to produce the desired combustible mixture.

suction line. Refrigerant line leading to the compressor.

suction pressure. Pressure in the suction line or evaporator of a refrigerating system.

sulfur dioxide (SO_2). One of several forms of sulfur in the air; an air pollutant generated principally from the combustion of fuels containing sulfur. A natural source of sulfur dioxide is volcanic gases.

sulfur smog. A fog composed of smoke particles, sulfur oxides, and high humidity. The major cause of damage comes from the reaction of sulfur trioxide with the water to form sulfuric acid (H_2SO_4) droplets.

summer degree hour. The product of 1 hour in time and the amount of the outside dry-bulb temperature at that hour is above a base temperature. Some correlation has been found between the number of degree-hours over 85°F and summer cooling load in residences.

summer valley. The depression that occurs in the summer months in the daily load of an electric system.

sump. A depression or tank that serves as a drain or receptacle for liquids for salvage or disposal.

sun. The luminous, celestial body around which the earth and other planets revolve, and from which they receive heat and light. Sunshine consists of a wide variety of electromagnetic waves. Its three main components are invisible heat waves, visible light rays of various colors, and invisible ultraviolet rays. Energy from the sun travels through space at a speed of 300,000 kilometers (or 186,000 miles) per second. Sunlight arriving at the edge of the earth's outer atmosphere carries energy at an approximately constant rate of 1.36 kilowatts per square meter (kW/m^2) [or

130 watts per square foot (W/ft^2)] of area covered. In terms of heat, this is equivalent to 428 British thermal units per square foot per hour (Btu/ft^2/hr). The average amount of energy reaching the earth's surface is equivalent to about 177 watts per square meter (or 16.4 watts or 58.5 Btu per square foot) per hour. Measurements over an entire year (including night and day, cloudy and clear conditions, winter and summer) in various locations in the United States have shown that about 13 percent of the sun's original energy arrives at ground level. With an estimated 5×10^9 more years of life in its present state, the sun may be considered, for practical purposes, an inexhaustible source of fuel.

Sun-a-therm®. A glazing material from AGS Industries Inc. which consists of twin-sheets of low-absorptance glass with the edges nearly hermetically sealed.

sun compass. A navigational compass that uses the sun and its calculated bearing to establish direction, especially in high altitudes.

sun effect. The quantity of heat from the sun tending to heat an enclosed space.

Sun-lite®. A glazing material from Kalwoll Corporation which is a translucent sheet of polyester and fiberglass.

sun observations. In surveying, fixation of longitude and/or latitude of a station, or orientation of a survey line, by use of a theodolite, to relate the position of the sun, sideral time, and the location of the theodolite.

sun path diagram (solar window). A circular projection of the sky vault, similar to a map, that can be used to determine solar positions and to calculate shading.

sun pillar. A vertical streak of light above the sun, usually seen at sunrise or sunset. It is caused by reflection from ice crystals.

sun position. The azimuth and elevation angles for specifying the direction antiparallel to the central ray from the sun.

sunrise. Defined in meteorology as the moment when the upper edge of the sun appears to rise above the apparent horizon on a clear day. See **sunset.**

suns. A dimensionless measure of the intensity of solar energy based on the amount of energy received from the sun (one sun) following its concentration by lenses or mirrors. Concentration ratios exceeding 10,000 suns have been achieved.

sunset. In meteorological terms, the moment when the upper edge of the sun appears to fall below the apparent horizon on a clear day. Effects of refraction cause the apparent position to be about 34 feet (10.4 meters) above the true position.

sunshine. The warmth and light given by the sun's rays.

sunspace. A living space enclosed by glazing; a solarium or greenhouse.

sun tempered. A building whose long walls and major glazing surfaces are oriented to the south. This maximizes beneficial sunlight warming the building in winter. Overhangs or shading devices shade glazing to minimize unwanted heat gain in summer. Solar tempering can be used to advantage in almost all climates.

suntime. Time measured on a basis of the sun's virtual motion. See **local solar time.** (See diagram p. 304.)

sun tracking. Following the sun with a solar collector to make the collector more effective.

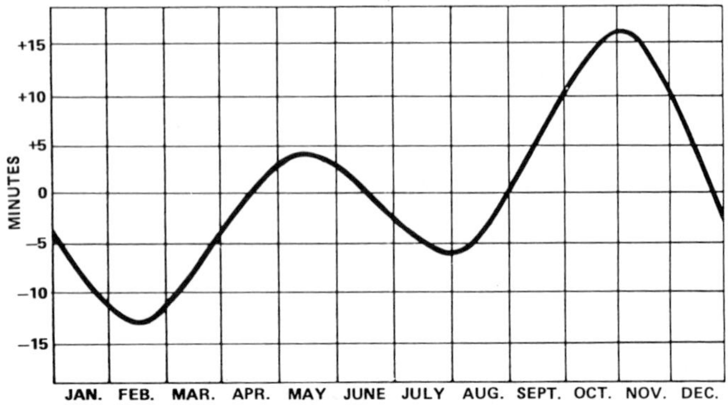

Equation of time diagram.

Sun window®. A system which uses a dark fluid to block solar radiation when it is not needed. The dark fluid is pumped through passages within the collector.

supercharge. Process of heating a storage material or room air beyond its heat capacity. Used in solarium or isolated gain designs as a means of increasing storage temperature.

supercharger. A blower that increases the intake pressure of an engine.

superconductivity. The abrupt and large increase in electrical conductivity exhibited by some metals as the temperature approaches absolute zero.

superconductor. A substance (metals, alloys, and compounds) exhibiting superconductivity; i.e., the disappearance or near disappearance of the electrical resistance of certain materials at temperatures below a transition zone (usually a few degrees above absolute zero).

superheated steam. Steam at a temperature higher than the boiling temperature corresponding to the pressure at which it exists.

supernatant. The liquid which collects on the top of a waste or sludge after the solids are allowed to settle. By removal of the supernatant the water plus dissolved organic matter can be separated from the solids.

supplementary heat. Extra heat provided by a conventional furnace when the available solar energy is insufficient to maintain the desired temperature.

surety bond. A type of insurance which satisfies the government's bonding requirements on distilled spirits production; obtainable from U.S. Treasury-authorized insurance companies, surety bonds usually carry an annual premium of 1 to 2 percent of face value.

surface area. The outside surface of an object that loses heat by convection or radiation. The bigger the surface area, the more heat lost.

surface combustion. Combustion of injected, properly proportioned fuel and air, on a surface or within a definite zone; when a mixture of air and gas or of air and oil vapor is forced through a porous wall, and ignited on the other side. Since necessary air is mixed with the gas, it will burn regardless of atmospheric conditions, even under water.

surface conductance. The heat transmitted from (or to) a surface to (or from) the fluid in contact with the surface in a unit of time per unit of surface area per degree tremperature difference between the surface and the fluid.

surface cooling. Air cooling by passing it over cold surfaces.

surface tension. Tensile forces on the thin surface film of an exposed liquid that tend to minimize the exposed area.

surface-to-volume ratio. The ratio of exposed surface of a building to occupied volume. A measure of exposure to harsh climate conditions causing unwanted heat loss and heat gain. (Lower numbers are desirable.) This ratio is especially useful in evaluating alternative building forms.

surface water. Water on the earth's surface exposed to the atmosphere as rivers, lakes, streams, oceans.

surfactant. An agent which modifies the physical, electrical, or chemical characteristics of the surface of a solid, or surface tensions of solids or liquids.

survival wind speed. The maximum wind speed a wind-energy conversion system in automatic, unattended operation (not necessarily generating) can sustain without damage to structural components or loss of the ability to function normally.

suspended matter. Particles of dust or other types of individual small granules or pieces sufficiently fine that they are suspended in air or carried by air, or fluid.

suspended solids. Sediment which is in suspension in water but which will physically settle out under quiescent conditions (as differentiated from dissolved material); solids that can be separated from a liquid by filtration.

swept area. Area as seen from the direction of the wind that a wind-energy conversion system rotor will pass over during one complete rotation.

swimming pool heating. The solar heating of both private and commercial-sized swimming pools is one of the most economical and widely available uses of solar energy. In its simplest form, solar pool heating can be accomplished by the use of a transparent "blanket," which collects heat during sunny periods and prevents heat loss at night or during cloudy days. This system provides low-cost pool heating particularly suited to dry and/or windy climates.

This "blanket" system can be supplemented by the use of solar collectors.

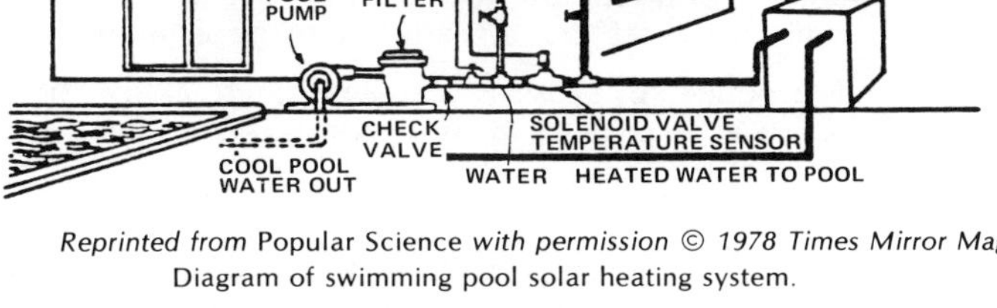

Reprinted from Popular Science *with permission © 1978 Times Mirror Magazines, Inc.*
Diagram of swimming pool solar heating system.

These may be plastic or metal, glazed or unglazed, mounted on a roof or on the ground. The collectors are put in line after the pump and filter (which are normal parts of the pool system anyway). This still represents a fairly inexpensive alternative to expensive fossil fuels.

swing check valve. A check valve which uses a hinged disc or clapper to limit the direction of flow. The pressure exerted by the fluid flowing through the valve forces the disc away from the seating surface. When the flow ceases, the clapper falls to its original position, preventing flow in the opposite direction.

swing joint. An arrangement of screwed fittings and pipe to provide for expansion in pipelines.

swivel joint. One employing a special fitting designed to be pressure tight under continuous or intermittent movement of the machine or part to which it is connected.

symbiosis. A mutually beneficial interdependency between two organisms, e.g., a leguminous plant and its infecting bacteria. Symbiosis can also refer to the net effect of two factors (chemicals, organisms, etc.) which exceeds the sum of the effects of each of the two factors in isolation.

synchronous generator. A type of electric generator which runs at synchronous (constant) speed and draws its excitation from an external or independent power source from the system it is feeding into. (It generates at variable power factor.)

synchronous inverter. Also called "line commutated inverter," inverts direct current (DC) to alternating current (AC) (see inverter) but must have another AC source (e.g., utility mains or AC gas generator) for voltage and frequency reference; AC is created synchronously (that is, in phase and at the same frequency as outside the AC source).

Courtesy DOE

Block diagram of synchronous inverter electrical system.

synergism. The action of two separate substances to produce an end effect which is greater than the sum of the effects of the two substances acting independently.

syntectic. Magmas produced by syntexis; also an isothermal reversible reaction in which a solid phase, on absorption of heat, is converted to two conjugate liquid phases.

synthetic fuels. A term now commonly reserved for liquid and gaseous fuels that are the product of a conversion process rather than mining or drilling; most frequently applied to cover liquids or gases derived from coal, shale, tar sands, waste, or biomass.

synthetic natural gas. A manufactured gas comprised chiefly of methane and roughly equivalent in heating value to natural gas. Synthetic natural gas can be synthesized from biomass or coal.

system loss. As applied in the electrical industry, the difference between the system net energy (power input and output), resulting from losses unaccounted for between the source of supply and the point of delivery.

system power coefficient. Power output of a wind-energy conversion system divided by the power available in its flow field at a specified wind speed.

system required reserve. The system reserve capacity needed as standby to ensure an adequate standard of service.

system-wide storage. An energy storage system accessible to, and chargeable by, any generating source in the system having available and/or excess capacity.

tachometer. An instrument of the direct-reading type, indicating the speed of a shaft or machine in revolutions per minute (rpm).

tail. Mechanical fin which orients a wheel to the direction of the wind.

tailvane. Vertical surface mounted on a wind-energy conversion system for control purposes. Some horizontal axis systems use this for yaw orientation and control; and some vertical axis systems use this to coordinate cyclic blade pitch.

tankage. The process of storing liquids in a tank; the capacity of tanks; the price paid for tank storage of liquids.

tap. An outlet on the side of a rainwater storage tank from which to draw stored water; analogous to the heat exchanger of a solar heating system.

tap temperature. Temperature at which fluid, such as hot water, is discharged from an outlet at the point of use.

tariff. A published volume of rate schedules and general terms and conditions under which a product or service will be supplied.

tax credits. An individual taxpayer may claim a credit against federal income tax for a portion of the cost of installing energy-conservation materials and renewable energy systems. The conservation credit is 15 percent of qualifying expenditures up to a maximum credit of $300. The renewable energy credit is 40 percent of the cost of installing solar, wind, or geothermal equipment up to a maximum credit of $4,000. Details on how to claim a credit and on what items qualify are contained in the Internal Revenue Service's Publication 903, "Energy Credits for Individuals." Many states now offer tax advantages for installing solar systems in addition to those offered by the Internal Revenue Service.

tax-free alcohol. Pure ethyl alcohol withdrawn free of tax for government, for hospital use, for science, or for humanitarian reasons; it cannot be used in foods or beverages; all purchasers must obtain BATF permits, post bonds, and exert controls upon storage and use of tax-free alcohol.

tax-paid alcohol. Pure ethyl alcohol which has been released from federal bond by payment of the federal tax of $21.00 per gallon at 200 proof or $19.95 per gallon at 190 proof.

technical fix. The use of technical solutions to problems which might be solved through institutional or behavioral changes. Sometimes used in a derogatory manner.

technology. The application of knowledge for practical purposes; for example, engineering designs to convert solar energy into more useful forms of energy such as electricity or space heating.

Tedlar®. A glazing material from Du Pont Co. which is a near-transparent sheet of polyvinyl fluoride.

tee. A pipe fitting, either cast or wrought, that has one side outlet at right angles to the run.

Teflon®. A glazing material of the Du Pont Co. which is a near-transparent sheet of fluorinated ethylene propylene.

Tempchek®. An insulating product of the Celotex Corporation consisting of a plate of urethane foam faced with asphalt saturated inorganic sheets.

temperature. A degree of hotness or coldness measured on a definite scale; as distinguished from heat. Heat is a form of energy; temperature, a measurement of its thermal effects.

temperature coefficient of a solar cell. The amount by which the voltage, current, or power from a solar cell will change with changes in the temperature of the cell.

temperature difference. One temperature subtracted from another temperature; analogous to depth difference.

temperature, equivalent. See **equivalent temperature.**

temperature-humidity index. An index that combines sensible temperature in degrees Fahrenheit and humidity in the air to arrive at a number that closely corresponds to the effective temperature. When the temperature–humidity index (THI) is 70, almost everyone is comfortable. When it is 71, 10 percent of the people will be uncomfortable; 72, 20 percent, etc., until, at 80, almost all people are uncomfortable.

temperature inversion. A layer of the atmosphere where the temperature increases with height. In this dictionary temperature inversion or inversion is also used to imply a layer where the potential temperature increases with height forming a stable layer, however, the reader should realize that the potential temperature can be increasing with height while the temperature is decreasing with height. The stable boundary layer is a potential temperature inversion, whereas the base of the lowest potential temperature inversion marks the top of an unstable or nearly neutral boundary layer.

temperature rise. The temperature difference between inlet and outlet of a solar collector; the temperature of a heat storage tank above the temperature to which it delivers heat.

temperature sensors. Devices that measure the temperature of the collectors and storage and send signals to the control unit to start or stop pumps or actuate valves.

temperature swing. The typical change in temperature between early morning and late afternoon in a passive solar heating system.

temperature zones. Areas controlled to maintain different temperatures within a house.

tempering valve. See **mixing valve.**

ten-megawatt-electric (10-MWe) pilot plant. The prototype solar thermal central receiver power system being constructed near Barstow, CA by the U.S. Department of Energy. The plant will have the capability of producing 10 megawatts of electricity (10 MWe) for use of the Southern California Edison system. See **Solar One.**

tensile structure. A structure formed by the stress or stiffening of its material.

ternary. This is a descriptive term meaning "three parts." In alcohol fuel production, when a third compound is added to an alcohol/water mixture to break the azeotropic mixture, a ternary azeotrope results and an anhydrous alcohol distillation becomes possible.

tertiary treatment. Wastewater treatment beyond the secondary or biological stage that includes removal of nutrients such as phosphorus and nitrogen, and a high percentage of suspended solids. Tertiary treatment, also known as advanced waste treatment, produces a high-quality effluent.

test reference year. A reference weather year for a particular location which is average in that extreme weather conditions have been eliminated.

tetraethyllead [$Pb(C_2H_5)_4$]. A volatile lead compound which is added in concentrations up to 3 milliliters per gallon (ml/gal) to motor gasoline to increase the antiknock properties of the fuel.

theoretical air requirement. Volume of air necessary to ensure the complete combustion of unit mass or volume of a fuel.

therm. A unit of heating value equivalent to 100,000 British thermal units (Btu); 1 therm equals 100 cubic feet (ft^3) of natural gas.

thermal batteries. Type of battery that requires heat to activate. The electrolyte employed is a mixture of anhydrous salts which conduct electricity when melted. The heat source can be ignited by either an electric match or a mechanical primer. When the battery is ignited, the heat source evolves sufficient heat energy to melt the electrolyte and so permits the battery to deliver a considerable amount of electrical power.

thermal capacitance. The ratio of the amount of heat added to the resulting rise in temperature in a unit mass of material, often called specific heat. Btu per pound per degree Fahrenheit (Btu/lb/°F.).

thermal capacity. Heat required to raise the temperature of a body 1°C.

thermal chimney. A vertical cavity through which heated air moves as a result of the stack effect. Used as a means of passive solar heat distribution or induced ventilation.

thermal conductance. A measure of the thermal conducting properties of a composite structure such as well, roof, etc., including the insulating boundary layers of air.

thermal conduction. The transfer of heat within a substance from points of higher to points of lower temperature.

thermal conductivity. A measure of the thermal conducting properties of a single material expressed in units of British thermal units per inch thickness (square foot) (hour) (degree Fahrenheit temperature difference).

thermal conductor. A material which readily transmits heat by means of conduction.

thermal contact. Two objects that can transfer heat easily from one to the other, usually by conduction.

thermal conversion. The transformation of heat into other forms of energy such as electricity or shaft power. Heat may be supplied from the sun either directly (solar collector) or indirectly (biomass).

thermal diffusivity. Thermal conductivity divided by the product of specific heat times unit weight.

thermal efficiency. The ratio of the electric power produced by a power plant to the amount of heat produced by the fuel; a measure of the efficiency of heat with which the plant converts thermal to electrical energy; also, relating to heat, a percentage indicating the available British thermal unit input that is converted to useful purposes.

thermal electric. Electric energy derived from heat energy, usually by heating a working fluid which drives a turbogenerator. See **solar thermal electric.**

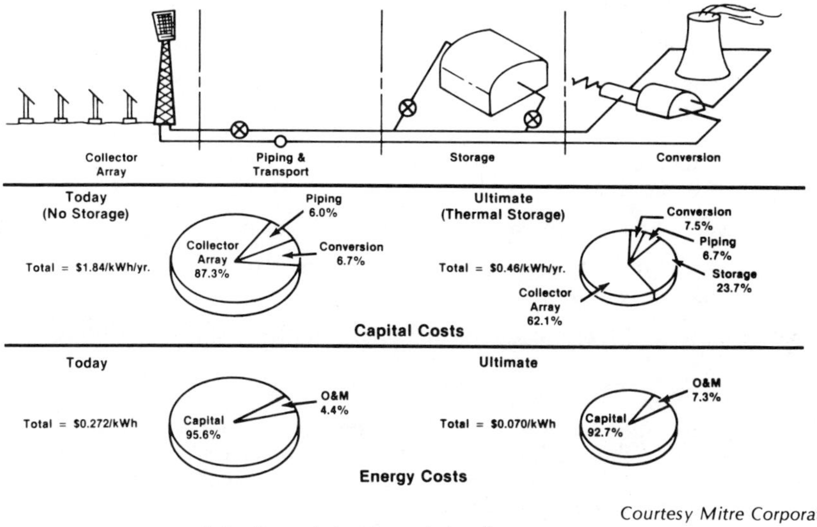

Courtesy Mitre Corporation

Solar thermal electric capital and energy costs

thermal energy. Heat energy.

thermal energy storage system. Any rechargeable unit capable of storing thermal energy for later use. Examples are storage as sensible heat in nitrate salt, sodium, rocks, water, or oil.

thermal envelope. The enclosure of a building which exhibits thermal resistance.

thermal equilibrium. In a system, the state at which there is no variation in temperature from one point to another.

thermal factors. A rating given to aluminum thermalized windows and doors, based on standards and test procedures formulated by the Architectural Aluminum Manufacturers Association. The number assigned to a window is based on temperature readings taken on various parts of the window using an inside temperature of 68°F (20°C), an outside temperature of 17.6°F (−8°C), and a 15-mile-per-hour wind. The number takes into consideration both a window's heat-transfer characteristics as well as its resistance to condensate formation. The higher the TF number, the better the window will be.

thermal insulation. A material having a relatively high resistance to heat transfer.

thermal lag. The time it takes to collect enough heat so that your solar storage unit climbs beyond its downpoint temperature after a period during which your system has collected no solar heat.

Thermalite®. An insulating product of the Northeast Energy Corporation, Econ Co. and Alpha Associates that is a honeycomb bubble plastic with adhesive backing.

thermal load. A demand for thermal energy (heat).

thermal mass. Any material used to store the sun's heat or the night's coolness. Water, concrete, and rock are common choices for thermal mass. In winter, thermal mass stores solar energy collected during the day and releases it during sunless periods (nights or cloudy days). In summer, thermal mass absorbs excess daytime heat and ventilation allows it to be discharged to the outdoors at night.

thermal performance. See **efficiency.**

thermal pollution. Degradation of water quality by the introduction of a heated effluent. Primarily a result of the discharge of cooling waters from industrial processes, particularly from electrical power generation. Even small deviations from normal water temperatures can affect aquatic life. Thermal pollution usually can be controlled by cooling towers.

thermal power. Energy generated through the use of heat energy.

thermal power plant. Any electric power plant which operates by generating heat and converting the heat to electricity.

thermal radiation. Electromagnetic radiation emitted by a warm body. See **infrared radiation.**

thermal ratchetting. The growth in diameter of tank walls with thermal cycling due to settling of solid material stored in the tank with rising temperature followed by overstressing of the walls when the temperature drops. This phenomenon is possible because of the differential thermal expansion of the stored material and tank wall (refers to solid media storage systems).

thermal resistance. Resistance or impediment to heat flow. A high thermal resistance means heat flows with difficutly, a low thermal resistance means heat flows easily; measure of a structure's degree of insulation.

thermal stratification. Process used in solar thermal storage systems whereby hot water (being less dense) floats on the top of more dense cold water.

thermal structure. The temperature variation with depth of seawater.

thermal water. The mineral-charged water that issues from a hot spring or geyser.

thermal wind. The increase in geostrophic wind with height due to horizontal temperature gradients. Its magnitude is about 3.6 kilometers per second (7.2 knots) per kilometer per 1°C per 100 kilometers.

Thermax®. An insulating product of the Celotex Corporation consisting of a plate of isocyonurate foam with a mixture of fiberglass. Both faces consist of aluminum foil.

thermic diode. A passive solar heating module that works on the thermosiphon principle. It has an inherent diode function to prevent heat loss at night.

thermionic conversion. The conversion of heat into electricity by boiling electrons from a hot metal surface and condensing them on a cooler surface. No moving parts are required.

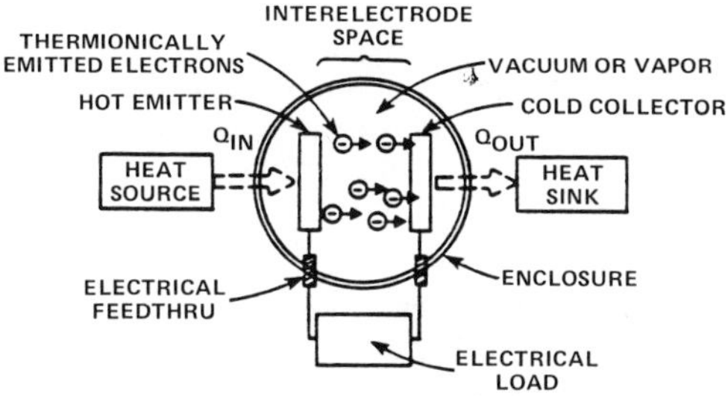

Courtesy SERI

Schematic diagram of thermionic converter.

thermionic engine. A conversion device in which electrical energy is produced directly from heat energy by the conversion of heat into electricity via boiling electrons from a hot metal surface and condensing them on a cooler surface.

thermistor. Sensing device which changes its electrical resistance according to temperature. Used in the control system to generate input data on collector and storage tank temperatures. See **sensor.**

Thermoanaerobactor ethanolicus. A microorganism that can convert sugar into ethanol. The bacteria has the ability to grow at temperatures as high as 172°F (77.8°C).

thermochemical conversion. The use of direct combustion, pyrolysis, gasification, or direct liquefaction to produce energy or energy products from biomass.

thermochemical conversion process. Any process which transforms an initial set of chemical reagents into a different product set of chemicals involving the application or deletion of heat energy.

thermochemical dissociation. A process for generating storable chemicals by using the sun's energy in a chemical reaction. A possible reaction could produce hydrogen and oxygen from water using only a high-temperature, solar-driven chemical reaction.

thermochemical heat. Thermal energy associated with a change in the chemical composition of a mixture of materials; for example, the heat released during combustion of a fuel. For application to energy storage, however, the chemical reactions must be controlled and reversible. That is, the products of the reaction can be recombined at will to recreate the starting mixture. The concepts of reversibility and controllability are illustrated by an automobile storage battery, although that device stores electric rather than thermal energy. In a storage battery, the input of electric energy (charging) causes chemical changes to occur and, once the battery is charged, the higher energy products within the battery remain a source of energy on demand. The battery may be connected to an electric motor or a light bulb, in which case the stored electric energy will be returned, accompanied by chemical changes within the battery the reverse of those which occurred during charging.

thermochemistry. The study of the heat changes accompanying chemical reactions.

thermocirculation. The convective circulation of fluid which occurs when warm fluid rises and is displaced by denser, cooler fluid in the same system.

thermocline. A specific configuration for sensible heat storage with a heated liquid such as water or oil. A single containment vessel is used to store both hot and cold fluid. Hot fluid remains at the top of the vessel and is added or withdrawn by displacement of (by) cooler fluid at the bottom of the vessel. The hot and cool fluid layers remain stratified because of the difference in density and because the containment vessel is filled with granular solid to impede mixing currents.

thermocline storage. The storage of thermal energy where the hot and cold media are in the same container (tank) using the thermocline principle which relies on a lower density hot fluid floating atop a higher density cooler fluid of the same type.

thermoconversion. The transformation of organic matter into other forms of energy such as electricity or fuels using heat. See **bioconversion.**

thermocouple. A thermoelectrical device consisting of two dissimilar wires with their ends connected together. A small voltage is generated when two junctions are at different temperatures; if one junction is kept at a reference temperature, the voltage generated in the other is a measure of the temperature of the other junction above the reference.

thermodynamic cycle. A method of using heat to give a cooling effect by evaporating (boiling) a liquid and then condensing (liquefying) it again.

thermodynamic efficiency. The highest fraction of input energy that can be converted into useful work by a heat engine. The Second Law of Thermodynamics establishes that maximum.

thermodynamics. A study of the transformation of energy into other manifested forms and of their practical application. The three laws are: first law, conservation of energy (energy may be transformed in an isolated system, but its total is constant); second law, heat cannot be changed directly into work at constant temperature by a cyclic process; third law, heat capacity and entropy of every crystalline solid become zero at absolute zero (0°K).

thermodynamics, laws of. The first law states that heat can be converted into work and work into heat; there is a constant relationship between the two quantities which is known as the mechanical equivalent of heat (1 calorie equals about 4.2 joules). The second law states that when a free exchange of heat takes place between two objects, the heat is always transferred from the warmer to the cooler object. See also **first law of thermodynamics; law of thermodynamics; second law of thermodynamics.**

thermoelectric conversion. The conversion of heat into electricity by the use of thermocouples.

thermoelectricity. The flow of electrical current when there is a difference in temperature between the junction of two different metal conductors joined together in a closed circuit.

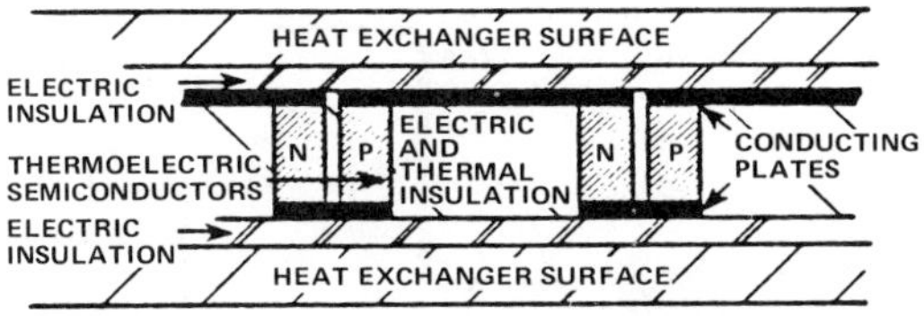

Courtesy SERI

Schematic diagram of thermoelectric generator.

thermoelectric metals. Metals or alloys used in thermocouples for measuring high temperatures.

Thermopane®. A glazing product by Libbey-Owens Ford Co. which is two sheets of glass with air space between the glass sheets. The edges are nearly hermetically sealed.

thermophilic. Organisms or processes which thrive at high temperature.

thermophotovoltaic cell. A device that concentrates sunlight on a small heat absorber made of metal or other suitable material, heating it to a high temperature. The secondary thermal radiation reemitted by the absorber is used as the energy source for a photovoltaic cell. The cell is chosen for maximum efficiency at the wavelength of the secondary radiation.

thermopile. A large number of thermocouples connected in series.

thermoplastic. A quality which allows a material to repeatedly soften when heated and harden when cooled. Typical of the thermoplastics family are the styrene polymers and copolymers, acrylics, cellulosics, polyethylenes, vinyls, nylons, and the various fluorocarbon materials.

Thermoply®. An insulating product of the Simplex Products Group that is a plate of very dense cardboard faced on one side with aluminum foil.

Thermos blanket®. An insulating product of King-Seeley Thermos Co. that is a vacuum-deposited aluminum film between polyethylene films.

Thermo-shade®. A glazing product from Solar Sunstill Inc. which is a near-transparent sheet, the transmittance of which varies closely with temperature.

thermosiphon. The principle that makes water circulate automatically between a solar collector and a storage tank above it, gradually increasing its temperature. A solar heating system that uses natural convection to transport heat from the collector to storage, by appropriately locating the storage in relation to the collector.

thermosiphoning. A method of circulating a fluid in which the warmer, less dense portion rises above the cooler. This method can be used in place of pumps to transfer solar-heated water or air.

thermosiphon system. System with natural thermal circulation—normally referring to water—in which hot water rises to the top of the system, displacing the colder water to the lowest point of the system. A heat exchanger in which the circulation of the working fluid is achieved by convection.

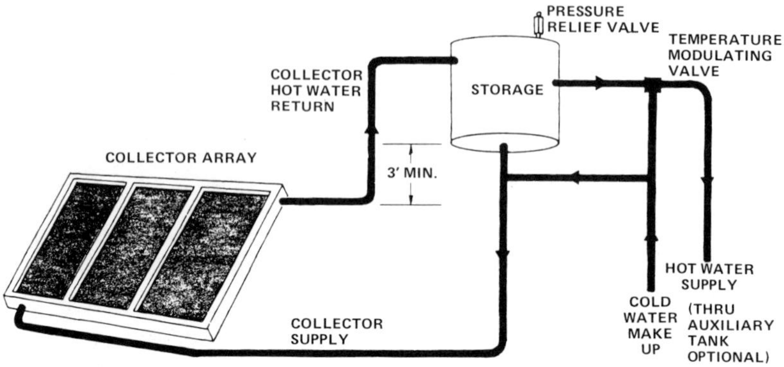

Courtesy National Copper Association

Pressurized thermosiphon domestic water heating system.

thermostat. The thermostat is the brain of a heating system. It must swiftly detect minute changes in room temperature and turn the heat source on and off accordingly. Moreover, it has to anticipate "overrun" of the heating system—the period when the blower of a warm-air system continues to operate after the burner has shut down.

thermostatic trap. The discharge from this type of trap is intermittent. Thermostatic traps are used to drain condensate from radiators, convectors, steam heating coils, unit heaters, and other similar equipment.

thin film. A layer of semiconductor material, for example polycrystalline silicon or gallium arsenide, typically a few hundredths of an inch or less in thickness, useful in making photovoltaic cells. Use of this material bypasses the costly steps of growing single crystal ingots and sawing them into wafers. Depending on the material, thin films may be produced in different ways, such as withdrawing a ribbon from a molten bath, slowly cooling a molten sheet on a substrate, or by spray coating.

thin stillage. The water-soluble fraction of a fermented mash plus the mashing water.

Thinsulate®. An insulating product of the 3M Company which is a bat of extrafine fibers of polyolefin.

Thomason solar house. A series of solar heated homes constructed by Harry H. Thomason in the 1950s and early 1960s in the Washington, DC area. Solar-generated heat is stored in a combination water-rock storage bin. The Thomason solar collector converts approximately 55 percent of the available sunlight into heat. (See diagram p. 317.)

three-phase current. Alternating current in which three separate pulses are present, identical in frequency and voltage, but separated 120 degrees in phase.

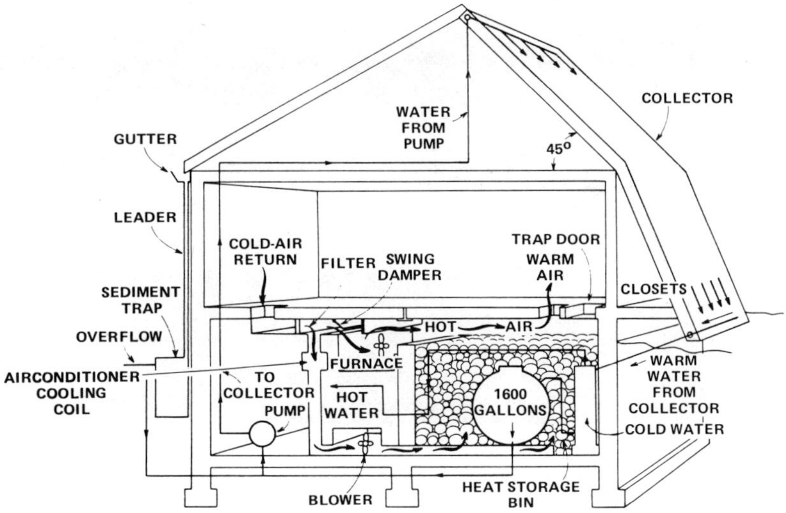

Courtesy A.J. Hand, Home Energy How-To, *Popular Science Book/Harper and Row*
Thomason's Solaris system is one of the simplest solar heating systems.

threshold dose. The minimum dose of a given substance necessary to produce a measurable physiological or psychological effect.

threshold velocity. The wind speed at which sand first begins to move or the minimum speed necessary to maintain sand motion once it has been set into motion.

throw. Distance from air supply opening measured in the direction of air flow, from the opening to the point where the air velocity is 50 ft/min.

throwaway option. Residential photovoltaic system designed to cover a share of total single-family residence load and peak power needs, but wasting all on-site photovoltaic potential not required for direct, simultaneous loads.

thrust. Force in the streamwise direction.

tidal power. Mechanical power, which may be converted to electrical power, generated by the rise and fall of ocean tides. The possibilities of using tidal power have been studied for many generations. The use of one or more tidal basins, separated from the sea by dams or barrages, and of hydraulic turbines through which water passes on its way between the basins and the sea. The world's largest tidal power plant is located on the estuary of the River Rance, in Brittany, France, completed in the 1960s. The disadvantages of tidal power generation are the very high capital cost of the dams or barrages, and possible ecological disturbance, but the major limitation imposed by the tidal cycle can be overcome by generating energy from both the filling and emptying of the basins to provide baseload power, and by using baseload power at times of low demand to pump water to higher reservoirs, from which it can be released to meet peak-load demands. (See photo p. 318.)

tide. Rise and fall of the surface of the sea due to the gravitational pull of the moon, generally taking place twice daily. In the open sea, this rise may not exceed 2

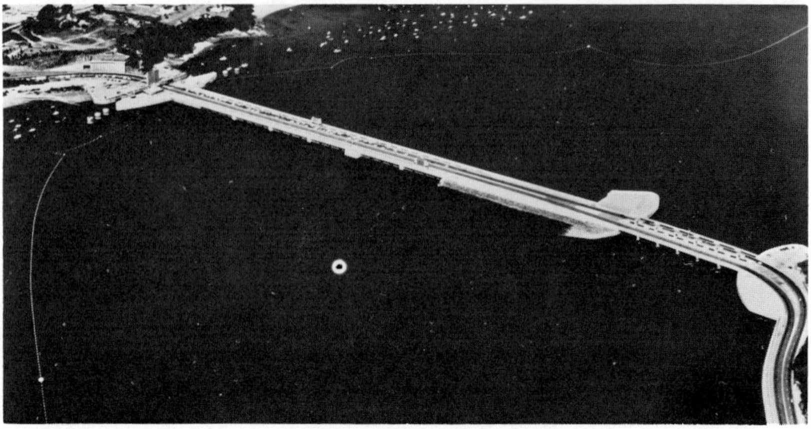

Courtesy DOE

Photograph of Rance River tidal power station.

feet (0.6 meters), whereas in the shallow seas bordering continents it may be more than 20 feet (6.1 meters), and in narrow tidal estuaries from 40 to 50 feet (12.2 to 15.2 meters). Since the moon travels in its own orbit in the same direction of the earth, a period of about 24 hours, 25 minutes will elapse between successive occasions when the moon is vertically above a given meridian. The interval between successive high tides will therefore be about 12½ hours.

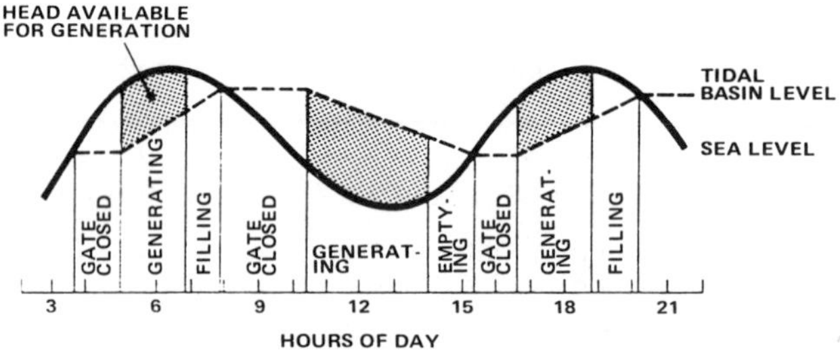

Courtesy NOAA

Illustration of the cycle of the tide.

tilt angle. The angle a solar collector makes with the horizontal. Besides facing south, collectors have a high angle for house heating, medium for hot water heating, and low for swimming-pool heating.

timber stand improvement. All intermediate cutting and thinning to improve the composition, structure, condition, and growth of a timber stand. See also **silviculture.**

time-lag heating. A process of heating a building's interior by using the heat loss properties of massive materials to delay the movement of solar heat. The period of time between the absorption of solar radiation by a material and its release into a space. Time lag is an important consideration in sizing a thermal storage wall or trombe wall.

Time of Day bill. Time-differentiated annual electric bill based on the time of the day. Utility peak-period pricing on a seasonal and/or weekday basis designed to discourage peak-load growth.

tip speed. Linear speed of the point on a wind-energy conversion system rotor farthest from the rotor axis.

tip-speed ratio. The trim aerodynamic blades of a wind generator extract more power from the wind at their tips than they do near the hub of the blade. This is expressed mathematically as the tip-speed ratio, which is the difference between the rotational speed of the tip of the blade and the actual velocity of the wind.

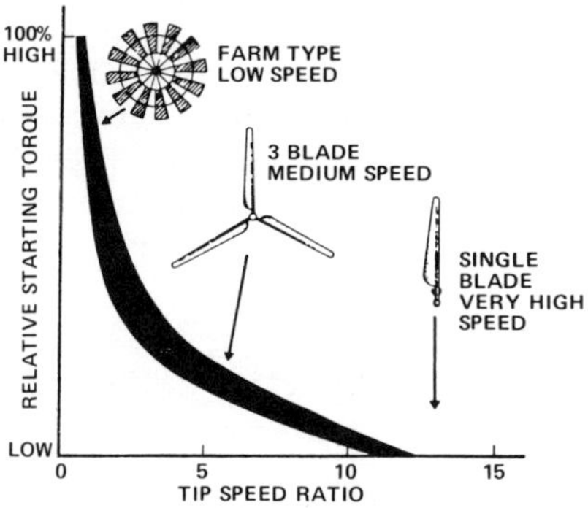

Courtesy DOE

Relative torque and tip speed for wind-energy conversion systems.

tip speed to wind speed. Ratio of the speed of the tip of a propeller blade to the speed of a windstream in which it is located: $R\Omega/V$.

tolerance. The relative capability of an organism to endure an unfavorable environmental factor.

ton. Any of various units of weight for large quantities: (a) a unit that equals 20 short hundredweight of 2000 pounds, used chiefly in the United States, Canada, and the Republic of South Africa; also called a short ton; (b) a unit that equals 20 long hundredweight or 2240 pounds, used chiefly in Britain; also called a long ton; (c) a unit of internal capacity for ships that equals 100 cubic feet (ft^3), also called a register ton; (d) a unit that approximately equals the volume of a long ton weight of seawater; used in reckoning the displacement of ships and it equals 35 cubic feet; also called a displacement ton; and (e) a unit of volume for cargo freight usually considered to be 40 cubic feet; also called a freight ton, a measurement ton.

tons of cooling. One standard commercial ton of refrigeration is defined as 288,000 Btu absorbed at a uniform rate during 24 hours.

tons of refrigeration. A common, commercial measure of refrigeration capacity, especially of air-conditioning apparatus cooling capacity, the equivalent of the heat required to melt 1 ton (2000 lb) of ice (heat of fusion is 144 Btu per lb) in 24 hours, hence, 288,000 Btu per day or 12,000 Btu per hour.

top assembly. The components of a water-pumping windmill, which include the wheel, gearbox, and tail.

topography. The configuration of a surface area including its relief, or relative elevations, and the position of its natural and man-made features.

topping. The distillation of crude oil to remove light fractions only.

topping cycle. A means to increase thermal efficiency of a steam-electric power plant by increasing temperatures and interposing a device, such as a gas turbine, between the heat source and the conventional steam-turbine generator to convert some of the additional heat energy into electricity. The higher temperature cycle in any energy conversion system where two (or more) separate cycles are used in cascade fashion (exhaust of one feeds input to another).

tornado-type wind turbine. An advanced wind-energy concept which would consist of a tall cylindrical tower with an open top, slotted side openings, and guide vanes to create a swirling tornadolike vortex flow. Outside air would be allowed to rush into the base of the tower and would be drawn upward through the vortex's low pressure core. Spinning rotor blades would drive a generator.

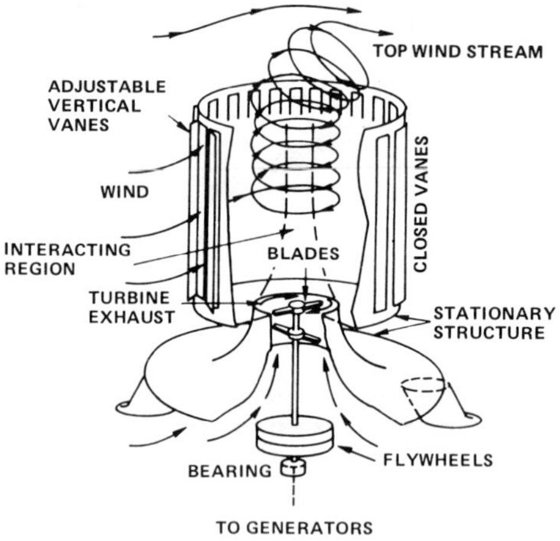

Courtesy DOE

Illustration of tornado-type wind turbine.

torque. A measure of force from windwheel, causing rotary motion of a power shaft.

total biomass volume. All above-ground parts of a plant.

total cooling effect. The difference between the total heat content of the airstream mixture entering an air conditioner per hour and the total heat of the mixture leaving per hour.

total cooling load. The sum of the sensible and latent heat components.

total energy. The total energy at any point in a moving fluid consists of the sum of the internal static, velocity, and potential energies at that point.

total energy house. A house designed to conserve energy.

total energy system. The productive use of thermal "waste" energy resulting from any electrical production process.

total gross energy consumption. Total energy inputs into the economy, including coal, petroleum, natural gas, and the electricity generated by hydroelectric, nuclear, and geothermal power plants. Gross consumption includes conversion losses by the electric power sector.

total heat. The sum of sensible heat and latent heat in a substance or fluid above a base point, usually 32°F or 0°C.

total incident insolation. Total incident insolation is the total energy received by a unit surface area for a specified time period.

total insolation. The sum of direct and diffuse insolation.

total (mean) sky cover. Fraction of the sky dome covered by clouds from sunrise to sunset.

total net energy consumption. Inputs into the final consuming sectors (household and commercial, industrial, and transportation), and consisting of direct fuels and electricity distributed from the electric power sector. Conversion losses in the electric sector constitute the difference between net and gross energy.

total pressure. The algebraic sum of the static pressure and velocity pressure at any particular point.

total solids. The residue remaining when the water is evaporated away from a waste and dried at 230°F (110°C).

tower. Subsystem of a wind-energy conversion system that holds the rotor, or other collection device, above the ground.

tower shadow. Aerodynamic wake created by airflow around a wind-energy conversion system tower.

toxicity. The quality or degree of being poisonous or harmful to plant or animal life.

toxic pollutants. A combination of pollutants including disease-carrying agents which, after discharge and upon exposure, ingestion, inhalation, or assimilation into any organism can cause death or disease, mutations, deformities, or malfunctions in such organisms or their offspring.

TRACE. A building energy analysis computer program developed by the Trane Company.

trace metals. Metals found in small quantities or traces usually due to their insolubility.

tracking collector. A collector that can rotate about one or two axes to face the sun; usually restricted to high-temperature concentrating collectors because of the complexity and cost of a tracking system. (See diagram p. 322.)

tracking system. Solar collectors which follow the sun. Also tracking mirrors follow the sun, keeping the sun's reflection focused on a stationary point.

trade winds. The wind system which blows from the subtropical highs toward the equatorial trough; a major component of the general circulation of the atmosphere.

trailing edge. On a wind-energy conversion system, the point on an airfoil farthest downstream when operating at zero angle of attack, or the area of a blade that is last to contact an element of airflow.

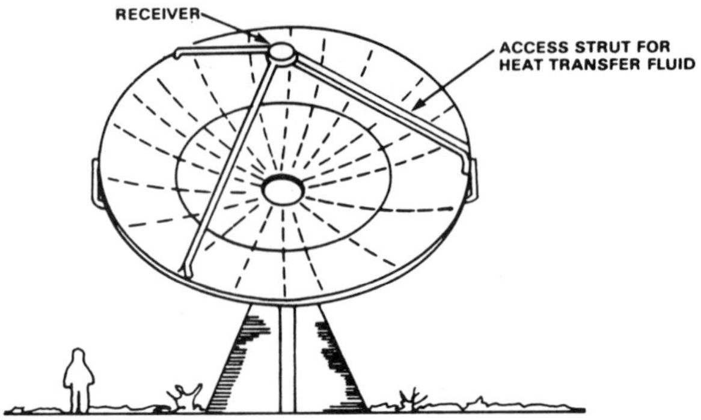

Courtesy DOE

Sketch of solar parabolic dish tracking collector.

transfer fluid. The transfer fluid is the medium such as air, water, or other fluid which passes through, or is in contact with the solar collector, and carries the thermal energy away from the collector. See **heat-transfer fluid.**

transfer medium. The substance that carries heat from the solar collector to storage, or from storage to the living areas.

transformer. A device which, through electromagnetic induction but without use of moving parts, transforms alternating or intermittent electric energy in one circuit into energy of similar type in another circuit, commonly with altered values of voltage and current.

translational motion. In a straight line.

translocation. Transfer of metabolites, nutritive material, etc., from one part of a plant to another.

translucent. The quality of transmitting light but causing sufficient diffusion to eliminate perception of distinct images.

transmission. The process by which solar radiation goes right through transparent materials such as glass, plastic, and air.

transmission level. The transmission level of the energy at any point in an energy-transmission system is the rate of flow of that energy, as expressed in terms of a specified reference rate of flow, and of the transmission loss by which the actual rate of flow must be reduced to equal the reference rate.

transmission line. A cable or wire through which electric power is moved from electric power generating plants to areas of use by consumers. It is usually operated at greater than 250,000 volts (V).

transmittance. The ratio of the radiant energy transmitted by the parallel-sided planar surface of a given material to the radiant energy incident on the upper surface of that material; it depends on the angle of incidence.

transmittance-absorptance product. The product of the transmittance of the transparent collector cover(s) and the absorptance of the collector plate.

transparent. A material, such as glass or plastic film, through which light can pass.

transparent frontal area. The area of the transparent frontal surface for flat-plate collectors.

transpiration. The process by which water vapor escapes from a living plant and enters the atmosphere.

transport. Heat flow by transporting or moving a substance which is hot. Often the substance moved is a gas or liquid (air or water); a type of thermal resistance.

trap. In plumbing, a device providing a liquid seal which will allow passage of sewage or waste matter but prevent the back passage of air.

tray. One of several types of horizontal perforated plates in a distillation column.

trickle collector. A solar collector in which water flows in channels covered by a single glazing from a pierced pipe header to a collecting trough. This collector concept was developed by H. E. Thomason.

trickle-down collectors. "Wet" solar collector where water runs down the surface of the absorber plate and collects in a gutter at the bottom for transfer to the heat-storage tank.

trickling filter. A device for the biological or secondary treatment of wastewater consisting of a bed of rocks or stones that support bacterial growth. Sewage is trickled over the bed enabling the bacteria to break down organic wastes.

trombe wall. A passive solar house using a glass-covered masonry wall with air holes at top and bottom of the wall. Heat is transferred to the house by conduction through the wall and thermosyphon through the air holes.

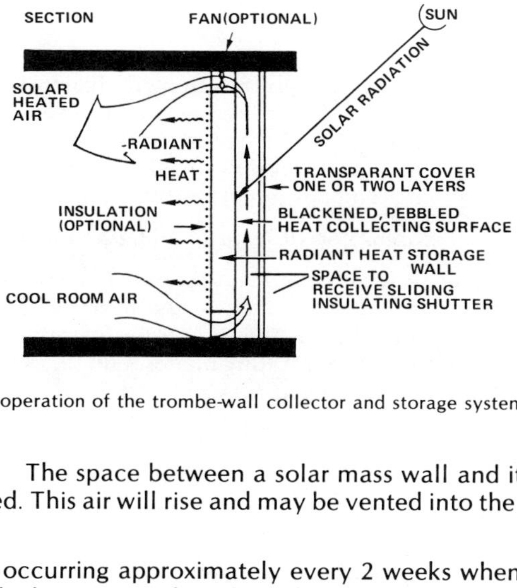

Courtesy SERI

The operation of the trombe-wall collector and storage system.

trombe wall cavity. The space between a solar mass wall and its exterior glazing in which air is heated. This air will rise and may be vented into the building's interior for distribution.

tropic tide. Tides occurring approximately every 2 weeks when the effect of the moon's maximum declination north and south of the equator is greatest.

troposphere. The portion of the atmosphere which extends out 7 to 10 miles (11.3 to 16.1 kilometers) from the earth's surface, and in which, generally, temperature decreases rapidly with altitude, clouds form, and convection is active.

true south. South with reference to the stars, not to a compass; opposite to the Polar Star, which lies to the true north of earth.

tubeaxial fan. An airfoil (propeller) or disc fan within a cylinder and including driving mechanism supports either for belt drive or direct connection.

tube-in-plate absorber. A metal absorber plate in which the heat-transfer fluid flows through passages formed in the plate itself.

tube-type collector. A collector in which the heat-transfer liquid flows through metal tubes that are fastened to the absorber plate by solder, clamps, or other means. See **collector.**

tubular centrifugal fans. Tubular centrifugal fans generally consist of a single width airfoil wheel arranged in a cylinder to discharge air radially against the inside of the cylinder. Air is then deflected parallel with the fan shaft to provide straight-through flow. Vanes are used to recover static pressure and straighten air flow.

Tuffaic® . A glazing product of Rohm & Haas consisting of a single sheet of polycarbonate.

Tuffak-twindow®. A glazing product of Rohm & Haas consisting of a double sheet of polycarbonate.

turbidimeter. A device used to measure the amount of suspended solids in a liquid.

turbidity. The condition of having the transparence or translucence disturbed, as when sediment in water is stirred up, or when dust, haze, clouds, etc., appear in the atmosphere because of wind or vertical currents. Also atmospheric haze resulting from aerosols and particulates.

turbine. A machine that converts the energy in a stream of fluid into mechanical energy by passing the stream through a system of fixed and/or moving fanlike blades, causing them to rotate. Turbines have wide uses in large-scale power generation (usually employing steam-driven turbines), small-scale power generation [e.g., Pelton wheel (q.v.)], jet aircraft propulsion (gas turbines deriving power from a stream of heated air), marine engines, etc. (See photograph p. 325.)

turbine generator. A combination of a turbine and an electric generator.

turboblower. A blower in which the rotating part is equipped with blades that rotate between stationary blades attached to the housing. The respective sets of blades are set at an angle such that, as the rotor turns, gases are pushed through the blades, and again by the next set of rotating blades, and finally discharged from the casing at the opposite end from which they entered.

turbulence. Rapid wind speed fluctuations; gusts are maximum values of wind turbulence; randomness in the winds.

turbulence intensity. Ratio of the first standard deviation of wind speed to the mean wind speed.

turbulent flow. This describes the alcohol/water flow, over packing or plates, as it descends the column, wherein the liquid velocity constantly changes in magnitude and direction. This condition must be achieved to get the greatest efficiency from a column. The packing or plates break the laminar flow into a turbulence, which allows all the alcohol molecules to have a chance to come to the surface of the liquid stream and escape as vapor.

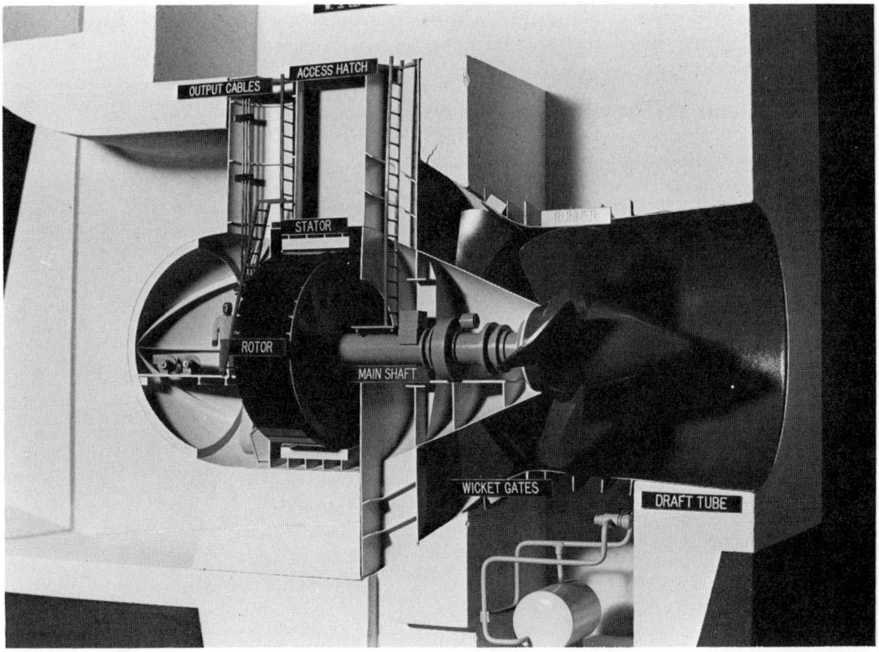

Courtesy DOE

Model of power generation 8-MW bulb turbine

turnaround efficiency. The resulting efficiency when energy is converted from one form or state to another form or state, and then reconverted to the original form or state.

turnbuckle. A device consisting of a link with screw threads at each end. This link is turned to bring the two objects connected by the turnbuckle closer together.

turning vane. Curved strips of short radii placed in a sharp bend or elbow in a rectangular duct to direct the air around the bend in streamlined flow.

turn out of the wind. To yaw a horizontal axis wind-energy conversion system so that its rotor axis is perpendicular to the wind direction.

Twindow. A glazing product of PPG Industries Inc. consisting of two sheets of glass with air space between. The edges are near-hermetically sealed.

twist. Variation in blade pitch of a wind-energy conversion system with respect to changes in radial station.

two-cycle engine. An engine in which only two strokes of the pistons, corresponding to one revolution of the crankshaft, are required to complete the cycle. In this cycle an explosion occurs on each downward stroke of the piston; the fresh charge is admitted and the exhaust gases expelled at or near the end of the stroke.

two-outlet heater. A water heater having one outlet going typically to the piping for a domestic hot-water system and the other outlet, through a pipe to a large storage tank.

two-pipe steam system. A heating system in which one pipe is used for supplying steam to a radiator and another pipe is used to return cooled water or condensate to boiler.

two-tank system. Solar water-heating system using a separate tank holding solar-heated water as a preheater for a smaller, conventionally fueled water heater. Two tanks must be used when the backup fuel is gas or oil.

Tyvek®. An insulating product of Du Pont Co. which is a sheet of very tough polyolefin fabric on one side of which aluminum has been vacuum deposited (spun-bonded).

U

ultimate analysis. The determination of the elements contained in a compound, i.e., carbon, hydrogen, oxygen, nitrogen, sulfur, and ash.

ultimate CO_2. The maximum theoretical percentage of flue gas CO_2 that it is possible to produce from the complete combustion of a fuel with the chemically correct fuel-air ratio.

ultrahigh voltages. Voltages greater than 765,000 volts (V).

ultraviolet. Electromagnetic radiation with wavelengths shorter than visible light and longer than x-rays.

ultraviolet light. Black light; invisible light rays from the portion of the spectrum that lies beyond the violet on the shorter wavelength side. Used to induce chemical reactions and produce fluorescence.

ultraviolet radiation. Electromagnetic radiation whose wavelength is shorter than that for visible light but longer than that for x-rays; radiant energy of wavelengths from 0.1- to 0.4-micron.

underground storage. The utilization of subsurface facilities for storing energy which has been transferred from its original location for the primary purposes of conservation, fuller utilization of distribution facilities, and more effective and economic delivery to markets.

understory. Trees and other vegetation growing under the main canopy of a forest stand.

unglazed. A collector with no glazing; convection heat losses are high without glazing since the absorbing surface is exposed to the wind.

unglazed collector. A collector without a cover plate.

union. The usual trade term for a device used to connect pipes. It commonly consists of three pieces: the thread end fitted with exterior and interior threads; the bottom end fitted with interior threads and a small exterior shoulder; and the ring which has an inside flange at one end while the other end has an inside thread like that on the exterior of the thread end. A gasket is placed between the thread and bottom ends, which are drawn together by the ring. Unions are very extensively used because they permit connections with little disturbance of the pipe positions.

union bonnet. Type of bonnet which is so constructed that the whole bonnet assembly, including the handwheel, stem, and disc assembly, may be quickly removed by unscrewing the bonnet union ring from the valve body.

union ring. The large nutlike ring which secures the union thread piece and the union shoulder piece together. It slips over and against the shoulder piece and screws onto the union thread piece.

union shoulder piece. That part of the union which is fastened to the pipe and retains the union ring.

union threaded piece. That part of the union which is fastened to the pipe and has external threads over which the union ring is screwed to effect a coupling.

Unipanel®. A trade name of the Solarex Corporation for a type of solar panel.

unit heater. A forced convection heating device of two types: an assembly of encased heating surface with fan and motor (or turbine) for connection to a source of steam or hot water; or an assembly same as the above plus a fuel burner for connection to a source of oil or gas (or supplied with coal).

unit ventilator. A unit heater connected to a source of ventilation air and frequently provided with an air filter.

universal curves. A family of solar-system performance curves that relate the fraction of demand met by solar energy to the solar building load ratio. These curves show correlations of solar-system performance produced by multiple, hourly computer simulations.

upright bucket trap. The discharge of condensate from this trap is intermittent. A differential pressure of at least 1 psi between the inlet and the outlet of the trap is normally required to lift the condensate from the bucket to the discharge connection. Upright bucket traps are commonly used to drain condensate and air from the blast coils, steam mains, unit heaters, and other equipment. This trap is well suited for systems that have pulsating pressures.

upstream. From a reference point, any point located nearer the origin of flow, that is, before the reference point is reached.

upwind. Horizontal axis wind-energy conservation system configuration where the wind intercepts the rotor plane prior to the tower. (See diagram p. 329.)

urban runoff. Storm water from city streets and gutters that usually contains a great deal of litter and organic and bacterial wastes.

useful energy gain. The energy collected by a solar collector which is not lost to the surroundings and can ultimately be used for space or water heating.

useful heat. The heat stored above room temperature in a house heating application. Heat stored below room temperature is not useful because it cannot warm the room (except as a preheat for a furnace); analogous to useful volume.

useful solar heat. Heat delivered by a solar system.

useful volume. The volume stored in a tank above the outlet tap; analogous to useful heat.

use pattern. The use pattern of the solar-energy system refers to the time when the system is needed. The daily use pattern for residences is both day and night, while offices and schools may be used only during daytime hours. The yearly use pattern,

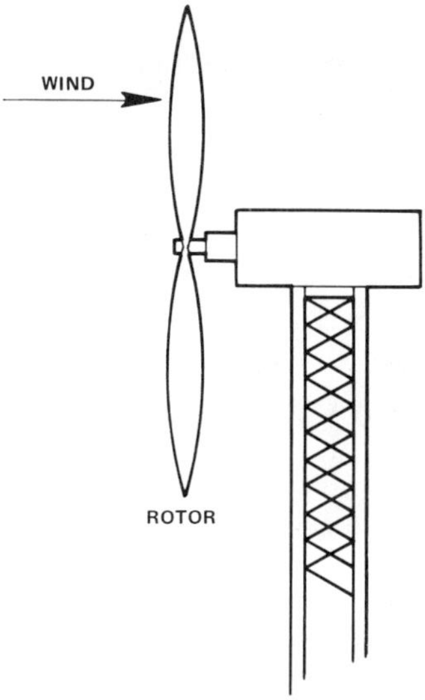

Courtesy DOE

Wind passes through wind rotor before hitting tower in upwind system.

is related to the function of the solar-energy system; for example, space heating is used only during the cold season, while domestic water heating is used all year. The use pattern largely determines the skyspace (q.v.) requirements of the solar-energy system.

utility electric. Device for measuring the amount of electricity used; the unit of measurement is the kilowatt hour. One kilowatt hour (kWh) is the equivalent of 1000 watts (W) of electricity used for 1 hour. Most electric meters have a set of dials read in multiples of 10, clockwise or counterclockwise. The dial on the far right indicates tens of kilowatt-hours; the next one, hundreds of kilowatt-hours; the next, thousands of kilowatt-hours. When the pointer is between two numbers, the lower of the two numbers is read.

utilization factor. The ratio of the maximum demand of a system, or part of a system, to the rated capacity of the system, or part of the system, under consideration.

U-value (coefficient of heat transfer). The number of Btu's that flow through 1 square foot (ft^2) of roof, wall or floor, in 1 hour, when there is a 1 °F state condition. The U-value is the reciprocal of the resistance or R-factor. The lower the U-value, the greater the insulating value of the wall or section.

V

vacuum. (1) A pressure less than atmospheric pressure, measured either from the base of zero pressure or from the base of atmospheric pressure. (2) A space devoid of matter.

vacuum distillation. The separation of two or more liquids under reduced vapor pressure; reduces the boiling points of the liquids being separated.

vacuum-evaporated film. A film generally formed on a sheet or plate by electrical evaporation of a metal or alloy in an evacuated chamber.

vacuum jacket. A vacuum around an evacuated tube collector that prevents convection heat loss.

vacuum system. A two-pipe steam-heating system equipped with vacuum pumps to permit maintenance of pressures below atmospheric within the radiators.

valance. A property of atoms determining the number of atoms with which they can combine in chemical reations.

valence. A property of atoms determining the number of atoms with which they can combine in chemical reactions.

valence state. Energy content of an electron in orbit about an atomic nucleus.

vaneaxial fan. An airfoil (propeller) or disc fan within a cylinder and equipped with air guide vane either before or after the wheel and including driving mechanism supports either for belt drive or direct connection.

vane ratio. In grilles, the ratio of vane depth to shortest opening width between two adjacent vane bars.

vapor. The gaseous phase of a material; for example, steam is the gaseous phase of water.

vapor barrier. A component of construction which is impervious to the flow of moisture and air and is used to prevent condensation in walls and other locations of insulation. A vapor barrier will prevent water vapor from condensing and collecting in home insulation. Vapor barriers can be Mylar or aluminum foil. (See diagram p. 331.)

vapor density. The relative density of a gas or vapor as compared with some specific standard, such as hydrogen.

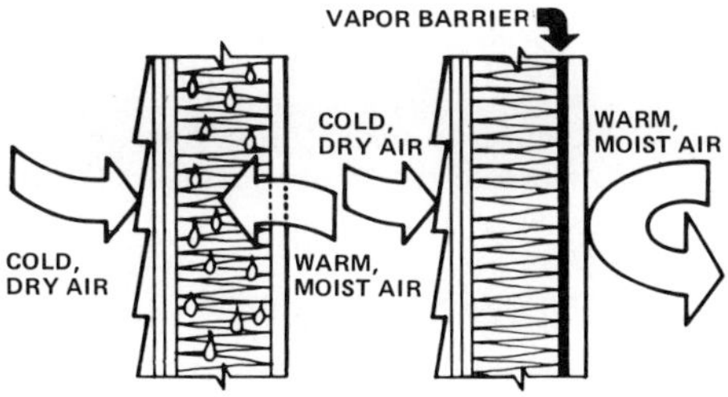

Courtesy USDA

Illustration of installation of vapor barrier.

vaporization. The act or process of changing a substance from the liquid to the gaseous state. One of three basic contributing factors to air pollution; the others are attrition and combustion.

vaporize. To change from a liquid or a solid to a vapor, as in heating water to steam.

vapor lock. The formation of vapor in a feed line by vapors generated from the fuel, resulting in an interruption of the flow of liquid fuel.

vapor plume. The stack effluent consisting of flue gas made visible by condensed water droplets or mist.

vapor pressure. The pressure at which a liquid and its vapors are in equilibrium at a definite temperature. If the vapor pressure reaches the prevailing atmospheric pressure, the liquid boils.

vapor system. A steam-heating system operating at pressures very near that of the atmosphere.

variance. Sanction granted by a governing body for delay or exception in the application of a given law, ordinance or regulation.

veering wind. A clockwise change in wind direction.

veiling reflection. Reflection of light from a tank, or work surface, into the viewer's eyes.

velocity. Rate of motion in a given direction.

velocity pressure. The pressure exerted by a moving fluid in the direction of its motion. It is the difference between the total pressure and the static pressure.

venetian blind. A set of horizontal opaque strips (or slate or vanes) that can be rotated at least 90° about horizontal axes so as to (a) form a contiguous, nearly coplanar set that greatly restricts passage of light and air or conversely (b) form an open set that freely permits such passage.

ventilation. The atmospheric air that is purposely allowed to enter an interior space to cool or freshen it; the principal air-conditioning process concerned with control of air circulation.

vent stack. A vertical vent pipe to provide air circulation to and from the drainage system.

vent system. Piping to provide a flow of air to or from a drainage system to protect trap seals from siphonage or back pressure.

venturi. A contraction in a pipeline or duct to accelerate the fluid and lower its static pressure. Used for metering and other purposes.

venturi meter. A form of the venturi tube arranged to measure the flow of a liquid in pipes. Small tubes are attached to the venturi tube at the throat and at the point where the liquid enters the converging entrance. The difference in pressure heads is shown on some form of manometer, and, from this difference and a knowledge of the diameters of the tubes, the quantity of flow is determined.

vertical-axis rotor-type wind turbines. A wind-energy turbine which is designed to accept wind from any direction and thus need not be turned into the wind.

vertical-axis wind turbine. A wind turbine for which the rotor rotates about a vertical axis. The concept of hub height does not exist for a vertical-axis machine so references to measurements at hub height should be substituted by measurements at the midpoint of the rotor.

Courtesy DOE

Vertical axis Darrieus wind-energy conversion system under test.

vermiculite. A loose-fill, mineral-insulation material that can be poured into small areas for complete coverage. A trade name for fluffy granules of expanded mica. This substance is noncombustible.

vertical multiple junction cell. A compound cell made of different semiconductor materials in layers one above the other like a club sandwich. Sunlight entering the top passes converts a separate portion of the spectrum into electricity, thus achieving greater total conversion efficiency of the incident light. Also called multiple junction cell.

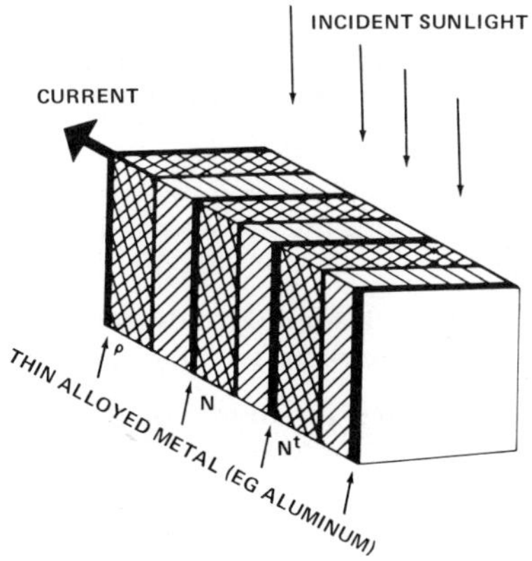

Courtesy SERI

Vertical multijunction photovoltaic cell.

violet cell. A new form of silicon solar cell which converts more sunlight to electricity from the violet/ultraviolet range of the light spectrum than conventional solar cells. The violet cell was developed by COMSAT's Dr. Lindmoyer.

viscosimeter. An instrument used to measure the viscosity of liquids.

viscosity. Any resistance to deformation that involves dissipation of energy by internal friction; in general, resistance to flow; that property of semi-fluids, fluids, and gases by virtue of which they resist an instantaneous change of shape or arrangement of molecules.

visible radiation. Radiant energy of wavelengths from 0.4 to 0.76 micron; the visible spectrum.

visible spectrum. That portion of the electromagnetic spectrum, the waves of which normally produce upon the human eye color sensations of red, orange, yellow, green, blue, violet or their intermediate hues, or of white light if the rays are combined.

volatile. Capable of being readily evaporated at a relatively low temperature.

volatile solids. The portion of the total solids which is driven off as a volatile gases when the sample is heated to 112° (600°C) for 1 hour. Often synonymous with organic matter content.

volatility. All substances can be made to vaporize under heat. Some materials will vaporize more easily than others, or with less heat. This tendency to vaporize can be described as the volatility of the substance. When one substance vaporizes sooner than another, it is said to be more volatile. Ethanol, for example, is more volatile than water. Therefore, when a mixture of these two is heated, there will be more alcohol than water in the vapor.

volt (V). The practical meter-kilogram-second (SI) unit of electrical potential difference and electromotive force that equals the difference of potential between two points in a conducting wire carrying a constant current of 1 ampere (A) when the power dissipated between these two points equals 1 watt (W). It equals the potential difference across a resistance of 1 ohm when 1 ampere of current is flowing through it.

voltage–current characteristic. A plot of the current output of a solar cell versus its terminal voltage. Usually several curves are given for different amounts of sunlight reaching the cell.

voltage regulation. The change in output voltage which occurs when the load is reduced from rated value to zero with the values of all other quantities remaining unchanged.

voltaic cell. A cell consisting of two electrodes and one or more electrolytes which, when connected in a closed circuit, will give out electrical energy.

volt-ampere. A unit of electrical measurement equal to the product of a volt and an ampere.

volume. The amount of space that a liquid occupies; analogous to heat.

volume flow. The movement of volume from one place to another; analoguous to heat flow.

volumetric efficiency. Volume of gas handled by a compressor divided by the displacement.

vortex. A flow of a fluid in which the steamlines form concentric circles. There are two kinds of vortex, forced and free, depending on whether torque is applied externally, e.g., a forced vortex can be made in a liquid contained in a vessel and stirred with a paddle; a free vortex occurs when the liquid is allowed to leave the cylinder through a small hole in the bottom, in which case the force is provided by gravity acting on the fluid. Vortices occur when a solid body moves through a fluid unless the body is designed (e.g., in an aerofoil) to avoid creating them.

W

wafer. A thin sheet of semiconductor material made by mechanically sawing it from a single crystal ingot.

wall orientation. The primary direction of an exterior vertical surface of a building facing within 45° of one of the directions (i.e., north, south, east, or west).

warm-air furnace. A vessel of steel or cast iron in which fuel is burned and over the surfaces of which air is passed and thus heated. In the heating trade designated as a furnace and not to be confused with a boiler.

warm-air heating system. A system whereby air is heated directly in a furnace and circulated to rooms to be heated by gravity or by a fan. The latter is termed forced warm-air heating.

warming-up allowance. An addition to the capacity of heating system (as calculated for heat loss) to provide for quick pick-up in the morning.

warm spring. A thermal spring.

waste heat. Heat which is at temperatures very close to the ambient and hence is not valuable for production of power and is discharged to the environment.

waste-heat boiler. One which uses heat of exit gases from furnaces to produce steam or to heat water.

waste pipe. A line carrying the discharge from any fixture except water-closets and the like, and which conveys the waste to the building drain or waste stack. If it is not directly connected to the building drain or soil pipe, it is an indirect waste pipe.

water-air heat exchanger. A device in which air is either heated or cooled by flowing water.

water budget. A budget of the incoming and outgoing water from a region, including rainfall, evaporation, runoff, seepage, and with perhaps special attention to ablation (evaporation) of snow, evapotranspiration from vegetation, dew, or other special aspects relevant for special interests (e.g., agriculture).

water cycle. See **hydrologic cycle.**

water distributing pipe. One carrying water from the water service pipe to the fixtures.

water flow velocity. The velocity of flow in any hot-water heating system will be that at which the resistance of the system is equal to the head produced by the pump. In other words, when the pump is started, the velocity will build up until the friction of flow in the system uses up the head produced by the pump at whatever speed the pump is run by the pump motor. The pump head produced depends on the pump design characteristics as well as speed. Performance curves are available from the pump manufacturer; they should be considered when selecting the pump.

Quiet operation is expected from hot-water heating systems. For residential and commercial buildings, it is common practice to limit velocities to 4 feet per second (fps), in small piping. In piping above 2-inch (50.8-mm) diameter, this velocity may be increased somewhat. In industrial buildings, friction loss in the piping is the limiting factor rather than the noise due to flow. Noises due to expansion can be prevented by designing the piping for necessary movement.

water heater. An appliance for supplying hot water for domestic or commercial purposes other than space heating.

water-heating. Using water as the heat transfer medium in a solar heating system; a hydronic system.

water hyacinth. A freshwater plant of tropical regions that has become a noxious weed in many parts of the world. Highly prolific, and reproducing mainly vegetatively, it can double its numbers in 8 to 10 days in water at a temperature of 50°F (10°C) or more, provided nutrients are present. It was introduced to North America in 1884 by visitors to the New Orleans Cotton Exposition, who brought specimens from Venezuela. NASA has experimented with the use of water hyacinth to provide fertilizer and feeding stuffs for livestock. May be candidate for energy biomass plant.

water line. The height of the normal water level in a boiler measured from the floor.

water main. The pipe that brings cold water into a house from a residential water system.

water pollution. The addition of sewage, industrial wastes, or other harmful or objectionable material to water in concentrations or in quantities sufficient to result in measurable degradation of water quality.

waterpower. The power of water derived from its gravity or its momentum as applied or applicable to the driving of machinery.

water quality criteria. The levels of pollutants that affect the suitability of water for a given use. Generally, water use classification includes: public water supply, recreation, propagation of fish and other aquatic life, agricultural use, and industrial use.

water quality standard. A plan for water quality management containing four major elements: the use (recreation, drinking water, fish and wildlife propagation, industrial, or agricultural) to be made of the water; criteria to protect those uses: implementation plans (for needed industrial-municipal waste-treatment improvements) and enforcement plans; and an antidegradation statement to protect existing high quality waters.

water rate. The weight of dry steam consumed by a steam engine for each horsepower per hour. The result is stated in either indicated horsepower (q.v.) or brake horsepower (q.v.).

water service pipe. Pipe from the water main to the building served.

watershed. The area drained by a given stream.

water-source heat pump. Heat pump that transfers heat from a water source to an indoor air-circulation system; frequently installed in commercial heating systems.

water storage wall. See **trombe wall.**

water supply system. The system for the collection, treatment, storage, and distribution of potable water from the sources of supply to the consumer.

water table. The upper level of groundwater.

water-to-air heat pumps. A heat pump which uses well or lake water or any other available water source as a heat sink or source, with the heat being delivered to the indoor area through the use of a direct refrigerant, is commonly referred to as a water-to-air type. The market for this type of unit is similar to the air-to-air type in the 2- to 50-ton range.

The water-to-air type heat pump has some of the same basic problems as a water-to-water type in regard to availability, quality, and disposal means for the water.

water-to-water heat pumps. The heat pump that uses water from wells, lakes, or other heat sources, with the heat being delivered to the indoor area through the use of a secondary refrigerant—in this case, water—is generally referred to as a water-to-water type.

Water-to-water heat pumps in this country have been used since the early 1930s, and, for the most part, have operated quite successfully. The growth of this type of system has been relatively slow and has been surpassed by the air-to-water type, which had its infancy, except on an experimental basis only, in the years following World War II.

water turbine. A prime mover coupled to an alternator and using a purely rotary motion to generate an alternating current. The main types of water turbines are: the Pelton wheel (q.v.) for high heads, the Francis turbine (q.v.) for low to medium heads, and the Kaplan turbine for a wide range of heads.

water vapor. A term used in psychrometry to denote the water in the air which actually is superheated steam.

water wall. An interior wall of water-filled containers constituting a one-step heating system which combines collection and storage.

water wheel. A wheel designed to be turned by the impact of flowing water, used to drive machinery, raise water, etc.

watt. The absolute meter-kilogram-second (SI) unit of power that equals 1 absolute joule per second. It is analogous to horsepower (hp) or foot-pounds per minute (ft-lb/min) of mechanical power. One horsepower is equivalent to approximately 746 watts (W).

watt-hour. The total amount of energy used in 1 hour by a device that uses 1 watt of power for continuous operation. Electrical energy is commonly sold by the kilowatt hour (1000 watt-hours).

wattmeter. An instrument for measuring electric power in watts, the unit of electrical energy, volts time amperes.

watt per square meter (W/m²). A measure of the energy in the wind passing through a square meter of area.

wavelength. The distance between similar points on successive waves. That of visible light varies between 300 angstrom units (violet) and 7600 angstrom units (red).

wave power. Using the sea's energy in advancing waves to generate power to drive a turbine.

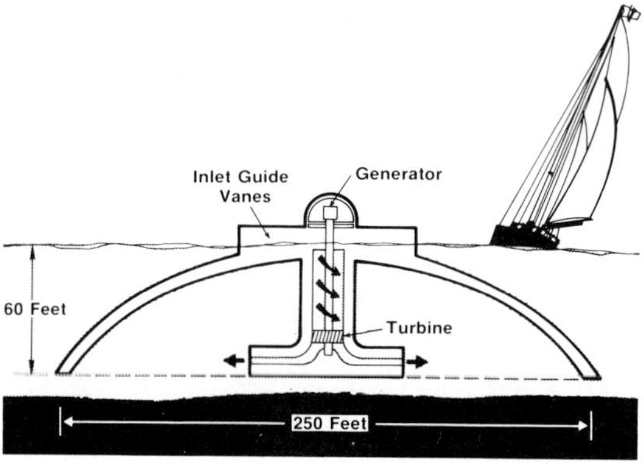

Courtesy Lockheed Corporation

Diagram of Lockheed's Dam-atoll wave turbine.

Weatherguard®. A glazing product from Technology Development Corp. which is a transparent sheet of vinyl plastic.

weathering. The group of processes, such as chemical action of air and rainwater, and of plants and bacteria, and the mechanical action of changes in temperature, whereby rocks, on exposure to the weather, change in character, decay, and finally crumble into soil.

Weatherizer®. A glazing product from Plasholite Inc. which is a transparent Mylar sheet said to be "semirigid."

weatherstripping. The use of thin spring metal, rolled vinyl, or foam-rubber tape to prevent infiltration of air and moisture around windows and doors.

wedge. The wedge-shaped device which fits into the seating surfaces of a gate valve and which is drawn out of contact with the seating surfaces to permit flow, or is pushed down into contact with the seating surfaces to close off flow through the valve.

welding fittings. Wrought or forged-steel prefabricated elbows, tees, reducers, saddles, and the like, beveled for welding to pipe.

welding neck flange. A flange with a relatively long neck beveled for butt welding to pipe.

well casing. The well driller usually installs a plastic or galvanized pipe in the well hole to support the well wall.

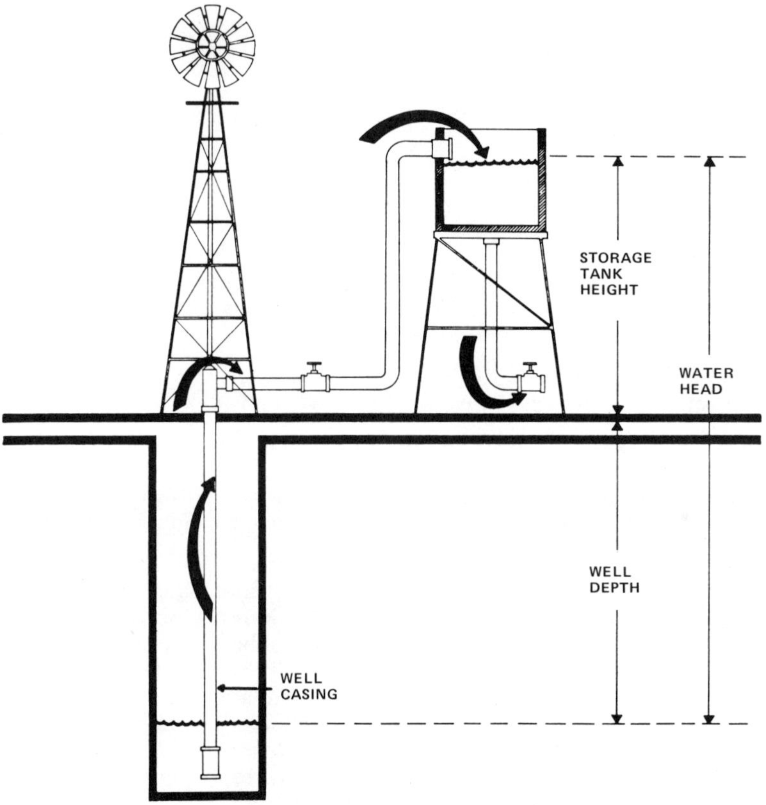

Water pumping wind system diagram showing well casing.

well seal. Mechanical plate inserted into the well casing. The drop pipe is inserted through well seal.

westerlies. Any winds with components from the west; also the dominant west-to-east motion of the atmosphere, centered over the middle latitudes of both hemispheres.

wet-bulb temperature. The temperature of the air is measured by a wet-bulb thermometer, and which is lower than the dry-bulb temperature (for all cases except when the air is saturated) in inverse proportion to the humidity.

wet heat. A slang expression for heating systems employing water or steam as the heating medium, as opposed to "dry" heat or warm-air systems.

wetlands. Swamps or marshes, especially as areas preserved for wildlife.

wet milling. A process similar to dry milling except that the various components of grain are separated in water.

wetting agent. A substance that lowers the surface tension of water and thus enables it to mix more readily; a chemical promoting adhesion of a liquid (usually water) to a solid surface.

wet vent. A soil or waste pipe that also serves as a vent.

wheel. The aerodynamic blade of a water-pumping multivane wind-energy conversion system. The diameter of the wheel is a major factor in the capacity of the wind machine.

wheeling. Transmission of electricity by a utility over its lines for another utility; also includes the receipt from and delivery to another system of like amounts (but not necessarily the same) energy.

whey. The watery part of milk separated from the curd in the process of making cheese; it is produced commercially in large quantities and can be used as a fertilizer, animal feed, or feedstock in the production of ethanol.

whole stillage. The undried "bottoms" from the beer well comprised of nonfermentable solids, distiller's solubles, and the mashing water.

wind chill. A measure of the quantity of heat that the atmosphere is capable of absorbing within an hour from an exposed surface. The term is based on a scale that correlates temperature and wind force. The term "wind chill" also used to describe the relative discomfort resulting from combinations of wind and temperature.

wind classification. A system designed to emphasize the forces mainly responsible for the characteristics of wind. Geostrophic winds have a balance between pressure gradient and the Coriolis force and blow along the isobars. Gradient winds are a modification in which the curvature of the isobars and of the flow is important, as in a cyclone. Isallobaric winds are caused by rapidly changing pressure patterns. Katabatic winds (cold downslope) and anabatic (warm upslope) winds are shallow and local, as are antitriptic (friction dominated) winds which are exemplified by katabatic ravine winds and cold outflows from storms. Land and sea breezes are antitriptic with time. The thermal wind is a geostrophic component. The ageostrophic component is due to acceleration or friction.

wind concentrator. A device or structure that concentrates a windstream.

wind-driven current. The ocean current which develops as a result of the changes in density distribution caused by the wind drift.

wind energy. A form of solar energy, since winds are caused by variations in the amount of heat the sun sends to different parts of the earth. Electricity is produced when a windmill catches the wind and revolves, rotating a turbine which powers an electric generator. A major drawback of wind energy is the large amount of wind needed to generate the 8000 watts (W) of electricity used by the average home. As with sunlight-produced energy, some backup resource is needed. (See table p. 341.)

wind-energy conversion system. A wide variety of machines have been designed to convert wind energy into useful power. The large-scale wind-energy conversion systems (WECS) currently being developed utilize a horizontal axis design. This design will require siting in high average wind speed areas. A machine that converts kinetic energy in the wind into a usable energy form. (See photograph p. 342.)

wind-energy conversion systems site. Location that is being considered for installation of a wind-energy conversion system or has a wind-energy conversion system installed.

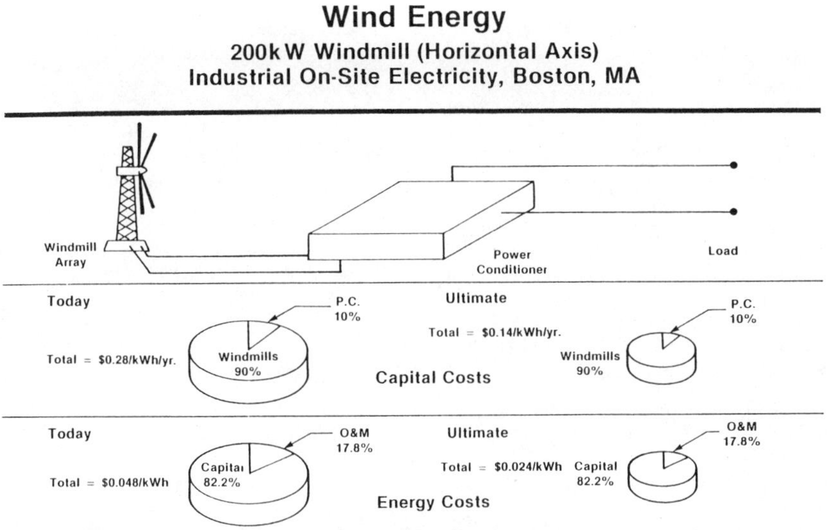

Courtesy Mitre Corporation

Wind-energy capital and equipment cost comparisons.

wind erosion. The removal of material from the land or from buildings by the action of the wind. The mechanisms include straightforward picking up of dust by the airflow, and the dislodging or abrasion of surface material by the impact of particles already airborne.

wind generators. Devices to extract energy from the wind to be used to generate electricity directly.

wind measurement. The measurement of wind may be achieved by the direct tracking of balloons, optically or by radar. Doppler radar may be used to determine the velocity of rain or other airborne objects, their horizontal motion being attributed to wind. The displacement of clouds as seen by satellite photography may be used, provided clouds can be satisfactorily identified as moving with the wind.

windmill. Archaic term for wind system; still used to refer to high-solidity rotor water-pumpers and older mechanical output machines.

windmill anemometer. An anemometer in which a windmill is driven by the air stream, and its rotation transmitted through gearing to dials or other recording mechanisms. In some instruments the rotating vanes and dials are in the same plane (both vertical), and in others the dial is horizontal. In the windmill type, the operation of air measurement involves readings of the dials at the beginning and end of a measured period, and a watch or clock is required. Windmill instruments may be fitted with an extension handle providing a form of remote control, and may be used to measure the air speed in an otherwise inaccessible spot.

windmills, rural. See **rural windmills.**

windpower. A power source derived from the use of wind-energy conversion systems which convert the wind's energy into electricity. The power contained in a moving wind stream is proportional to the first power of both the air density and the area of the wind stream and proportional to the third power of the wind velocity.

Courtesy SCE (Southern California Edison)

Horizontal and vertical axis wind-energy conversion systems being tested by Southern California Edison.

windpower plants. Wind turbines supplying electric power to a grid.

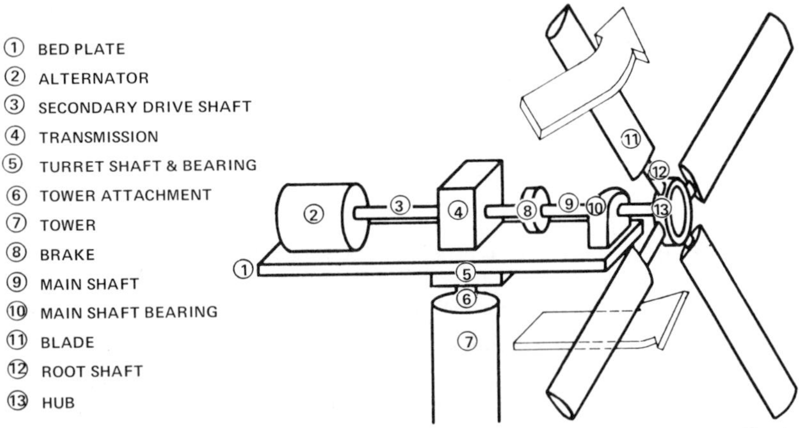

Courtesy DOE

Key elements of horizontal axis wind-energy conversion system.

windpower profile. How the windpower changes with height above the surface of the ground or water; the wind power profile is proportional to the cube of the wind speed profile. See **wind speed profile.**

wind pressure. The pressure on a structure due to wind, which increases with wind velocity approximately in accordance with the formula $p = 0.003v^2$, where p is the pressure in pounds per square feet of area affected and v is the wind velocity in miles per hour.

wind profile. The variation of wind characteristics (e.g., mean speed, direction, turbulence level) with altitude.

wind resource assessment. The interpretations of the spatial and temporal distributions of the wind climatology and wind resource interpolated between permanent or long-term data stations using historical data from those stations. The resolution of these assessments is as fine as can be responsibly made from sparse data, but is too coarse to be used in wind-energy conversion system (WECS) site selection except to provide a preliminary indication of the magnitude and distribution of the wind resource for an area.

wind rose. A graph showing the average wind speed from each of the standard compass points and the percent of time the wind blows from each direction. (Usually monthly or yearly wind averages and 16 equally spaced directions are used.)

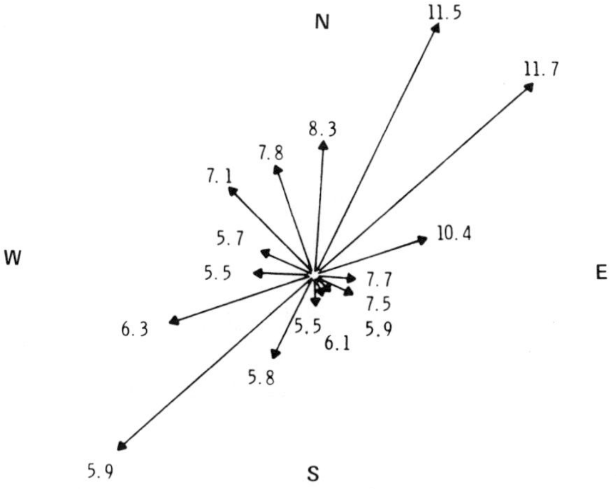

Courtesy DOE

Typical wind rose diagram.

wind shadow thermals. The strength of the wind and the terrain determine the intensity of turbulence. Where the wind is reduced (e.g., on the lee side of a hill) the depth of air warmed by the ground heated in sunshine is less and the maximum temperature reached greater. Glider pilots find that intense thermal upcurrents rise from such places from time to time. A special case is a field of ripe wheat: the temperature within the crop is higher than in a green crop, and a gust of wind bearing down the stalks releases a body of very warm air.

wind shear. In general, the rate of change of the wind vector with distance in a direction perpendicular to the wind direction. Usually, the term implies the change of wind with height, and near the ground it is often used to designate the vertical gradient of horizontal velocity.

wind speed. Wind speed as a rate of flow undisturbed by obstacles.

wind-speed duration curve. A graph which shows the distribution of wind speeds as a function of the cumulative hours that the wind speed exceeds a given wind speed in a year.

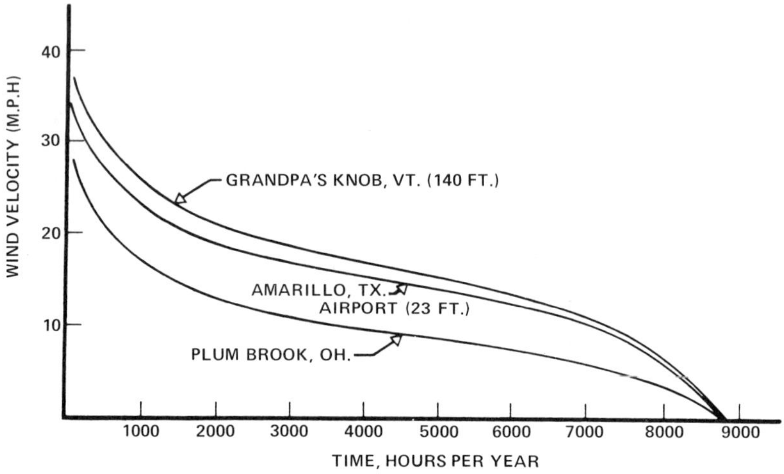

Courtesy DOE

Wind duration curves.

wind-speed frequency curve. The probability density function curve for wind speed, expressed in probability units per speed interval or in hours/year/speed interval.

wind-speed profile. How the wind speed changes with height above the surface of the ground or water.

wind system weight. Total weight of a wind-energy conversion system including everything above ground level. (This includes tower, machine, wiring, etc., but excludes footings, guy anchors, etc.)

wind turbine generator. A wind system which produces electrical power. Preferred use is wind-energy conversion system. (See diagram p. 345.)

wind turbine or wind machine. See **wind-energy conversion system.**

wind variation. The wind varies in time and place. In the wake of a building, tree, or cliff it may fluctuate by 50 percent or more in a very few seconds and vary in direction by 180°. With the passage of storms and with time of day and season it varies over minutes, hours, days, and months. Even one year may differ markedly from another at the same place. These variations have to be taken into account in defining "the wind," its average or mean, and in determining the exposure of the measuring instrument. (See diagram p. 345.)

wind velocity. Wind speed and direction in an undisturbed flow (a vector quantity).

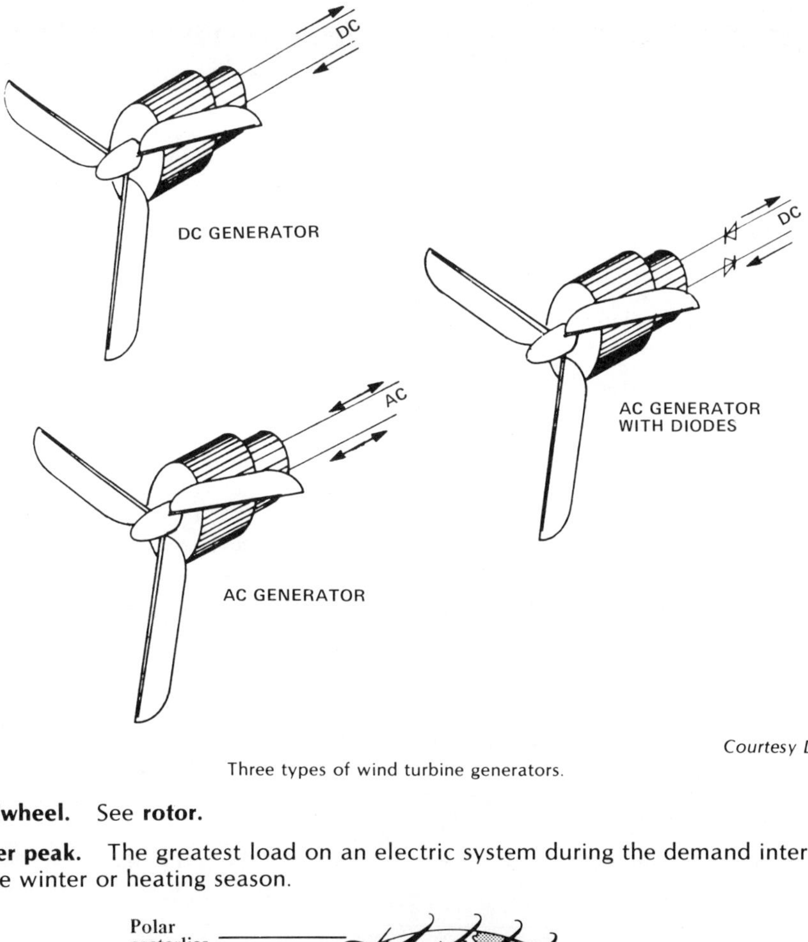

Courtesy DOE

Three types of wind turbine generators.

windwheel. See **rotor.**

winter peak. The greatest load on an electric system during the demand interval in the winter or heating season.

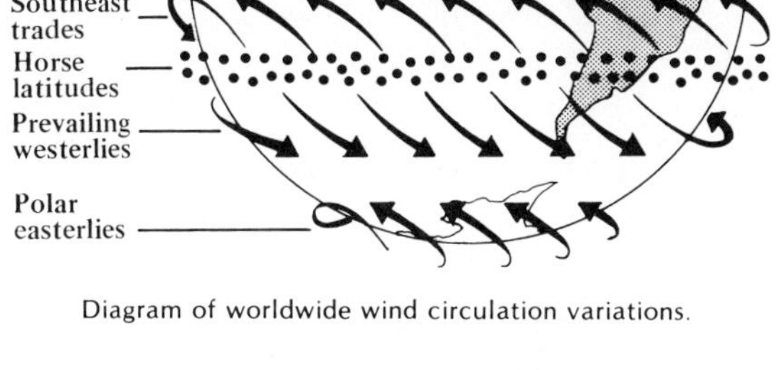

Courtesy DOE

Diagram of worldwide wind circulation variations.

wiped joint. A lead joint in which the molten solder is poured upon the joint, after scraping and fitting the parts together, and the joint is wiped up by hand with a cloth pad while the metal is plastic.

wood alcohol. See **methanol.**

wood-burning stoves. Heating furnace which burns wood. Five of the basic stove types include: simple box stove, ashley airtight stove, base-burning (Riteway) stove, down-draft (Vermont) airtight stove, and front-end (Scandinavian Jotul) combustion stove.

wood gas. Gas produced during production of charcoal by heating wood in the absence of air and usually used as a fuel at the production site.

wood heating efficiency. The quality of producing the maximum heat with the least wood.

work. Energy transferred from one body to another in such a way that a difference in temperature is not directly involved. Work is measured in terms of the amount of a force multiplied by the distance it has traveled. The practical unit is the foot-pound (force in pounds weight multiplied by distance in feet).

working capital. Capital which is used for initial training, inventory, and labor.

working fluid. Fluid in electrical generation plants that is heated by the energy source and then expands through the turbine without leaving the system.

working pressure. Normal operating gauge pressure in a device or system.

wort. The raw product (feedstock) that has been prepared and is ready for inoculation with yeast; unfermented mash.

wrought pipe. This term refers to both wrought steel and wrought iron. Wrought in this sense means worked, as in the process of forming furnace-welded pipe from skelp, or seamless pipe from plates or billets. The expression "wrought pipe" is thus used as a distinction from cast pipe. When wrought-iron pipe is referred to, it should be designated by its complete name.

wye. A pipe fitting with a side outlet at other than 90° to the main axis.

X

xenobiotics. Matter that can enter the environment only by human activities.

Y

yaw. Rotation of a horizontal axis wind-energy conversion system about its yaw axis.

yaw angle. Angle between the mean wind direction and the wind-energy conversion system rotor axis.

yaw axis. A vertical axis about which a horizontal axis wind-energy conversion system changes its directional orientation.

yaw control. A wind-energy conversion system control system function that effects yaw angle or yaw rate.

yaw orientation. A change made in yaw angle due to a wind-energy conversion system control action.

yaw rate. Yawing angular rate.

year. Measure of time. Solar year (successive intervals between transits of first point of Aries) is 365.2422 mean solar days; civil year is 365.2425; and sidereal year 365.2564 days.

yeast. Single-cell microorganism (fungi) that produce alcohol and CO_2 under anaerobic conditions and acetic acid and CO_2 under aerobic conditions; the microorganism that is capable of changing sugar to alcohol by fermentation. The yeast often used for fermentation ethanol is called *Saccharomyces cerevisiae*.

yeast starter. This is used to create vigorous fermentation action and reduce fermentation time to a minimum.

yield point. The stress at which a material exceeds its elastic limit. Below this stress the material will recover its original size on removal of the stress; above this stress, it will not.

yoke vent. A pipe extending upward from a waste stack to a vent stack to prevent pressure changes or differences in the stacks.

Z

Zer-O-Cel®. An insulating product of the National Gypsum Co. which is a sheet of urethane foam.

zero differential escalation. A condition where the general inflation rate and the escalation of a specific commodity (such as fuel) are identical.

Zero Perm®. An insulating product by Alumiseol Corporation consisting of aluminum foil between 1/2 mil Mylar sheets. Zero Perm is often used as a vapor barrier.

zinc (Zn). A lustrious, bluish-white metallic element in group II of the periodic system.

zinc-bromine batteries. An advanced battery system in early research stage.

zinc-chlorine battery. The zinc-chlorine battery operates at ambient temperatures and has an aqueous electrolyte. The reactants are zinc and chlorine gas. A unique feature of this system is the method used to store chlorine when the battery is charged. The chlorine gas is chilled in the presence of water, which is part of the electrolyte, to form an ice, known as chlorine hydrate. This solid form of storage provides added safety in case of accidental exposure. To discharge the battery, the chlorine ice is heated to revaporize the chlorine which is then consumed by reaction at one of the electrodes.

zone heat. A central heating system which is arranged so that different temperatures can be maintained in two or more areas of the building being heated.

zoning. The division of a building or group of buildings into separately controlled spaces where different conditions can be simultaneously maintained.

Zonolite®. An insulating product of Alumiseol Corporation consisting of a sheet of ¼-inch hardboard faced with Zero Perm.

zooplankton. Planktonic animals that are a supply or food for fish.

zymase. In the last century, prior to understanding how yeast enzymes functioned, the term zymase was applied to the complete enzymatic system of fermentation. Although several enzymes and coenzymes were later found to actually produce the fermentation process within the yeast cell, the term is still used to describe the process, as if it were a single function. Therefore, zymase can be thought of as that enzyme complex present in yeast which converts sugars to alcohol and carbon dioxide.

zymosis. See **fermentation.**

METRIC CONVERSION TABLES

The following table translates various units into the international system of units (or SI units). This conversion is accomplished by multiplying by the appropriate numbers from Column 1.

Some users may also wish to use the table to better understand the SI units. To change *out* of SI units, multiply by numbers from Column 2.

Column 1 has the dimensions of (SI units)/(non-SI units), while Column 2 is just the inverse. For convenience, Column 2 gives less accuracy.

The numbers in parentheses in this table are the exponents of 10 in scientific notation. For example, there are 6.1 (9) joules in a barrel of petroleum, which is 6.1×10^9, or 6,100,000,000 joules. Also, there is 6.4516 (-4) in.^{2} in a square meter, which is 6.4516×10^{-4}, or 0.00064516 in.^{2}.

This table should rarely be used for other conversion purposes, although it is possible to do so. For example, to determine the number of square feet in an acre, one can use values from either column as follows:

1. Using column 2:

$$(\text{ft}^2/\text{acre}) = (\text{ft}^2/\text{m}^2) \div (\text{acres}/\text{m}^2) = 10.8 \div 2.47(-4) = 43{,}725 \text{ (approximate)}$$

2. Using column 1:

$$(\text{ft}^2/\text{acre}) = (\text{m}^2/\text{acre}) \div (\text{m}^2/\text{ft}^2) = 4046.873 \div 0.09290304 = 43{,}560.01 \text{ (more accurate)}$$

Clearly, the second method will be preferable in most cases.

Variables and Units (ordered by size)	Multiplier to Obtain SI-Units $\left(\frac{\text{SI Units}}{\text{Non-SI Units}}\right)$ Column (1)	Approximate Inverse $\left(\frac{\text{Non-SI Units}}{\text{SI-Units}}\right)$ Column (2)
Mass Density (kg/m^3)		
pounds/cubic foot (lb/ft^3)	16.01846	0.0624
pounds/gallon (lb/gal)	119.8264	8.34 (-3)
ounces/cubic inch (oz/in.^{3})	1.729994 (3)	5.78 (-4)
pounds/cubic inch (lb/in.^{3})	2.767990 (4)	3.61 (-5)
Mass Per Unit Time (kg/s)		
pounds/hour	1.259979 (-4)	7.94 (3)
pounds/minute	7.559873 (-3)	132.3
pounds/second	0.4535924	2.20
tons (short)/h	0.2519958	3.97
Force (newton, N)		
poundal	0.1382550	7.23
ounce-force (ozf)	0.2780139	3.60
pound-force (lbf)	4.448222	0.2248
ton-force	8.896444 (3)	1.124 (-4)
Pressure (pascal, Pa)		
poundals/ft^2	1.488164	0.672
pound force/ft^2	47.88026	0.0209

Variables and Units (ordered by size)	Multiplier to Obtain SI-Units (SI-Units / Non-SI Units) Column (1)	Approximate Inverse (Non-SI Units / SI-Units) Column (2)
millimeters of mercury (torr)	1.33322 (2)	7.50 (–3)
inches of mercury (60°F)	3.37685 (3)	2.96 (–4)
pound force/in.² (psi)	6.894757 (3)	1.450 (–4)
atmospheres	1.013250 (5)*	9.87 (–6)
Energy (joules, J)		
electron volt	1.60219 (–19)	6.24 (18)
erg (dyne-cm)	1.0 (–7)*	1.0 (7)*
foot-poundal	0.04214011	23.7
watt-second	1.0*	1.0
foot-pound force	1.355818	0.738
calorie	4.1868*	0.239
Btu	1055.056	9.48 (–4)
watt-hour	3600.*	2.78 (–4)
kilocalorie	4186.8*	2.39 (–4)
ft³ of natural gas	1.06 (6)	9.4 (–7)
kilowatt-hour	3.6 (6)*	2.78 (–7)
liter of gasoline	3.4 (7)	2.9 (–8)
m³ of natural gas	4.0 (7)	2.5 (–8)
therm	1.055056 (8)	9.48 (–9)
gallon of gasoline	1.29 (8)	7.8 (–9)
gigajoule	1.0 (9)*	1.0 (–9)*
MBtu	1.055056 (9)	9.48 (–9)
barrel of petroleum	6.1 (9)	1.64 (–10)
ton of coal	25.4 (9)	3.9 (–11)
tonne of coal	28. (9)	3.6 (–11)
tonne of petroleum	44. (9)	2.3 (–11)
tonne of natural gas	56. (9)	1.8 (–11)
Length (meters, m)		
inch (in.)	0.0254*	39.37
foot (ft = 12 in.)	0.3048*	3.28
yard (yd = 3 ft)	0.9144*	1.09
mile (mi = 1760 yd)	1609.344*	6.21 (–3)
Area (m²)		
square inches (in²)	6.4516 (–4)*	1550.
square feet (ft²)	0.09290304*	10.8
square yards (yd²)	0.8361274	1.20
acre (a)	4046.873	2.47 (–4)
square miles (mi²)	2.589988 (6)	3.86 (–7)
Volume (m³)		
cubic inches (in³)	1.638706 (–5)	6.10 (4)
gallons (gal)	3.785412 (–3)	264.
cubic feet (ft³)	2.831685 (–2)	35.3

* Figure is exact.

Variables and Units (ordered by size)	Multiplier to Obtain SI-Units (SI-Units / Non-SI Units) Column (1)	Approximate Inverse (Non-SI-Units / SI-Units) Column (2)
cubic yards (yd^3)	0.764555	1.31
acre-feet	1233.489	8.90 (–5)
Mass (kg)		
ounce (oz)	2.834952 (–2)	35.3
pound (lb)	0.4535924	2.20
ton	907.1847	1.10 (–3)
Time (s)		
minute, mean solar (min)	60*	0.0167
sidereal	59.83617	0.0167
hour, mean solar (h)	3600.*	2.78 (–4)
day, mean solar	8.640 (4)*	1.157 (–5)
year (365 days)	3.153600 (7)*	3.17 (–8)
sidereal	3.155815 (7)	3.17 (–8)
Velocity (m/s)		
feet/hour (ft/h)	8.466667 (–5)	1.181 (4)
feet/minute (ft/min)	5.080 (–3)*	1.969 (2)
inches/second (in./s)	0.0254*	39.37
feet/second (ft/s)	0.3048*	3.28
kilometers/hour (km/h)	0.2777778	3.6*
miles/hour (mi/h)	0.44704*	2.24
knots	0.514444	1.944
Volume Per Unit Time (m^3/s)		
gallons/day	4.381264 (–8)	2.28 (7)
cubic inches/minute (in^3/min)	2.731177 (–7)	3.66 (6)
gallons/minute	6.309020 (–5)	1.585 (4)
cubic feet/minute (ft^3/min)	4.719474 (–4)	2.120 (3)
cubic feet/second (ft^3/s)	2.831685 (–2)	35.3
Energy Surface Density (J/m^2)		
Btu/ft^2	1.135653 (4)	8.81 (–5)
cal/cm^2	4.1868 (4)*	2.39 (–5)
kWh/m^2	3.6 (6)*	2.78 (–7)
$MBtu/ft^2$	1.135653 (10)	8.81 (–11)
Energy Volume Density (J/m^3)		
Btu/ft^3	3.725895 (4)	2.68 (–5)
kWh/m^3	3.6 (6)*	2.78 (–7)
cal/cm^3	4.1868 (6)*	2.39 (–7)
$MBtu/ft^3$	3.725895 (10)	2.68 (–11)
Power (watts, W)		
erg/sec	1.0 (–7)*	1.0 (7)*
ft-lbf/h	3.766161 (–4)	2.655 (3)

*Figure is exact.

Variables and Units (ordered by size)	Multiplier to Obtain SI-Units (SI Units / Non-SI Units) Column (1)	Approximate Inverse (Non-SI Units / SI-Units) Column (2)
ft-lbf/min	2.259697 (–2)	44.25 (1)
Btu/h	2.930711 (–1)	3.41
ft-lbf/s	1.355818	0.738
cal/s	4.184*	0.239
gigajoules/year	31.68753	0.0316
horsepower (550 ft-lbf/s)	745.6999	1.341 (–3)
horsepower (electric)	746.*	1.340 (–3)
Btu/s	1055.056	9.48 (–4)
petroleum (tonnes/yr)	1394.	7.17 (–4)
ton of refrigeration	3.516800 (3)	2.84 (–4)
petroleum (bbl/day)	7.06 (4)	1.42 (–5)
quads/year	3.343212 (10)	2.99 (–11)
Power Density (W/m^2)		
ergs/($cm^2 \cdot s$)	1.0 (–3)*	1.0 (3)*
Btu/($ft^2 \cdot h$)	3.154591	0.317
Btu/($ft^2 \cdot min$)	1.892755 (2)	5.28 (–3)
calories/($cm^2 \cdot min$)	6.973333 (2)	1.434 (–3)
watts/cm^2	1.0 (4)*	1.0 (–4)*
Btu/($ft^2 \cdot s$)	1.135653 (4)	8.81 (–5)
Heat Capacity (J/kg·°K)		
cal/gm·°C	4.1868 (3)*	2.39 (–4)
Btu/lb·°F	4.1868 (3)*	2.39 (–4)
Thermal Conductance ($W/m^2 \cdot °K$)		
Btu/($h \cdot ft^2 \cdot °F$)	5.678263	0.1761
Temperature °C		
°Fahrenheit (°F)	Subtract 32; Multiply by 5/9	Multiply by 9/5; add 32

*Figure is exact.

FOR MORE INFORMATION

This appendix provides the name, address, phone number, and area of interest of major solar support organizations as noted below:

I. Private Organizations
II. National Government
III. Regional

This information is based on the Solar Energy Information Data Base, Solar Locator publication.

I. PRIVATE ORGANIZATIONS

Alternate Energy Resources Organization (AERO)
435 Stapleton Building
Billings, Montana 59101
406/259-1958
Areas of Interest: Promotes local self-sufficiency through small-scale solar heating, wind power, low-head hydro, geothermal, and wood and other biomass energy resources; lobbies for state conservation and energy legislation; assists communities involved in renewable energy projects.

Alternate Sources of Energy
Route 2
Milaca, Minnesota 56353
612/983-6892
Areas of Interest: Practical applications of solar technology, community self-reliance.

American Society of Heating, Refrigerating and Air Conditioning Engineers, Inc. (ASHRAE)
345 East 47 Street
New York, New York 10017
212/644-7931
Areas of Interest: Professional association; solar heating and cooling of buildings.

American Wind Energy Association (AWEA)
1050 17th St., NW
Suite 1100
Washington, D.C. 20036
202/775-8910
Areas of Interest: Design, manufacture, installation, and operation of wind energy conversion systems (electrical and mechanical).

Arizona Public Service Company (APS)
P. O. Box 21666
Phoenix, Arizona 85036
602/254-6283 (General Energy Information)
602/271-2418 (Solar)
Areas of Interest: Solar thermal electric development and utilization, photovoltaic development and applications, industrial process heat applications, irrigation pumping applications; medium- and low-temperature solar applications which include swimming pools, space heating and cooling; domestic solar water-heating financing for residential customers.

Bio-Energy Council
1625 Eye Street NW, Suite 825-A
Washington, D.C. 20006
202/833-5656
Areas of Interest: Development and use of solar energy stored in plant matter.

Biomass Energy Institute, Inc.
P. O. Box 129, Postal Station C
Winnipeg, Manitoba R3M 3S7
Canada
Areas of Interest: Biomass energy production, harvesting, and conversion.

California Institute of Public Affairs
P. O. Box 10
Claremont, California 91711
714/624-5212
Areas of Interest: Compiles and publishes directories and other reference materials on domestic and international institutions active in environmental and natural resource issues and undertakes regional studies.

Center for Renewable Resources
1001 Connecticut Avenue, 5th Floor
Washington, D.C. 20036
202/466-6350
Areas of Interest: All forms of solar energy.

Citizens Energy Project
1110 6th Street NW, Suite 300
Washington, D.C. 20001
202/387-8998
Areas of Interest: Solar energy technology, with emphasis on passive design, alcohol fuels, and renewable energy systems suited for community-scale use.

Congressional Information Service (CIS)
600 Montgomery Building
Washington, D.C. 20014
202/654-1550
Areas of Interest: Private publisher of major statistical publications of the federal government (1960 to present), working papers and bills submitted before Congress (1970 to present).

Consumer Action Now (CAN)
355 Lexington Avenue
New York, New York 10017
Areas of Interest: Educational and lobbying organization concerned with energy conservation and solar energy issues, and government policies affecting energy consumers; encourages women to take more active energy roles.

The Energy Institute, Inc.
5716 Jonathan Mitchell Road
Fairfax Station, Virginia 22039
703/250-5136
Areas of Interest: Renewable energy sources including solar, wind, biomass, methane, solar thermal, bioconversion, and alcohol fuels.

Energy Research and Information Foundation
3500 Kingman Boulevard
Des Moines, Iowa 50311
515/277-0968
Areas of Interest: Solar heating (passive and active); solar architecture, greenhouses, solar grain drying, wind energy.

Environment Information Center (EIC)
292 Madison Avenue
New York, New York 10017
212/949-9494
Areas of Interest: Private publisher and clearinghouse of energy and environmental information.

Environmental Action Coalition
235 East 49 Street
New York, New York 10019
212/929-8481
Areas of Interest: Educational information packages covering all aspects of energy.

Environmental Action Reprint Service (EARS)
Box 545
La Veta, Colorado 81055
303/742-3221
Areas of Interest: Clearinghouse offering more than 300 publications from outside sources covering: solar heating and cooling, wind power, solar cells, methane generation, water power, wood energy. Topics include architecture and design, system plans and blueprints, teaching aids, directories, food and energy policy, and community self-reliance.

Environmental Information Center of the Florida Conservation Foundation, Inc.
935 Orange Avenue
Winter Park, Florida 32789
305/644-5377
Areas of Interest: Organizes research and distributes information on Florida environmental issues. Promotes local use of solar energy.

Farallones Institute
1516 Fifth Street
Berkeley, California 94710
415/525-1150
Areas of Interest: Operates two centers for research, education, and demonstration of environmentally sound, energy-conserving technologies and lifestyles appropriate to urban and rural settings.

Friends of the Earth, Energy Department
124 Spear Street
San Francisco, California 94105
Areas of Interest: Technical and sociopolitical aspects of adopting "soft" energy paths, domestically and abroad.

Institute for Ecological Policies

2408 18th Street NW
Department A
Washington, D.C. 20009
Areas of Interest: Promotion of solar energy and other technologies which further community self-sufficiency.

Institute for Local Self-Reliance
1717 18th Street NW
Washington, D.C. 20009
202/232-4108
Areas of Interest: Promoting urban self-reliance through decentralized energy sources, intensive agriculture, and waste recovery.

International Biomass Institute
1522 K Street NW, Suite 600
Washington, D.C. 20005
202/783-1133
Areas of Interest: Biomass and other renewable energy systems which promote community self-sufficiency.

International Solar Energy Society (ISES)
P. O. Box 52
Parkville
Victoria, Australia 3052

ISES American Section
American Technological University
P. O. Box 1416
Killeen, Texas 76541
817/526-1300

Regional Chapter Coordinator
G. Lee Salmon
P. O. Box 238
Loveland, Colorado 80537
303/669-4811
Areas of Interest: Applications of solar energy.

Iowa Corn Promotion Board
200 West Towers
1200 35th Street
West Des Moines, Iowa 50265
515/225-9242
Areas of Interest: Production of alcohol fuels, gasohol.

League of Women Voters (LWV)
1730 M Street NW
Washington, D.C. 20036
202/296-1770
Areas of Interest: Community organizing, energy issues.

Montana Power Company
40 East Broadway
Butte, Montana 59701
406/723-5421, ext. 2650
Areas of Interest: Active and passive solar heating, and wind conversion.

National Council of Churches Energy Project
475 Riverside Drive, Room 572
New York, New York 10027
212/870-2385
Areas of Interest: Social and economic implications of energy.

New Alchemy Institute
237 Hatchville Road
East Falmouth, Massachusetts 02536
Areas of Interest: Studies environmentally benign and biologically integrated food production, shelter, and energy technologies.

New York State Electric and Gas Corporation
4500 Vestal Parkway East
Binghampton, New York 13902
607/729-2551
Areas of Interest: Solar and wind energy.

Oregon Solar Institute
215 SE Ninth Avenue, Suite 21
Portland, Oregon 97214
503/232-4741
Areas of Interest: Private, nonprofit organization seeking to develop low-cost solar heating and photovoltaic systems; public education activities in Pacific Northwest.

Pacific Gas and Electric Company (PG & E)
Energy Conservation and Services Department
Solar Group
245 Market Street, Room 1700
San Francisco, California 94106
415/781-4211
800/792-8000 (PG & E service area only)
Areas of Interest: Residential, commercial, industrial, and agricultural uses of solar heating (active and passive), photovoltaics, and wind energy systems; has built and is currently monitoring solar demonstration homes.

Portland General Electric Company (PGE)
121 SW Salmon Street
Portland, Oregon 97204
503/226-8462

503/226-8478
Areas of Interest: Technical and economic feasibility of solar heating and hot water within its service area; has built and is currently monitoring a series of residential solar demonstration projects (new and retrofit).

Solar Bookstore
2239 East Colfax
Denver, Colorado 80206
303/321-1645
Areas of Interest: Solar and alternative energy sources.

Solar Energy Industries Association (SEIA)
1001 Connecticut Avenue NW
Washington, D.C. 20036
202/293-2981
Areas of Interest: Trade and lobbying organization of the solar industry.

Solar Energy Information Services (SEIS)
P. O. Box 204
18 Second Avenue
San Mateo, California 94401
415/347-2640
Areas of Interest: Private publisher which researches, compiles, and markets solar energy information.

Solar Energy Institute of North America (SEINAM)
1110 6th Street NW
Washington, D.C. 20001
202/667-6611
Areas of Interest: Professional and consumer organization fostering use and development of economical solar systems.

SUNRAE
5679 Hollister Avenue
Goleta, California 93017
Areas of Interest: Citizen lobbying organization which promotes applications of solar energy in California.

Total Environmental Action, Inc. (TEA)
Church Hill
Harrisville, New Hampshire 03450
603/827-3374
Areas of Interest: Passive and hybrid solar heating, and solar system design.

Union Electric Company
P. O. Box 149
St. Louis, Missouri 63166
Areas of Interest: Measures solar radiation.

U.S. League of Savings Associations
111 East Wacker Drive
Chicago, Illinois 60601
Areas of Interest: Energy conservation in homes.

Volunteers in Technical Assistance (VITA)
3706 Rhode Island Avenue
Mt. Rainier, Maryland 20822
Areas of Interest: Private nonprofit organization seeking to apply appropriate energy technologies (including small-scale solar) to solve energy problems of lesser developed areas.

PERIODICALS

This section includes a partial listing of major solar energy or solar-related magazines. Most items are available on a subscription basis or can be located at many public or university libraries. Subscription rates quoted are current as of the date of publication and may be subject to change without notice.

No attempt was made to list the many newsletters issued by regional or local solar energy or solar-related citizen organizations; check with the organization to find out whether they publish a newsletter of their activities. This section also does not include technical journals in solar-related fields (such as architecture or power system design). Consult the **Solar Energy Update** or other energy indexes or abstract journals (available in many libraries), or ask a reference librarian to locate the names of technical journals pertaining to your solar interests.

Adobe News
Box 702
Los Lunas, New Mexico 87031
Bimonthly, $8/year. Magazine with in-depth articles and information on earthen-brick buildings, emphasis on providing shelter in the southwestern United States and desert regions.

Advanced Solar Energy Technology
1609 West Windrose
Phoenix, Arizona 85029

Monthly, $60/year. Technical newsletter focusing on new developments and patents.

AERO Sun-Times
Alternate Energy Resources Organization
435 Stapleton Building
Billings, Montana 59101
Monthly, $12/year (or free with $15 annual membership). "How-to-do-it" and technical articles on low-cost renewable energy systems.

Alternate Sources of Energy Magazine
Route 2
Milaca, Minnesota 56353
Bimonthly, $10/year. Practical and imaginative use of alternate energy resources including plans, projects, current events, new product evaluations, and book reviews.

Biomass Refining Newsletter
Document Distribution Service
Solar Energy Research Institute
1617 Cole Boulevard
Golden, Colorado 80401
Semiannual, free. Technical update on research (sponsored by the U.S. Department of Energy) on the production of liquid fuels through fermentation processes.

Canadian Renewable Energy News
P. O. Box 4869, Station "E"
Ottawa, Ontario, Canada K1S 5B4
Monthly, $7.50/year. Covers renewable energy and appropriate technology news from Canada. Canadian perspective on U.S. energy scene.

Cascade
Cascadian Regional Library
1 West Fifth Avenue
Eugene, Oregon 97440
Ten issues, $12/year, free to community groups or co-ops in Pacific Northwest. Information on alternative energy, food, gatherings, and other aspects of community cooperation.

Coevolution Quarterly
Box 428
Sausalito, California 94965
Quarterly, $12/year. Information and commentary relating to adoption of self-reliant lifestyles, including those of community and other small-scale energy systems.

Commerce Business Daily
Superintendent of Documents
U.S. Government Printing Office
Washington, D.C. 20402
Daily, $105/year (first-class postage), $80/year (second-class postage). Daily listing of U.S. Government procurement invitations, contract awards, subcontracting leads, foreign business opportunities, and contracts funded by the U.S. Department of Energy.

Critical Mass Journal
P. O. Box 1538
Washington, D.C. 20013
Monthly, $7.50/year (individuals). Covers nuclear and alternative energy issues.

Earth Shelter Digest
WEBCO Publications
479 Fort Road
St. Paul, Minnesota 55102
Bimonthly, $15/year. Case studies and topical articles on earth-sheltered homes. Written for homeowners, builders, and architects.

Energy
Business Communications Inc.
9 Viaduct Road
P. O. Box 2070C
Stamford, Connecticut 06906
203/325-2208
Quarterly, $52/year. Current events, economics, engineering, interpretation, and analysis of issues facing all energy technologies (including solar).

Energy and Alternatives
P. O. Box 681
Salem, Oregon 97308
Quarterly, $4/year. "In search of tomorrow's energy for today's consumer." Practical workings and discussion of government activities affecting the use of renewable energy technologies.

Energy Consumer
Office of Consumer Affairs
U.S. Department of Energy
Room 8-G 031
Washington, D.C. 20585
Free. How to become involved or take advantage of federal energy education, grant, or assistance programs intended for consumers.

Energy Daily
300 National Press Building

Washington, D.C. 20045
Daily, $525/year. Newsletter covering general energy technology news, discussing in some detail, policy decisions and major technological developments and breakthroughs.

Energy Digest
Alternative Sources of Energy, Inc.
Route 2
Milaca, Minnesota 56353
Monthly, $6/year (charter rate). Regional newsletter for the upper Midwest, covering renewable energy and appropriate technology issues.

Energy Insider
U.S. Department of Energy
Room 8-G 031
Washington, D.C. 20585
Monthly, subscription prices to be determined. Newspaper of the U.S. Department of Energy, including calendar of upcoming speeches and conferences.

Energy Research and Technology
National Science Foundation
RANN Document Center
1800 G Street NW, Room 1241
Washington, D.C. 20550
Free. Solar energy research and development information.

Energy Report to the States
Energy Policy Project
National Conference of State Legislatures
1405 Curtis Street, 23rd Floor
Denver, Colorado 80202
Biweekly, $15/year, free to state legislators and staff. Newsletter, providing updates and analyses of state energy legislation activities.

Energy Research Digest
P. O. Box 17162
Washington, D.C. 20041
Biweekly, $150/year. Newsletter, covering new developments in various energy technologies. Highlights new publications available from NTIS.

Energy Review
Environmental Studies Institute
2074 Alameda Padre Serra
Santa Barbara, CA 93103
Bimonthly, $55/year. Summaries of books, reports, articles, conferences, grants, and audiovisual materials in all areas of energy.

Energy Today
Trends Publishing Company
National Press Building
Washington, D.C. 20045
Biweekly, $165/year. Covers all aspects of all energy sources and technologies, ranging from policy and economics to current technical developments. Reviews current literature.

Environmental Study Conference Weekly Bulletin
Environmental Study Conference
U.S. Congress
3334 House Annex 2
Washington, D.C. 20515
Weekly, $150/year. Tracks and analyzes proposed energy and environmental legislation being considered by the U.S. Senate or House of Representatives.

High Country News
Box K
Lander, Wyoming 82520
Biweekly, $10/year. Environmentally oriented newspaper covering the Rockies and Northern Plains. Reports on the people of the region, natural history, conservation. Western coal activities, solar power, agriculture, and changing land-use patterns.

In Review
Document Distribution Service
Solar Energy Research Institute (SERI)
1617 Cole Boulevard
Golden, Colorado 80401
Monthly, free. Update on SERI's research, development and demonstration activities. Highlights significant new SERI reports.

Inside DOE
McGraw-Hill Inc.
1211 Avenue of the Americas
New York, New York 10020
Weekly, $320/year. In-depth reports on the U.S. Department of Energy.

Mother Earth News
P. O. Box 70
Hendersonville, North Carolina 28739
Bimonthly, $10/year. How to lead a low-consumption lifestyle, in harmony with Mother Nature. Good source of information from the grassroots level.

Monthly Energy Review
Superintendent of Documents

U.S. Government Printing Office
Washington, D.C. 20402
Monthly, $23/year ($33/year via first-class mail). Summary report of U.S. energy supply, imports, and consumption levels by energy source.

New Alchemy Newsletter
New Alchemy Institute
237 Hatchville Road
Woods Hole, Massachusetts 02536
Quarterly, $25/year (with membership). Newsletter, covering all aspects of research carried out at the Institute including agriculture, aquaculture, and bioshelters.

New Roots
New England Appropriate Technology Network, Inc.
P. O. Box 548
Greenfield, Massachusetts 01301
Bimonthly, $8/year (individuals), $12/year (institutions). Covers the movement toward self-reliance in the Northeast. Topics include aquaculture, waste recycling, food co-ops, and appropriate technology.

Outlook (formerly Acorn)
Governor's State University
Park Forest South, Illinois 60466
Ten issues, $6/year (individuals), $12/year (institutions). Covers small-scale technology and self-reliance; includes nontechnical news on local projects and people.

People and Energy
Department A
2408 18th Street NW
Washington, D.C. 20009
Bimonthly, $15/year (reduced rates available for low-income persons). Covers and analyzes developments in renewable energy policies and programs, reviews current literature. Write for free sample copy.

Popular Science
Times Mirror Magazines, Inc.
380 Madison Ave.
New York, New York 10017
Monthly, $10/year. Magazine covering homes, new products, and do-it-yourself solar projects; includes regular energy news features.

RAIN: Journal of Appropriate Technology
2270 NW Irving Street
Portland, Oregon 97210
Monthly, $10/year. Focuses on small-scale, home-built and community projects.

Small Farm Energy Project Newsletter
Center for Rural Affairs
P. O. Box 736
Hartington, Nebraska 68739
Bimonthly, free. Provides information for farms on renewable energy sources, such as waste recycling, composting, energy conservation, wind, solar, and biofuels.

Solar Age
Solar Vision, Inc.
Church Hill
Harrisville, New Hampshire 03450
Monthly, $20/year. Official publication of the American Section of the International Solar Energy Society; articles, reviews, interviews, and news of all solar energy technologies and related topics.

Solar Collector
Florida Solar Energy Center
Education and Information Division
300 State Road 401
Cape Canaveral, Florida 32920
Quarterly, free. News of activities and new publications of the Florida Solar Energy Center, including testing programs and conferences. Also includes updates on Florida solar legislation.

Solar Events Calendar
Document Distribution Service
Solar Energy Research Institute
1617 Cole Boulevard
Golden, Colorado 80401
Quarterly, free. Listing of solar conferences, symposia, workshops, and other formal meetings. Includes section listing calls-for-papers.

Solar Energy
Pergamon Press Inc.
Maxwell House
Fairview Park
Elmsford, New York 10523
Monthly, $176/year ($30/year for individuals working with institutions already subscribing). Technical journal of the International Solar Energy Society, containing scientific and engineering papers.

Solar Energy Digest
CWO 4 WB Edmondson

P. O. Box 17776
San Diego, California 92117
Concise summaries of solar energy developments, including ongoing research and literature reviews.

Solar Energy Intgelligence Report
Business Publishers Inc.
P. O. Box 1067
Silver Spring, Maryland 20910
Weekly, $90/year. Current domestic and international developments, including proposed legislation, research and commercialization programs, technical developments, business news, literature reviews, new patents, and grant opportunities.

Solar Energy Research and Development Report
U.S. Department of Energy
Washington, D.C. 20545
Periodic update (free) on funding opportunities, conferences, and government assistance programs.

Solar Energy Update
National Technical Information Service (NTIS)
5285 Port Royal Road
Springfield, Virginia 22161
Monthly, $27.50/year. Includes indexes and abstracts of technical reports and other solar publications available through NTIS.

Solar Engineering
Solar Engineering Publishers, Inc.
8435 North Stemmons Freeway
Suite 880
Dallas, Texas 75247
Monthly, $15/year. Official magazine of the Solar Energy Industries Association, emphasizing the technical aspects of solar energy.

Solar Greenhouse Digest
P. O. Box 3218
Kingman, Arizona 86401
Bimonthly, $7/year. Focuses on solar greenhouse projects across the U.S., including do-it-yourself information. Also reviews current legislation and literature.

Solar Heating and Cooling
Gordon Publications Inc.
20 Community Place
Morristown, New Jersey 07960
Bimonthly, $8/year. The practical aspects of solar heating and cooling, focusing on already installed systems and readily available components. Subscription includes Annual Buyers Guide, published every March.

Solar Law Reporter
P. O. Box 5400
Denver, Colorado 80217
Bimonthly, $12/year. A publication of the Solar Energy Research Institute covering important legal issues surrounding the adoption of solar energy. Includes regular section tracking new solar legislation.

Solar Ocean Energy Liaison (formerly OTEC Liaison)
1303 South Michigan Avenue
Chicago, Illinois 60605
312/427-3000
Monthly, $95/year. Newsletter featuring current developments, government procurements and contract awards, conferences, and events relating to Ocean Thermal Energy Conversion technology.

Solar Utilization News
Alternate Energy Institute
P. O. Box 3100
Estes Park, Colorado 80517
303/586-5636
Monthly, $8/year (individuals), $15/year (libraries). Current events, technical developments, do-it-yourself information, system descriptions, and literature reviews.

Sun Times
Solar Lobby
1001 Connecticut Avenue NW, 5th Floor
Washington, D.C. 20036
Monthly, $15/year. Timely reports on policy, legislative and grassroots solar developments.

Sun Up
P. O. Drawer S
Yucca Valley, California 92284
Monthly, $6.50/year. Solar industry and new equipment news, government solar-incentive developments. Official newsletter of the California Solar Energy Association included with all subscriptions.

Synerjy
P. O. Box 4790
Grand Central Station
New York, New York 10017

Semiannual, $10/year (individuals), $18/year (institutions). Comprehensive bibliography of books, periodicals, journal articles, and research reports covering all forms of solar energy, hydropower, energy storage, and electricity; also lists publishers and book distributors. Each issue lists only new materials (back issues available).

Tranet
P. O. Box 587
Rangeley, Maine 04970
Quarterly, $15/year (with membership). Newsletter and directory of individuals and groups around the world actively developing appropriate technologies. Includes reviews of current literature.

The Neighborhood Works
Center for Neighborhood Technology
570 West Randolph Street
Chicago, Illinois 60606
Biweekly, $25/year (individuals and nonprofit groups), $40/year (government organizations and libraries). Covers community development and self-reliance for urban neighborhoods; topics include energy conservation, alternative energy, and co-ops.

Wind Energy Report
Wind Publishing Co.
P. O. Box 14
Rockville Centre, New York 11571
Monthly, $95/year. International newsletter focusing on technical aspects of wind energy conversion. Includes reviews of current literature.

Windletter
American Wind Energy Association
1621 Connecticut Avenue NW
Washington, D.C. 20009
Bimonthly, $10/year (free to association members). Newsletter of the American Wind Energy Association.

Wind Power Digest
115 E. Lexington
Elkhart, Indiana 46516
Attn: Mr. Michael Evans
Quarterly. Comprehensive coverage of ongoing developments in the wind energy field from both consumer and industry perspectives. Articles, interviews, and reviews of the latest wind energy happenings. Also publishes Wind Access Catalogue as an annual supplement to regular subscription.

Wind Technology Journal
P. O. Box 7
Marston Mills, Massachusetts 02648
Quarterly, $20/year ($15/year to American Wind Energy Association members). Technical journal covering state-of-the-art research.

Wood 'n Energy
5 South State Street
Concord, New Hampshire 03301
Quarterly, $3/year. Covers all aspects of harvesting and production of energy from wood, from wood-burning stoves to the production of wood-alcohol fuels.

II. NATIONAL GOVERNMENT

ACTION
806 Connecticut Avenue, N.W.
Washington, D.C. 20525
800/424-8580
Areas of Interest: Umbrella agency which includes VISTA (domestic) and the Peace Corps (international); programs enlist volunteers who teach skills to promote economic self-sufficiency to low-income persons in the U.S. and lesser developed areas of the world. Skills may include the construction of small-scale solar energy systems. Write or call for more information on ACTION programs.

Community Services Administration (CSA)
Energy Office
1200 19th Street N.W., Room 334
Washington, D.C. 20506
Areas of Interest: Strives to increase economic self-sufficiency of low-income and elderly persons. Solar-related programs include demonstration projects, grants and low-interest loans, technical training and assistance, encouragement of new small businesses, and home weatherization.

Contact the National Center for Appropriate Technology for up-to-date information on community action programs and appropriate energy projects (including small-scale solar systems) nationwide.

Consumer Information Center
Pueblo, Colorado 81009
Areas of Interest: Since 1970, the

Center has published the Consumer Information Catalog (quarterly, free) listing 200 consumer publications.

Environmental Study Conference (ESC)
U.S. Congress
3334 House Annex 2
Washington, D.C. 20515
202/225-2988
Areas of Interest: Informal bipartisan legislative research organization studying environmental and energy issues.

Library of Congress
Reference Section
Science and Technology Division
10 First Street S.E.
Washington, D.C. 20540
202/287-5580
Areas of Interest: Solar and wind energy; other renewable energy sources.

National Aeronautics and Space Administration (NASA)
Washington, D.C. 20546
Areas of Interest: Conducts broad range of research and development programs applying aerospace technologies to the solution of energy problems on earth. Areas of solar research include space heating and cooling, photovoltaics and wind, satellite power stations, and energy storage technologies.

National Bureau of Standards (NBS)
Solar Technology Group
Building 225, Room B-150
Washington, D.C. 20234
Areas of Interest: Active and passive solar heating and cooling. Monitors passive solar buildings, tests flat-plate collectors, has developed solar heating and cooling performance criteria (Minimum Property Standards), and is monitoring selected solar heating and cooling demonstration projects (residential and commercial buildings) for U.S. Departments of Energy, and Housing and Urban Development; evaluates energy-related inventions.

National Center for Appropriate Technology (NCAT)
P. O. Box 3838
Butte, Montana 59701
406/494-4572
Areas of Interest: Utilization of renewable energy sources by low-income persons. Design, application, and performance monitoring of passive and active solar systems, wind, small-scale hydropower, biofuels (including wood and methane), home energy and resource conservation, housing rehabilitation, solar greenhouse design and operation, community gardening.

National Climatic Center
Federal Building
Asheville, North Carolina 28801
704/258-2850 Ext. 683
Areas of Interest: Climate data including temperature, solar radiation, wind direction and speed.

National Energy Information Center
Office of Energy Information Services
Energy Information Administration
U.S. Department of Energy,
EI-71, MS-240
1726 M Street N.W.
Washington, D.C. 20461
202/634-5610
Areas of Interest: Collects and analyzes statistics on supply, demand, allocation, and consumption of all energy resources; energy policy issues.

National Technical Information Service (NTIS)
5285 Port Royal Road
Springfield, Virginia 22161
703/557-4600
703/557-4650 (Document Orders)
703/557-4780 (Title Information)
703/557-4642 (NTISearch)
Areas of Interest: Clearinghouse for over 600,000 technical reports produced from U.S. government-sponsored research, translations of selected foreign research documents, computer models, patents, and other materials.

Superintendent of Documents
U.S. Government Printing Office (GPO)
Washington, D.C. 20402
202/783-3238 (Publications Information)

Areas of Interest: Wide-interest publications produced from U.S. government-sponsored work. Over 25,000 titles available.

Technical Information Center (TIC)
U.S. Department of Energy
P. O. Box 62
Oak Ridge, Tennessee 37830
Areas of Interest: Research and development sponsored by the U.S. Department of Energy.

U.S. Department of Agriculture (USDA)
Washington, D.C. 20450
Areas of Interest: Solar-related interests include space heating of farm residences and service buildings (technical assistance and FmHA-backed loans), alcohol fuel production, utilization of energy from wood and biomass residues, crop drying, and weatherization of farm buildings.

Cooperative Extension Service Program: USDA provides major funding and technical coordination of state and county extension service activities nationwide. Encompassing food production, home economics, and urban youth activities, this program offers hands-on technical assistance in all solar areas mentioned above. Cooperative Extension Service agents may be found at county seats.

U.S. Department of Energy (DOE)
Washington, D.C. 20545
Areas of Interest: Carries out national energy policies and programs, including the research, development, and application of solar energy technologies. Sponsors National Solar Heating and Cooling Demonstration Program for commercial buildings.

U.S. Department of Housing and Urban Development (HUD)
Washington, D.C. 20410
Areas of Interest: Solar heating and cooling of residences; sponsors National Residential Solar Heating and Cooling Demonstration Program. As part of demonstration program, works with National Bureau of Standards to implement solar heating and cooling system performance standards (Minimum Property Standards).

III. REGIONAL GOVERNMENT, OTHER GOVERNMENT-RELATED ORGANIZATIONS, AND UNIVERSITIES

Arizona State University Library
Solar Energy Collection
Tempe, Arizona 85281
602/965-7608
Areas of Interest: Solar heating and cooling, solar thermal power generation, and photovoltaic applications

California Office of Appropriate Technology
1530 Tenth Street
Sacramento, California 95814
Areas of Interest: Development of less costly and less energy-intensive technologies including waste disposal, recycling, natural resource conservation, building design, agriculture, and transportation.

Center for Energy Policy and Research
New York Institute of Technology (NYIT)
Old Westbury, New York 11568
516/686-7578
Areas of Interest: Energy conservation and renewable energy technologies; energy policy research and legislation.

Center for Solar Energy Applications
San Jose State University
Washington Square
San Jose, California 95192
408/277-2939
Areas of Interest: Dissemination of solar energy information to the public.

Colorado Energy Research Institute
2221 East Street
Golden, Colorado 80401
303/279-2881
Areas of Interest: Policy research in energy technologies, including solar and energy conservation.

Energy Resources Center
U.S. Department of Energy
111 Pine Street
San Francisco, California 94111
414/556-7328
Areas of Interest: U.S. Department of Energy information resources, grants

and other programs, energy statistics.

Florida Solar Energy Center
300 State Road 401
Cape Canaveral, Florida 32920
305/783-0300
800/432-0575 (State Solar Hotline)
Areas of Interest: Tests solar equipment; conducts educational programs, workshops.

Hawaii Natural Energy Institute (HNEI)
2540 Dole Street
Honolulu, Hawaii 96822
808/948-8890
Areas of Interest: Research, development, and demonstration of renewable natural energy resources with commercialization as the goal.

Institute of Energy Conversion
Energy Information Services
University of Delaware
1 Pike Creek Center
Wilmington, Delaware 19808
302/994-0915
302/995-7155
Areas of Interest: Energy conservation for home, solar "total energy" systems (space heating, hot water, photovoltaics), thermal energy storage.

Johnson Environmental and Energy Center
University of Alabama
P. O. Box 1247
Huntsville, Alabama 35807
205/895-6257
800/572-7226 (Alabama only)
Areas of Interest: Solar energy commercialization, tests collectors for private industry, compiles utility rate and other economic data, collects solar radiation data, evaluates solar system and building designs for federal demonstration programs.

Massachusetts Solar Action Office
73 Tremont Street, Room 800
Boston, Massachusetts 02108
617/727-7297
Areas of Interest: State solar policy advisory agency; works with state and local officials on solar-related regulatory matters; monitors technical performance of systems installed in conjunction with HUD solar hot water initiative.

NASA Langley Research Center
Technology Utilization Office
Hampton, Virginia 23665
Areas of Interest: Energy-efficient building design and construction; has erected Technology Utilization House (prototype energy-conserving home utilizing solar heating).

NASA Lewis Research Center
Technology Utilization Office
21000 Brook Park Road
Cleveland, Ohio 44135
216/433-4000 ext. 6833
Areas of Interest: Terrestrial applications and testing of photovoltaic cells, large horizontal axis wind turbines, electrochemical and flywheel energy storage.

NASA Marshall Space Flight Center
Technology Utilization Office
Code AT01
Marshall Space Flight Center
Alabama 35812
Areas of Interest: Testing solar heating and cooling systems.

New Mexico Energy Institute at New Mexico State University
P. O. Box 3 EI
Las Cruces, New Mexico 88003
505/646-1745
Areas of Interest: Administers state solar, wind, and geothermal programs.

New Mexico Energy Institute at University of New Mexico
117 Richmond Drive NE
Albuquerque, New Mexico 87106
Areas of Interest: Administers state energy conservation, synfuels, nuclear, and socioeconomic energy studies.

New Mexico Solar Energy Institute
Box 3 SOL
Las Cruces, New Mexico 88003
505/646-1846
Areas of Interest: Coordinates state and federal solar research, development, and demonstration activities within New Mexico; develops solar equipment standards in cooperation with private industry; tests solar heating and cooling equipment; disseminates information on solar heating and cooling, wind, photovoltaics, and biomass energy.

Pacific Northwest Laboratories
Wind Energy Program
P. O. Box 999, Building HS-1
Richland, Washington 99352
Areas of Interest: Wind energy characteristics and site validation.

Palo Alto Resource Conservation Program
250 Hamilton Avenue
Palo Alto, California 94301
415/329-2241
Areas of Interest: Swimming pool solar heating, building code modifications, energy conservation and solar feasibility for buildings, information on local solar activities.

Regional Energy/Environment Information Center
Denver Public Library
1357 Broadway
Denver, Colorado 80203
303/837-5994
Areas of Interest: Energy conservation and environmental information.

Rocky Flats Plant
Wind Energy Program
Rockwell International Energy Systems Group
P. O. Box 464
Golden, Colorado 80401
Areas of Interest: Small wind energy conversion systems; tests commercially available wind systems.

Sandia Laboratories
Advanced Energy Projects
Division 4715
Albuquerque, New Mexico 87185
Areas of Interest: Vertical axis wind energy turbines, passive solar energy.

Smithsonian Science Information Exchange, Inc. (SSIE)
170 M Street NW, Room 300
Washington, D.C. 20036
202/381-4211
Areas of Interest: Research from all fields of science, domestic and abroad.

Solar CAL
1111 Howe Avenue, Suite 315
Sacramento, California 95825
916/920-7921
800/952-5670 (State Solar Hotline)
Areas of Interest: State solar policy advisory agency.

Solar Data Center
Trinity University
Box 500
715 Stadium Drive
San Antonio, Texas 78284
Areas of Interest: Collects detailed local solar radiation and climate date; constructed local solar heating and cooling demonstration project; developing test facility for passive solar systems.

Solar Energy Applications Laboratory
Colorado State University
Fort Collins, Colorado 80523
303/491-8618
Areas of Interest: Solar energy applications research.

Solar Energy Center
Department of Mechanical Engineering
University of Florida
Gainesville, Florida 32611
Areas of Interest: Solar cooling, test collectors for private industry, energy-conserving building design.

Solar Energy Center
Department of Architecture
University of Oregon
Eugene, Oregon 97403
Areas of Interest: Solar heating and hot water systems; solar resource assessment.

Solar Energy Research Institute (SERI)
1617 Cole Boulevard
Golden, Colorado 80401
303/231-1415
Areas of Interest: Manages and coordinates research, and development, and works with all levels of government to promote the application of the solar energy technologies.

Southern Energy/Environment Information Center
One Exchange Place
2300 Peachford Road, Suite 1230
Atlanta, Georgia 30338
404/455-8841
Areas of Interest: All aspects of energy, with concentration on state legislation and policy matters, and technical aspects.

Technology Applications Center
Institute of Applied Research Services
University of New Mexico

Albuquerque, New Mexico 87131
505/277-3622
Areas of Interest: Solar energy utilization for industry; remote sensing of natural resources.

Tennessee Valley Authority (TVA)
Information Office
400 Commerce Avenue
Knoxville, Tennessee 37902
800/251-9242 (TVA service area outside Tennessee)
800/362-9250 (Tennessee only)
Areas of Interest: Renewable energy sources including wind, small hydro, biomass, active and passive solar heating and hot water systems.

Underground Space Center
University of Minnesota
11 Mines and Metallurgy
221 Church Street SE
Minneapolis, Minnesota 55455
Areas of Interest: Earth-sheltered building design (including passive solar applications), heat transfer studies; monitors several underground dwellings.

University of Wisconsin Solar Energy Laboratory
Engineering Research Building
1500 Johnson Drive
Madison, Wisconsin 53706
608/263-1586
Areas of Interest: Developed F-Chart and TRNSYS computer simulations for solar heating and cooling applications for U.S. Department of Energy.

Western Regional Information Service Center (WRISC)
Lawrence Berkeley Laboratory
Building 50, Room 130
Berkeley, California 94720
415/486-6307
Areas of Interest: Technical and nontechnical information on all energy sources.

BIBLIOGRAPHY

Agricultural and Industrial Process Heat

Brown, K. C., et al. *End Use Matching for Solar Industrial Process Heat.* Golden, CO: Solar Energy Research Institute, October 1979. SERI/T.

Bryant, H. S.; Bowser, R. P.; Schioler, O. T.; and Wittanberg, L. J. "Construction and Initial Operation of the Miamisberg Salt-gradient Pond." *Proceedings of the International Solar Energy Congress,* 1979, Atlanta, GA.

Energy Research and Development Agency, Division of Solar Energy. *Solar Program Assessment: Environmental Factors.* Washington, DC: Energy Research Development Agency: March, 1977. Includes the following reports: "Solar Heating and Cooling of Buildings," ERDA 77-47/1; "Solar Agricultural and Industrial Process Heat," ERDA 77-47/1; "Solar Agricultural and Industrial Process Heat," ERDA 77-47/2; "Photovoltaics," ERDA 77-47/3; "Solar Thermal Electric," ERDA 77-47/4; "Solar Total Energy Systems," ERDA 77-47/5; "Wind Energy Conversion," ERDA 77-47/6; "Fuels from Biomass," ERDA 77-47/7; and "Ocean Thermal Energy Conversion," ERDA 77-47/8, 1977.

Gupta, B., and Kreith, F. (eds). *Proceedings of the Solar Thermal Concentrating Collector Symposium,* June 14-15, 1978. Golden, CO: Solar Energy Research Institute, August 1978. SERI/TP-34-048.

Intersociety Energy Conversion Engineering, Proceedings of the 14th Conference, August 5-10, 1978, Boston, MA. Washington, DC: American Chemical Society, Vol. 1, p. 49. IEEE 79 CH/477-9 Energy.

Jayadev, T. S., et al. *Conversion System Overview Assessment.* Golden, CO: Solar Energy Research Institute, January, 1979. SERI/TR-35-078.

Jayadev, T. S., and Henderson, J. "Modelling and Optimization of Solar Ponds." International Solar Energy Society Meeting, 1979, Atlanta, GA.

Solar Industrial Process Heat Conference, Proceedings. Vol. 1, October 18-20, 1978, Denver, CO. Golden, CO: Solar Energy Research Institute, October 1978. SERI/TP-49-065.

Solar Energy Research Institute. *Putting the Sun to Work in Industry.* Golden, CO: Solar Energy Research Institute, February 1979. SERI/SP-34-175.

Tabor, H. "Solar Ponds." *Solar Energy,* **7** (4): 189-94 (1963).

U.S. Department of Energy, Assistant Secretary for Conservation and Solar Applications and Assistant Secretary for Environment. *Environmental Development Plan (EDP), 1977.* Washington, DC: U.S. Department of Energy, March, 1978. Includes the following reports: "Solar Agricultural and Industrial Process Heat," DOE/EDP-0002; "Fuels from Biomass," DOE/EDP-0005; "Ocean Thermal Energy Conversion," DOE/EDP-0006; and "Wind Energy Conversion," DOE/EDP-0007.

U.S. Department of Energy. *National Program Plan for Research and Development in Solar Heating and Cooling for Building, Agricultural and Industrial Applications.* Washington, DC: U.S. Department of Energy, August, 1978. DOE/CS-0008.

U.S. Department of Energy. Research, Development, and Demonstration Panel. "Agricultural and Industrial Process Heat." In: *Domestic Policy Review of Solar Energy.* Washington, DC: U.S. Department of Energy, October, 1978, p. C-5. Report No. TID-28837. Available from: NTIS, U.S. Department of Commerce, 5285 Port Royal Road, Springfield, VA.

U.S. Department of Energy. *Solar Energy for Agricultural and Industrial Process Heat, Program Summary.* Washington, DC: U.S. Department of Energy, September 1978. DOE/CS-0053.

Weinberger, H. "The Physics of the Solar Pond." *Solar Energy,* **8** (2): 45-56 (April, 1964).

Basic Solar Research

Buchsbaum, S. J. (ed.) *Report of the Office of Science and Technology Policy Working Group on Basic Research in the Department of Energy.* Washington, DC: U.S. Department of Energy, June, 1978.

"Carter Budget Tilts Back to Basics for Research." *Science,* **199:** 507-10 (February 3, 1978).

Metz, W. D. "Report of the Office of Science and Technology Policy Working Group on Basic Research in the Department of Energy." *Science,* **202** (4365): 293-94 (October 20, 1978).

National Science Board, National Science Foundation. *Basic Research in the Mission Agencies, Agency Perspectives on the Conduct and Support of Basic Research.* Washington, DC: National Science Board, National Science Foundation, 1978.

New York Times. 1-4, December 16, 1977. Abstract from Solar Energy Information Bank Data Base.

Stevens, D. K. *DOE Summary of Materials Sciences Program,* FY 78. Washington, DC: U.S. Department of Energy, September, 1978.

U.S. Department of Energy. *Program Information, Basic Energy Sciences.* Washington, DC: U.S. Department of Energy, April, 1978.

U.S. Department of Energy, Office of Basic Energy Sciences. *Materials Sciences Programs Fiscal Year 1978.* Washington, DC: U.S. Department of Energy, September, 1978. DOE/ER-0013.

U.S. Department of Energy. *Research Projects in the Physical Sciences.* Washington, DC: U.S. Department of Energy, April, 1978. DOE/ER-0003(78).

U.S. Department of Energy. *Summaries of FY 1977 Research in the Chemical Sciences.* Washington, DC: U.S. Department of Energy, February, 1978.

U.S. Department of Energy, Washington, DC: Office of Science and Technology. *Policy Working Group on Basic Research.* U.S. Department of Energy, Office of Energy Research, 1978.

Biomass

A Guide to Commercial Scale Ethanol Production and Financing, Golden, CO: Solar Energy Research Institute, 1980.

Anderson, Russell E. *Biological Paths to Self-Reliance—A Guide to Biological Solar Energy Conversion.* New York: Van Nostrand Reinhold, 1979.

Benneman, J. R. *Biofuels: A Survey.* Palo Alto, CA: Electric Power Research Institute, June, 1978. ER-746-SR.

Bennington, J. *Solar Energy: A Comparative Analysis to the Year 2020.* McLean, VA: MITRE Corporation, March, 1978.

Bio-Energy Council. *Bio-Energy Directory.* Washington, DC: The Bio-Energy Council, June, 1978.

Biomass Energy Institute. *Biomass Energy Success Stories: A Portfolio Illustrating Current Economic Uses of Renewable Biomass Energy.* Winnipeg, Canada: Biomass Energy Institute, Inc., March 1978. HCP/T0285-01.

"Brazil Hot After Gasohol." *The Denver Post,* August 23, 1978.

Bureau of Mines Circular No. 8549. *Energy Potential from Organic Wastes: A Review of the Quantities and Sources.* Washington, DC: U.S. Bureau of Mines, 1972.

Burwell, C. C. "Solar Biomass Energy: An Overview of U.S. Potential." *Science,* **199** (4333): 1041-8 (March 10, 1978).

Capturing the Sun Through Bioconversion. Proceedings of a Conference, March 10-12, 1976, Washington, DC. Washington, DC: Center for Metropolitan Studies, 1976.

Clark, Wilson. "Solar Bioconversion." In *Energy for Survival.* Garden City, NY: Anchor Press/Doubleday, 1974.

Designing an Intensified National Biomass Energy Effort. Bio-Energy Seminar, October 18, 1977. Available from: The Bio-Energy Council, 1337 Connecticut Avenue N.W.,Suite 204, Washington, DC 20036.

Energy Research and Development Agency, Division of Solar Energy. *Solar Program Assessment: Environmental Factors.* Washington, DC: Energy Research and Development Agency, March, 1977. Includes the following reports: "Solar Heating and Cooling of Buildings," ERDA 77-47/1; "Solar Agricultural and Industrial Process Heat," ERDA 77-47/2; "Photovoltaics," ERDA 77-47/3; "Solar Thermal Electric," ERDA 77-47/4; "Solar Total Energy Systems," ERDA 77-47/5; "Wind Energy Conversion," ERDA 77-47/6; "Fuels from Biomass," ERDA 77-47/7; and "Ocean Thermal Energy Conversion," ERDA 77-47/8.

Evaluation of the Use of Agricultural Residues as an Energy Feedstock: A Ten-Site Survey, Stanford Research Institute, Vol. 2, January 1978, TID-27904; and *Cost Analysis of Aquatic Biomass Systems,* July 25, 1978; C00/4000-78/1.

Executive Office of the President, Council on Environmental Quality. *Solar Energy—Progress and Promise.* Washington, DC: Council on Environmental Quality, April, 1978.

Federal Operations Panel. *Domestic Policy Review of Solar Energy: A Response Memorandum to the President of the United States.* Washington, DC: U.S. Department of Energy, February, 1979. TID 28834. (Formerly TID 22834.)

Fuel From Farms—A Guide to Small Scale Ethanol Production, Golden, CO: Solar Energy Research Institute, 1980.

Golueke, Clarence G., and McGauhey, P. H. "Waste Materials." In: *Annual Review of Energy.* Jack M. Hollander (ed.) Palo Alto, CA: Annual Review, Inc.

Horner, and Shifrin, St. Louis. *Energy Recovery from Waste.* Washington, DC: U.S. Environmental Protection Agency, 1972.

Hunt, V. Daniel. *The Gasohol Handbook.* New York: Industrial Press, 1981.

Jackson, Frederick R. *Energy from Solid Waste.* Park Ridge, NJ: Noyes Data Corporation, 1974.

Klass, D. L. "Energy from Biomass and Wastes: 1978 Update." *Energy From Biomass and Wastes:* Institute of Gas Technology Symposium, August 14-18, 1978, Washington, DC. Chicago, IL: Institute of Gas Technology, Copyright September, 1978.

Lipinsky, E. S. "Fuels From Biomass: Integration With Food and Materials Systems." *Science,* **199** (4329): 644-51 (February 10, 1978).

Madewell, J. R. "Quads From Biomass." *Energy From Biomass and Wastes,* Institute of Gas Technology Symposium, August 14-18, 1978, Washington, DC. Chicago, IL: Institute of Gas Technology.

"Marine Biomass." *Energy Users Report,* (No. 263), Washington, DC: Bureau of National Affairs, Inc., August 24, 1978.

Metz, William D. "Report of the OSTP Working Group on Basic Research in the Department of Energy." *Science,* **202** (4365): 293-94 (October 20, 1978).

National Research Council, Canada. Research Plans and Publications Section. *Wood and Charcoal as Fuel for Vehicles.* 3d Ed. Ottawa, Canada, 1944.

Overview of Research and Development Activities in Biomass. Booz-Allen & Hamilton, Inc., August, 1978. Prepared for the Gas Research Institute.

Poole, A. *Energy From Biomass: A Conceptual Overview; Final Report.* Oak Ridge, TN: Institute for Energy Analysis, September, 1977.

Ramsev, E. M. *Biomass Conversion Industry—A Description of the Market Chain.* Golden, CO: Solar Energy Research Institute, August 9, 1978.

Rensselaer Polytechnic Institute. *Second Annual Fuels From Biomass Symposium,* Troy, NY, June 20-22, 1978.

"Resource Applications: Moving Proven Energy Into the Marketplace." *Energy Insider,* **1** (23): 4 (August 21, 1978).

Salo, D. J., et al. "Land Suitability and Availability." In: *Silvicultural Biomass Farms,* Vol. 3. McLean, VA: MITRE Corporation, May, 1977, MTR 7347.

Salo, D. J., et al. *Near-Term Potential of Wood as a Fuel.* McLean, VA: MITRE Corporation, August, 1978. HCP/T41-1-02.

Schooley, F. A. *Mission Analysis for the Federal Fuels From Biomass Program. Vol. 1: Summary and Conclusions.* Menlo Park, CA: Stanford Research International, December, 1978.

Science and Public Policy Program, University of Oklahoma, Organic Farms. *Energy Alternatives: A Comparative Analysis.* Washington, DC: U.S. Government Printing Office, May 1975.

Stanford Research International. *Crop, Forestry and Manure Residue Inventory: Continental United States.* Vols. 1-8. Menlo Park, CA: Stanford Research International, June, 1976. TID-27162.

U.S. Department of Energy, Office of Alcohol Fuels. *Alcohol Fuels Program Plan.* Washington, DC: U.S. Department of Energy, March, 1981.

U.S. Department on Agriculture, Forest Service. *A National Framework for Energy From Biomass.* Washington, DC: U.S. Department of Agriculture, April 28, 1978.

U.S. Department of Energy, Assistant Secretary for Conservation and Solar Applications, and Assistant Secretary for Environment. *Environmental Development Plan (EDP), 1977.* Washington, DC: U.S. Department of Energy, March, 1978. Includes the following reports: "Solar Agricultural and Industrial Process Heat," DOE/EDP-0002; "Fuels from Biomass," DOE/EDP-0005; "Ocean Thermal Energy Conversion," DOE/EDP-0006; and "Wind Energy Conversion," DOE/EDP-0007.

U.S. Department of Energy. *Solar Energy: A Status Report.* Washington, DC: U.S. Department of Energy, June, 1978. DOT/ET-0062.

U.S. Department of Energy, Domestic Policy Review of Solar Energy Integration Group. *Status Report on Solar Energy domestic Policy Review.* Washington, DC: U.S. Department of Energy, August 25, 1978. DOE/ET-0021/1.

U.S. Department of Energy, Division of Solar Technology. *Fuels From Biomass Program: Program Summary.* Washington, DC: U.S. Department of Energy, January, 1978. DOE/ET-0022/1.

U.S. Department of Energy, Solar Working Group. *Solar Energy Research and Development: Program Balance: a*

Review. Washington, DC: U.S. Department of Energy, February, 1978. DOE/IR-004.

Witwer, J. R., et al. *A Comparative Evaluation of Solar Alternatives: Implications for Federal RD&D*. Menlo Park, CA: Stanford Research International, January, 1978. TID-28533/2.

Wood Energy Institute, Society for the Protection of New Hampshire Forests. *Wood'N Energy*. For information contact: Society for the Protection of New Hampshire Forests, 5 South Street, Concord, NH 03301.

Concentrator Collectors

Baldwin, A. J. *Performance and Durability of Acid-Etched Solar Glass*. Minneapolis, MN: Honeywell, Inc., Energy Resources Center, 1980.

Bowmar, S. H.; Brown, C. T.; and Teague, H. L. "Advanced Component Test Facility." *Review of Advanced Solar Thermal Power Systems*. November 15-17, 1978. Denver, CO. DOE/JPL-1060-78/5.

Butler, B. L. "Solar Materials R&D Overview for Solar Thermal Power Systems." *11th National SAMPE Technical Conference,* November 13-15, 1979, Boston MA. Available from Society for the Advancement of Material and Process Engineering, P.O. Box 613, Azusa, CA 91702.

Collares-Pereira, M., and Rabl, A. *Simple Procedure for Predicting Long Term Average Performance of Nonconcentrating and of Concentrating Solar Collectors*. Argonne, IL: Argonne National Laboratory, June, 1978. ANL-78-67.

Dudley, V. E., and Workhoven, R. M. *Summary Report: Concentrating Solar Collector Test Results, Collector Module Test Facility*. Albuquerque, NM: Sandia Laboratories, March, 1979. SAND 76-09771.

Duffie, J. A., and Beckman, W. A. *Solar Energy Thermal Processes*. New York: John Wiley & Sons, 1974.

Gupta, B. P., and Kreith, F. (eds.) *Proceedings of the Solar Thermal Concentrating Collector Technology Symposium*. Golden, CO: Solar Energy Research Institute, August, 1978. SERI/TP-34-048.

Harrison, D., and McCulloch, W. H. *Midtemperature Solar Systems Test Facility (MSSTF) Project Test Results*. Albuquerque, NM: Sandia Laboratories, November, 1978. SAND 78-1088.

Kreith, F., and Kreider, J. F. *Principles of Solar Engineering*. New York: McGraw-Hill Book Company, 1978.

Leonard, James A. *Linear Concentrating Solar Collectors—Current Technology and Applications*. Albuquerque, NM: Sandia Laboratories, March, 1979. SAND 76-09771.

Solar Industrial Process Heat Conference Proceedings, October 18-20, 1978. Golden, CO: Solar Energy Research Institute, January, 1979. SERI/TP-49-065.

Solar Thermal Test Facilities Users Association, Proceedings of Annual Meeting: Technical Sessions. April 11-12, 1978, Golden, CO.

Spectrolab Inc. *System Design Report for a 10 Kilowatt Photovoltaic Concentrator System,* May, 1978. SAND 78-7025.

Teagan, W. P., and Cunningham, D. R. *Conceptual Design and Analysis of a Compound Parabolic Concentrator Collector Array*. Cambridge, MA: Arthur D. Little, Inc., August, 1977. ANL-K77-3855-1.

Electrical and Mechanical Storage

Baylin, F. *An Assessment of Battery Technologies*. Unpublished SERI report, October, 1978. For further information contact: Frank Baylin, Solar Energy Research Institute, Golden, CO.

Electric Power Research Institute. *An Assessment of Energy Storage Systems Suitable for Use by Electric Utilities*. Palo Alto, CA: Electric Power Research Institute, July, 1976. EPRI Report EM-264.

Ma, F. S., and Isaksen, L. *Impacts of Dispersed Supply Management on Distribution Systems*. Systems Control, Inc., June, 1978.

Metz, William D. "Energy Storage and Solar Power: An Exaggerated Problem," *Science,* **200** (4349): 1471-73, June, 1978.

Systems Control, Inc. *Network Reliability of Dispersed Storage*. Palo Alto, CA: Electric Power Research Institute, July, 1978. EPRI Report RP-917.

U.S. Department of Energy. *Applied Research on Energy Storage and Conversion for Photovoltaic and Wind Energy Systems*. Washington, DC: U.S. Department of Energy, January, 1978. Report HCP/T-22221-1.

U.S. Department of Energy, Assistant Secretary for Energy Technology. *Solar, Geothermal, Electric and Storage Systems Program Summary Document*. Washington, DC: U.S. Department of Energy, March, 1978, and January 20, 1979.

U.S. Department of Energy. *Solar, Geothermal, Electric and Storage Systems, Program Summary Document, FY 1980.* Washington, DC: U.S. Department of Energy, January 20, July, 1979. DOE/ET-0102.

Energy, General

Clark, Wilson. *Energy for Survival—The Alternative to Extinction.* Garden City, NY: Anchor Press/Doubleday, 1975.

Dubin, Fred, and Long, Chalmers. *Energy Conservation Standards.* New York: McGraw-Hill Book Company, 1978.

Energy—Global Prospects 1985-2000. Workshop on Alternative Energy Strategies. New York: McGraw-Hill Book Company, 1977.

Energy in America's Future. Baltimore, MD: The Johns Hopkins University Press, 1979.

Energy—The Next Twenty Years. Cambridge, MA: Ballinger Publishing Company, 1979.

Energy Policy Project of the Ford Foundation. *A Time to Choose: America's Energy Future.* Cambridge, MA: Ballinger Publishing Company, 1974.

Federal Energy Administration. *Project Independence Report.* Washington, DC: Government Printing Office, 1974. GPO Stock No. 4118-00029.

Hammond, Allen L.; Metz, William D.; and Maugh II, Thomas H. *Energy and the Future.* Second Printing. Washington, DC: American Association for the Advancement of Science, 1973.

Hollander, Jack M. (ed.) *Annual Review of Energy.* Palo Alto, CA: Annual Reviews, 1981.

Hunt, V. Daniel. *Handbook of Conservation and Solar Energy.* New York: Van Nostrand Reinhold, 1982.

Kahn, Herman. *The Next 200 Years.* New York: William Morrow & Company, 1976.

Maddox, John. *Beyond the Energy Crisis: A Global Perspective.* New York: McGraw-Hill Book Company, 1975.

National Research Council. *Energy in Transition 1985-2010.* Final Report of the Committee on Nuclear and Alternative Energy Systems. San Francisco CA: W. H. Freeman & Company, 1979.

Science and Public Policy Program. University of Oklahoma, Norman, OK. *Energy Alternatives, A Comparative Analysis.* Washington, DC: Council on Environmental Quality, 1975.

Yergin, Daniel, et al. (eds.) *Energy Future—Report of the Energy Project at the Harvard Business School.* New York: Random House, 1979.

Federal Solar Program Controversies

Bennington, G., et al. *Solar Energy: A Comparative Analysis to the Year 2020.* McLean, VA: The MITRE Corporation, METREK Division, March, 1978. MTR-7579.

"DOE Skimps on Basic Research." *Industrial Research/Development,* September, 1978.

Executive Office of the President, Council on Environmental Quality. *Solar Energy: Progress and Promise.* Washington, DC: Government Printing Office, April, 1978. GPO No. 041-011-00036-0.

Katzenberg, R. "Solar Policy Review." *Wind Power Digest,* Winter 1978.

Lovins, A. B. *Soft Energy Paths: Toward a Durable Peace.* Cambridge, MA: Ballinger Publishing Company, 1977.

Morrison, D. C. *Energy, Appropriate Technology, and International Interdependence.* Prepared for the Annual Meetings of the Society for the Study of Social Problems, San Francisco, California, September 1978. For further information, contact the author at the Woodrow Wilson International Center for Scholars, Smithsonian Institution Building, Washington, DC 20560.

Nash, H. (ed.) *The Energy Controversy.* San Francisco, CA: Friends of the Earth, 1979.

Odland, R., et al. *Decentralized Energy Studies: Progess Report No. 2.* Golden, CO: Solar Energy Research Institute, April 1979. SERI/PR-53-220.

Perry, H., and Streiter, S. *Multiple Paths for Energy Policy: A Critique of Lovins' Energy Strategy.* New York: National Economic Research Associates, Inc., 1978.

Ray, D. L., Chairman, U.S. Atomic Energy Commission. *The Nation's Energy Future.* A Report to Richard M. Nixon, President of the United States. Washington, DC: Atomic Energy Commission, December 1, 1973. Wash-1281.

Simard, R. L. *A Critique of Amory Lovins' Soft Energy Paths.* Oak Ridge, TN: Oak Ridge National Laboratory, 1978. ORNL/TM-6233.

Solar Energy Research Institute. *Public Meeting on Solar Technology,* August 8-9, 1978, Washington, DC. Golden, CO: Solar Energy Research Institute.

SERI/TP-54-205.

Solar Lobby. *Blueprint for a Solar America.* Washington, DC: Solar Lobby, January 1979.

U.S. Congress, Office of Technology Assessment. *Application of Solar Technology to Today's Energy Needs.* Washington, DC: Government Printing Office, June, 1978. GPO No. 052-033-00539-5.

U.S. Department of Energy, Solar Working Group. *Solar Energy Research and Development: Program Balance.* Washington, DC: U.S. Department of Energy, February, 1978.

U.S. Department of Energy. *Solar Energy: A Status Report.* Washington, DC: U.S. Department of Energy, June, 1978. DOE/ET-0062.

U.S. Department of Energy. *Domestic Policy Review of Solar Energy: A Response Memorandum to The President of the United States.* Washington, DC: U.S. Department of Energy, TID-22834.

United States General Accounting Office. *The Magnitude of the Federal Solar Energy Program and the Effects of Different Levels of Funding.* Washington, DC: United States General Accounting Office, February 2, 1978. EMD-78-27.

Whitwer, J. G., et al. *A Comparative Evaluation of Solar Alternatives: Implications for Federal R&D.* Menlo Park, CA: Stanford Research International, January, 1978. SRI Project 6375.

Financing Solar Energy

American Insurance Association. *Special Hazards Bulletin—Solar Energy Systems,* 1977. Z-70.

Isakson, N. "Making the Most of Energy in Real Estate." *Society of Real Estate Appraisers Seminar,* Spring, 1978. Sponsored by U.S. Department of Energy.

Smith, P. K. *Information Requirements of the Solar Community—A Management Summary.* Golden, CO: Solar Energy Research Institute, August, 1978.

U.S. Department of Energy, Domestic Policy Review of Solar Energy Integration Group. *Status Report on Solar Energy Domestic Policy Review.* Washington, DC: U.S. Department of Energy, August 25, 1978. Public Review Copy.

International Projects

Ashworth, John H. *Renewable Energy Sources For the World's Poor: A Review of Current International Development Assistance Programs.* Golden, CO: Solar Energy Research Institute, May, 1979. Appendix. SERI/TR-51-195.

Bracken, R. B. *Status of Collaborative Programs in International Solar Energy.* Golden, CO: Solar Energy Research Institute, February, 1978. SERI-18.

Campbell, R. *Export Potential for Photovoltaic Systems.* Richland, WA: Pacific Northwest Laboratory, April, 1979. Prepared for U.S. Department of Energy. Available from: NTIS, Springfield, VA. DOE/CS-0078.

Corcoran, W. L. "Commercialization Activities—International Solar Commercialization Working Group (ISCWG)." Sponberg, R. C., and Corcoran, W. L. (eds.) *DOE/Solar Export Opportunities Workshop,* January 8-10, 1979, Atlanta, GA. Golden, CO: Solar Energy Research Institute, 1979. pp. 95-111.

Dematrops, J. *European Industrial Fnergy Conservation Technologies: A Survey.* Washington, DC: Department of Energy, Conservation and Solar Applications, Fall, 1979. Available from: Technical Information Center (TIC), Oak Ridge, TN.

Diachok, D. *Addendum to Proposal for Cooperative Research Grant: Further Development of Two-Sided Solar Cells.* Golden, CO: Solar Energy Research Institute, February, 1979, SERI/SP-43-135.

Diachok, D. *Addendum to Proposal for Cooperative Research Grant: Methane Production from Anaerobic Breakdown of Urban Solid Waste.* Golden, CO: Solar Energy Research Institute; February, 1979. SERI/SP-43-156.

Gopalakrishnan, N. K. (Tata Energy Research Institute, Bombay.) "Global Survey of Solar Energy R&D." *Urja.* **3** (2): 43-50 (January 21, 1978).

International Energy Agency. *Annual Report on Energy Research, Development and Demonstration; Activities of the IEA, 1977-1978.* Paris: Organization for Economic Cooperation and Development, 1978.

National Academy of Sciences, Board on Science and Technology for International Development. *Energy for Rural Development: Renewable Resources and Alternative Technologies for Developing Countries.* Washington, DC: National Academy of Sciences, Commission on International Relations, 1976. Report No. 18.

National Research Council, Panel on

Methane Generation. *Methane Generation from Human, Animal and Agricultural Wastes.* Washington, DC: National Academy of Sciences, 1977. Available from NTIS, Springfield, VA. PB-276 469.

North Atlantic Treaty Organization, Committee on the Challenges of Modern Society (CCMS). *Solar Energy Pilot Study.* College Park, MD: University of Maryland, Department of Mechanical Engineering, CCMS Projects Office, October, 1978. CCMS No. 83/UMD-4908-13.

Saudi Arabian Ministry of Finance and National Economy, U.S. Department of Energy and U.S. Department of Treasury. *Saudi Arabian-United States Joint Commission on Economic Cooperation: Project Agreement.* Washington, DC: Government Printing Office, October, 1978. 4 p. 261-155/2026.

United States Agency for International Development. *Mali Renewable Energy Project Paper.* Washington, DC: United States Agency for International Development, June 1, 1978. Appendix.

Labor

Butt, S. *Solar Market Capture and Market Penetration.* Washington, DC: Solar Energy Industries Association, October, 1976.

Carlough, E. J. Solar Energy Industries Association Annual Meeting, Boston, MA, October 4, 1978.

Chapman, D. "Energy Conservation, Employment, and Income." *Operations Research Society of America,* May 9-11, 1977.

Coates, J. F. "Population and Education: How Demographic Trends Will Shape the U.S." *The Futurists,* **XII** (1), February, 1978.

Environmentalists for Full Employment. *Joint Economic Committee Holds Jobs and Energy Hearings: Energy and Labor Departments Confess.* Washington, DC: Environmentalists for Full Employment, Spring, 1978. p. 1.

Federal Energy Administration. *Project Independence Blueprint Task Force Report on Solar Energy.* Washington, DC: U.S. Department of Commerce, November, 1974. PB 248 507.

Federal Energy Administration. *Project Independence Blueprint Task Force Report on Nuclear Energy.* Washington, DC: U.S. Department of Commerce, November, 1974. PB 248 504.

Impacts Panel. *Domestic Policy Review of Solar Energy.* Washington, DC: U.S. Department of Energy, October, 1978. TID #28835.

Mason, B., and Armington, K. *Direct Labor Requirements for Select Solar Energy Technologies: A Review and Synthesis.* Golden, CO: Solar Energy Research Institute, August, 1978.

Orsak, C. G., et al. *Assessment of the Need for Developing and Implementing Technical and Skilled Worker Training for the Solar Energy Industry.* Corsicana, TX: Navarro College, January, 1978. DOE No. EG-77-S-04-3869.

Schurr, Sam H., and Darmstadter, J. "The Energy Connection." *Resources Bulletin,* (53). Aspen Seminar on Perspectives on Energy and Society, August 18, 1976. Washington, DC: Resources for the Future, Fall 1976.

Shulze, W. D., and Ben-David, S. *The Economics of Solar Home Heating: A Study Prepared for the Use of the Joint Economic Committee.* Washington, DC: Congress of the United States, March 13, 1977.

Solar Energy Research Institute, Market Development Branch, Technology Commercialization Division. *Organized Labor and Solar Energy.* Golden, CO: Solar Energy Research Institute, February, 1979. SERI/TR-62-148.

Solar Energy Research Institute, Market Development Branch, Technology Commercialization Division. *Organized Labor and Solar Energy.* Golden, CO: Solar Energy Research Institute, February, 1979. SERI/TR-62-148.

Spewak, P. C. *Labor Requirements for Solar Heating, Cooling, and Hot Water.* McLean, VA: MITRE Corporation, September, 1977. MITRE Working Paper No. 12569.

U.S. Congress, Office of Technology Assessment. *Solar Technology to Today's Energy Needs,* June, 1977.

U.S. Congress. Subcommittee on Energy of the Joint Economic Committee of Congress. *Creating Jobs Through Energy Policy.* Washington, DC, March 15-16, 1978.

U.S. Department of Energy, News Release, May 25, 1978.

Materials Research

American Society of Mechanical Engineers. *Journals of Solar Energy Engineering.* F. Kreith (ed.). New York: American Society of Mechanical Engineers, 1979.

Buchsbaum, S. J. (ed.) Report of the Office

of Science and Technology Policy Working Group on Basic Research in the Department of Energy. June, 1978.

Oak Ridge National Laboratory. *Reference Guide to the Technical Literature of the Metals and Ceramics Division of Oak Ridge National Laboratory, April, 1979-June, 1979.* Oak Ridge, TN: National Laboratory, September, 1979. ORNL/MC-27/2.

Pacific Northwest Laboratories. *A Methodology for Identifying Materials Constraints to Implementation of Solar Energy Technologies.* Pacific Northwest Laboratories for the Department of Energy, July, 1978. PNL-2711, UC-63.

Research Projects in the Physical Sciences. April, 1978. DOE/ER-0003.

Seraphin, B. O. (ed.) *Solar Energy Materials.* Amsterdam, Netherlands: North Holland Publishing Company, **1** (1 and 2) (February, 1979).

Solar Energy Research Institute. *Solar Optical Material Program Activity Committee, Progress Report, July 1, 1978 to January 31, 1979.* Golden, CO: Solar Energy Research Institute, July, 1979. SERI/PR-31-137.

U.S. Department of Energy. *Materials Sciences Program Fiscal Year 1978.* Office of Basic Energy Sciences, September, 1978. DOE/ER-0013.

U.S. Department of Energy. *Solar Heating and Cooling Research and Development Project Summaries.* May, 1978. DOE/CS-0010.

U.S. Department of Energy. *Solar Thermal Power Systems Program.* January, 1978. DOE/ET-0018/1.

Ocean Energy

Clegg, Peter. *New Low-Cost Sources of Energy for the Home; with Complete Illustrated Catalog.* Charlotte, VT: Garden Way Pub., 1975.

Creager, William Pitcher; Justin, Joel D.; and Hinds, Julian. *Engineering for Dams.* New York: J. Wiley & Sons; London, Chapman & Hall Ltd., 1945.

Davey, Norman. *Studies in Tidal Power.* London: Constable & Co., Ltd., 1923.

Gillett, Colin Anson. *Can the Tides be Harnessed?* Raetihi, N.Z., 1958.

Handbook of Homemade Power. By the Staff of the Mother Earth News. New York: Bantam Books, 1974.

Macmillian, Donald Henry. *Tides.* New York: American Elsevier Pub. Co., 1966.

Morton, M. Granger. (ed.) Ocean Thermal Energy Conversion. *Energy and Man: Technical and Social Aspects of Energy.* New York: IEEE Press, 1975.

Paton, Thomas Angus Lyall, and Brown, J. Guthrie. *Power from Water.* London: L. Hill, 1961.

Rash, Don E., et al. *Energy Under the Ocean: A Technology Assessment.* Norman, OK: University of Oklahoma Press, 1973.

Reynolds, John. *Windmills and Watermills.* London: H. Evelyn, 1970.

Ross, David. *Energy From the Waves.* Oxford, England: Pergamon Press, 1979.

Tidal Power: Proceedings of the International Conference on the Utilization of Tidal Power, Nova Scotia Technical College, 1970. Gray, T. J., and Gashus, O. K. (eds.) New York: Plenum Press, 1972.

Wilson, Paul N. *Water Turbines.* London: H. M. Stationery Office, Obtainable in the U.S.A. Palo Alto, CA: Pendragon House, 1974.

Ocean Thermal Energy Conversion

Bender, E. "The OTEC Gamble." *Sea Technology,* 6 (August, 1979).

Booda, L. "OTEC Leads the Way—Ocean Energy Challenges Technology Grows." *Sea Technology,* 10-17 (August, 1978).

Cohen, R. "An Overview of the U.S. OTEC Development Program." *Proceedings of the Energy Technology Conference,* November, 1978, Houston, TX.

Cohen, R., and Dunning, F. S., Jr. "On Island Strategy for OTEC Commercialization." Solar Energy and Conservation Symposium-Workshop, December 11-13, 1978, Miami, FL. Miami, FL: University of Miami, 1978.

Energy Research and Development Agency, Division of Solar Energy. *Solar Program Assessment: Environmental Factors.* Washington, DC: Energy Research Development Agency, March, 1977. Includes the following reports: "Solar Heating and Cooling of Buildings," ERDA 77-47/1; "Solar Agricultural and Industrial Process Heat," ERDA 77-47/2; "Photovoltaics," ERDA 77-47/3; "Solar Thermal Electric," ERDA 77-47/4; "Solar Total Energy Systems." ERDA 77-47/5; "Wind Energy Conversion," ERDA 77-47/6; "Fuels from Biomass," ERDA 77-47/7; and "Ocean Thermal Energy Conversion," ERDA 77-47/8.

MITRE Corporation, METREK Division. *OTEC Commercialization Analysis.* McLean, VA: The MITRE Corporation, January, 1979, MTR-7963.

Office of Technology Assessment. *Renewable Ocean Energy Sources.* Washington, DC: Office of Technology Assessment, U.S. Congress, May 1978. OTA-0-63.

U.S. Department of Energy, Assistant Secretary for Conservation and Solar Applications and Assistant Secretary for Environment. *Environmental Development Plan (EDP), 1977.* Washington, DC: U.S. Department of Energy, March, 1978. Includes the following reports: "Solar Agricultural and Industrial Process Heat," DOE/EDP-0002; "Fuels from Biomass," DOE/EDP-0005; "Ocean Thermal Energy Conversion," DOE/EDP-0006; and "Wind Energy Conversion," DOE/EDP-0007.

U.S. Department of Energy, Assistant Secretary for Energy Technology. *Solar, Geothermal, Electric and Storage Systems Program Summary Document FY 80.* Washington, DC: U.S. Department of Energy, January 20, 1979.

U.S. Department of Energy. *Energy Insider,* **2** (1) (January 8, 1979).

U.S. Department of Energy. *Ocean Currents Energy Conversion and Ocean Waves Energy Conversion.* U.S. Department of Energy Program Information Handout. Available from: Wilbur Sherwood, Program Manager, (202) 376-4835.

U.S. Department of Energy. *Ocean Systems, Program Summary.* Washington, DC: U.S. Department of Energy, December, 1978. DOE/ET-0083.

U.S. Department of Energy. *Ocean Systems Program Summary.* Washington, DC: U.S. Department of Energy, February, 1979. DOE/ET-0083.

Watt, A. D.; Mathews, F. S.; and Hathaway, R. E. *Open Cycle Ocean Thermal Energy Conversion—A Preliminary Engineering Evaluation.* Golden, CO: Colorado School of Mines, Engineering Physics Department, December 31, 1977. ALO/3723-76/3.

Other Conversion Technologies

Armas, B.; Combesure, C.; Dusseau, J. M.; Darolles, J. M.; Robert, J. L.; and Pistonlet, B. "Preparation of Refractory Thermoelectric Materials by CVD—Application to Solar Energy Conversion." Electrochemical Society Fall Meeting, 1976, pp. 646-47.

Briot, R.; Gonnard, P.; and Fetiveau, Y. "Conversion Thermoélectrique par Transitions de Phase avec des Matériaux Ferroélectriques." *Revue de Physique Appliqué.* **13:** 366 (July, 1978).

Cole, T. "Thermoelectric Energy Conversion with the Sodium Heat Engine." Solar Thermal Test Facilities Users Association Annual Meeting, April, 1978. Golden, CO.

Elliott, G. R. B. *Night Storage and Backup Generation with Electrochemical Engines.* Los Alamos, NM: Los Alamos Scientific Laboratory, 1978. Report No. LA-VR-78-606.

Fitzpatrick G. O., and Britt, E. J. "Thermionic Energy Conversion for Solar Applications." Solar Thermal Test Facilities Users Association Meeting, April, 1978, Golden, CO.

Jayadev, T. S.; Benson, D.; Henderson, J.; and Finegold, J. *Assessment of Solar Conversion Technologies—Solar Thermoelectrics, Vol. I.* Golden, CO: Solar Energy Research Institute, August, 1979. SERI/TR-035-078.

Neill, J. M.; Elsner, N. B.; and Sonn, D. L. *Thermoelectric Application to Solar Power.* November, 1978. GA-C15168.

Pisharody, R. K., and Garvey, L. P. "Modified Silicon-Germanium Alloys with Improved Performance." 13th Intersociety Energy Conversion Engineering Conference, August, 1978.

Sotirova, M.; Sotirov, S.; and Andreev, A. "Production and Properties of Thermoelectric Substances Based on Bismuth Telluride-Selenide Compounds." *Friedberg-Forschungs, Chemical Abstracts No. 143604.* Vol. 84, p. 570, 1978.

U.S. Department of Energy. *Thermionic Energy Conversion Technology Development.* U.S. Department of Energy, Program Plan, January, 1978.

Weber, N. "A Thermoelectric Device Based on Beta-Alumina Solid Electrolyte." *Energy Conversion,* **14** (1): 2 (1974).

Yater, J. C. "Reversible Thermoelectric Power Conversion of Energy Fluctuations." Second International Conference on Thermoelectric Energy Conversion, March, 1978, University of Texas—Arlington.

Passive Solar Heating and Cooling

Feldman, S., and Anderson, B. *The Impact of Active and Passive Solar Building Designs on Utility Peak Loads.* Washington, DC: Energy Research and Development Administration, September 30, 1977. ERDA Contract No. EG-77-G-01-4029.

Holtz, Michael J.; Place, W.; and Kemmerud, R. *A Classification Scheme for the Common Passive and Hybrid Heating and Cooling Systems.* Golden, CO: Solar Energy Research Institute, August 1979, Prepared for the Department of Energy, TP-63-218.

Judkoff, R. *A National Plan for Passive Design and Analysis Tool Development.* Golden, CO: Solar Energy Research Institute, Passive Technology Branch, 1979.

Mazeria, E. *The Passive Solar Energy Book.* Emmaus, PA: Rodale Press, 1979.

McFarland, R. D., and Balcomb, J. D. "The Effect of Design Parameter Changes on the Performance of Thermal Storage Wall Passive Systems." San Jose, CA: A paper prepared under the auspices of the U.S. Department of Energy for the Third National Passive Solar Conference, January 12-13, 1979.

Morse, F., and Maybaum, Michael. *Commercialization Strategy of Passive Solar Heating.* Washington, DC: U.S. Department of Energy, November, 1978.

U.S. Department of Energy. *Commercialization Task Force Reports,* Final Draft, Vol. IV, September 1979.

U.S. Department of Energy. *Interim Report: National Program Plan for Passive/Hybrid Heating and Cooling.* Washington, DC: U.S. Department of Energy, June 1979. DOE/CS-0089.

Yanda, B., and Fisher, R. *The Food and Heat Producing Greenhouse: Design, Construction, Operation.* Santa Fe, NM: John Muri Publications, 1976.

Photoconversion

Allen, M. J. "Direct Conversion of Radiant into Electrical Energy Using Plant Systems." *Living System as Energy Converters,* New York: Elsevier/North Holland Biomedical Press, 1977.

Bolton, J. R. (ed.). *Proceedings of the Fourth International Congress on Photosynthesis,* 1977, pp. 621-34.

Bolton, J. R. "Solar Fuels." *Science,* **202** (4369): 705-11 (November 17, 1978).

"DOE Skimps on Basic Research." *Industrial Research/Development.* **20** **(9)** 50 (September, 1978).

Metz, W. D. "Report of the OSTP Working Group on Basic Research in the Department of Energy." *Science,* **202** (4365): 293-94 (October 20, 1978).

Milne, T.; Connoly, J.; Inman, R.; Reed, T.; and Seibert, M. *Research Overview of Biological and Chemical Conversion Methods and Identification of Key Research Areas for SERI.* Golden, CO: Solar Energy Research Institute, September, 1978. SERI/TR 33-067.

Photovoltaics

Aerospace Corporation. *Overview of Photovoltaic Market Studies.* Washington, DC: U.S. Department of Energy, May, 1978.

Barnett, A. M.; Bragagnolo, J. A.; Hall, R. B.; Phillips, J. E.; and Meakin, J. D. "Achievement of 9.159% Efficiency in Thin Film CdS/Cu_2S Solar Cells." 13th Institute of Electrical and Electronics Engineers (IEEE) Photovoltaics Specialists Conference—1978, June 5-8, 1978, Washington, DC. New York: Institute of Electrical and Electronics Engineers (IEEE), 419-420, 1978.

BDM Corporation. *Photovoltaic Power Systems Market Identification.* Vol. I, May, 1977; Vol. II, November, 1977; Vol. III, November, 1978. Washington, DC: U.S. Department of Energy. HCP/M 2533-01/1.

BDM Corporation. *Photovoltaic Incentive Option.* Washington, DC: August, 1978.

Booz-Allen & Hamilton. *Assessment of Photovoltaics Industry, Market and Technologies,* Washington, DC: U.S. Department of Energy, August, 1975.

Campbell, R. *Export Potential for Photovoltaic Systems.* Richland, WA: Pacific Northwest Laboratory, April, 1979. Prepared for U.S. Department of Energy. Available from NTIS, Springfield, VA. DOE/CS-0078.

Costello, D. *Photovoltaic Venture Analysis,* Golden, CO: Solar Energy Research Institute, Inc., 1978. SERI/TR-52-040.

DeMeo, E. A., and Bos, P. B. *Perspectives on Utility Central Station Photovoltaic Application.* Electric Power Research Institute (EPRI), Special Report. Palo Alto, CA: Electric Power Research Institute (EPRI), January, 1978. ER-589-SR.

Hauser, J. "Multijunction Photovoltaic Devices." Photovoltaics Advanced Materials Review Meeting, October, 1978, Vail, CO. Golden, CO: Solar Energy Research Institute. SERI/TP-49-105.

Intertechnology/Solar Corporation. *Photovoltaic Energy Technology Market Analysis.* Washington, DC: U.S. Department of Energy, January, 1978.

Kazmerski, L. *Photovoltaic Research Report—FY 78.* Golden, CO: Solar Energy Research Institute, March, 1979. SERI/TR-32-150.

Kirpich, A. C., et al. "Regional Analysis of

Residential Photovoltaic System Concepts," 13th Photovoltaic Specialists Conference, June 1978, Washington, DC. New York: Institute of Electrical and Electronics Engineers, p. 1171.

Leonard, S. L. "Central Station Power Plant Applications for Photovoltaic Solar Energy Conversion," 13th Photovoltaic Specialists Conference, June 1978, Washington, DC. New York: Institute of Electrical and Electronics Engineers, p. 1190.

Lindmayer, J. "Characteristics of Semicrystalline Silicon Solar Cells." 13th Institute of Electrical and Electronics Engineers (IEEE) Photovoltaics Specialists Conference—1978, June 5-8, 1978, Washington, DC. New York: Institute of Electrical and Electronics Engineers (IEEE), 1978. pp. 1096-1100.

Masden, G., and Backus, C. E. "Increased Photovoltaic Conversion Efficiency Through Use of Spectrum Splitting and Multiple Cells." 13th Institute of Electrical and Electronics Engineers (IEEE) Photovoltaics Specialists Conference—1978, June 5-8, 1978, Washington, DC. New York: Institute of Electrical and Electronics Engineers (IEEE), 1978. pp. 853-858.

Maycock, P. D. "The Development of Photovoltaics as a Power Source of Large Scale Terrestrial Application." 13th Institute of Electrical and Electronics Engineers (IEEE) Photovoltaics Specialists Conference—1978, June 5-8, 1978, Washington, DC. New York: Institute of Electrical and Electronics Engineers (IEEE), 1978, pp. 9-15.

Moon, R. L.; James, L. W.; Vanderplas, H; Yep, T. O.; Chai, Y. G.; and Antypas, T. A. "Multigap Solar Cell Requirements and the Performance of A1GaAs and Si Cells on Concentrated Sunlight." 13th Institute of Electrical and Electronics Engineers (IEEE) Photovoltaics Specialists Conference—1978; June 5-8, 1978, Washington, DC. New York: Institute of Electrical and Electronics Engineers (IEEE), 1978.

Noland, M. *SERI Annual Operating Plan Fiscal Year 1978.* Golden, CO: Solar Energy Research Institute, October, 1977.

Noland, M. *SERI Annual Operating Plan Fiscal Year 1979,* Vol. III, pp. 10-22; Vol. II, pp. 5-17. Golden CO: Solar Energy Research Institute, May, 1978. SERI-25.

Ratten, E. J. "Survey of Near Term Civilian Application in the U.S." *Mission Analysis of Photovoltaic Solar Energy Conversion,* Vol. II. March, 1977. SAN/1101/PAS-1/2.

Sears, D. R.; Marrifield, D. V.; and Perry, M. M. *Environmental Impact Statement for a Hypothetical 1000 MWe Photovoltaic Solar-Electric Plant.* Lockheed Missles and Space Company, Inc., August, 1977. EPA-600-77-085. Prepared for U.S. Environmental Protection Agency, Office of Research and Development, Cincinnati, OH.

Solar Energy Research Institute. "Plan for Development of Preliminary Performance Criteria and Test Standards for Photovoltaic Systems." *Annual Review of Solar Energy.* Golden, CO: Solar Energy Research Institute, November, 1978, pp. 97-100. SERI/TR-54-066.

Terrestrial Central Station Array Analysis Support Study, Bechtel National, Inc., May 1978. DOE/JPL954848-78/1.

U.S. Department of Energy. *National Photovoltaic Program Plan,* U.S. Department of Energy, March, 1978. DOE/ET-0035 (78).

U.S. Department of Energy, Office of Technology Assessment. *Application of Solar Technology to Today's Energy Needs.* Vol. I. Washington, DC: U.S. Department of Energy, Office of Technology Assessment, 1978. pp. 406-407.

U.S. Department of Energy. *Photovoltaic Power Systems Market Identification and Analysis.* U.S. Department of Energy. HCP/M2533-01/2.

U.S. Department of Energy. *Photovoltaics Systems, Program Summary.* Washington, DC: U.S. Department of Energy, December, 1978. DOE/ER-0075.

U.S. Public Law 95-590. *National Solar Photovoltaic Energy Research, Development, Demonstration Act of 1978.*

Social Impacts of Solar Energy

Berry, Brian, "The Counterurbanization Process; How General?" *Human Settlement Systems—International Perspective on Structure, Change, and Public Policy.* Hansen, Niles N. (ed.) Cambridge, MA: Ballinger Press, 1978.

Brunner, Ronald D. *Energy Conservation and Renewable Resources: A Decentralized Approach.* Institute of Public Policy Studies, University of Michigan, December, 1978.

Cambel, A. B.; Coates, V. T.; Hower, C. R.; Kay, P.; and Nutt, H. V. *Social Assessment of Onsite Solar Energy Technologies.* Washington, DC: George Washington University, June, 1978.

Cook, Jeffrey. *Solar Consumers: An Investigation Toward Commercialization.* Arizona State University, June, 1977.

Farhar, B. C.; Weis, P.; Unseld, C. T.; and Burns, B. *Public Opinion about Energy: A Literature Review.* Golden, CO: Solar Energy Research Institute, 1979. SERI/TR-51-155d.

Farhar, B. C.; Unseld, C. T.; Caputo, R.; and Easterling, J. *Citizen Participation in the Domestic Policy Review of Solar Energy.* Golden, CO: Solar Energy Research Institute, January, 1979. SERI/TR-53-126d.

Gallup Organization, Inc. "Energy Crisis—The Public Concerned but not Well Informed." *The Gallup Report, 1.* June, 1977.

Henderson, Hazel. *Creating Alternative Futures.* Berkley Windhaven: 1978.

Hornick, William F., and Enk, Gorden A. *Value Issues in Technology Assessment.* Rensselaerville, New York: The Institute on Man and Science, 1978. Prepared for the U.S. Department of Energy.

Leonard-Barton, Dorothy. *The Diffusion and Adoption of Solar Equipment Among California Homeowners: Report on a Pretest Study.* Institute for Communication Research, Stanford University, December, 1978.

Martino, Joseph P.; ;Lenz, Ralph C., Jr.; and Chen, Juei-Lin. *Technology Assessment: An Appraisal of the State of the Art.* Dayton Research Institute, University of Dayton, 1978.

National Academy of Sciences. *Interim Report of the National Research Council Committee on Nuclear and Alternative Energy Systems.* Washington, DC: National Academy of Sciences, January, 1977.

National Academy of Sciences. "Sociopolitical Impacts of Nuclear and Alternative Energy Systems." *Final Report of the CONAES Panel on Risk/Impact.* National Academy of Sciences, 1979. Chapter 8.

Perlman, R., and Warren, R. L. *Families in the Energy Crises: Impacts and Implications for Theory and Policy.* Cambridge, MA: Ballinger Press, 1977.

Quinn, J., and Ohi, J. *A Compendium of Decentralized Energy Studies, Programs and Projects.* Golden, CO: Solar Energy Research Institute, 1979, SERI/RR-53-316.

Real Estate Research Corporation. *Selling the Solar Home: Some Preliminary Findings.* Washington, DC: U.S. Government Printing Office, 1978. Prepared for the U.S. Department of Housing and Urban Development (HUD).

Roper Organization. *Roper Reports.* New York: Roper Organization, February, March, November, 1977.

RUPI, Inc. *Federal Incentives for Solar Homes, An Assessment of Program Options.* Washington, DC: Division of Energy, Building Technology and Standards Office of Policy Development and Research, U.S. Department of Housing and Urban Devleopment (HUD), November, 1977.

San Diego Gas and Electric Company. *Solar Background Attitude and Awareness Study.* San Diego, CA: San Diego Gas and Electric Company, October, 1976.

U.S. Department of Commerce, Social and Economic Statistics Administration. Bureau of the Census. *We the Americans: Our Cities and Suburbs.* Washington, DC: May, 1978.

U.S. Department of Energy, Office of Science and Technology. *Policy Working Group on Basic Research.* Washington, DC: U.S. Department of Energy, Office of Energy Research, 1978.

Unseld, C. T., and Crews, R. *Residential Solar Energy Users: A Review of Empirical Studies and Related Literature.* Golden, CO: Solar Energy Research Institute, 1979. TR-53-245d.

Unseld, C. T.; Morrison, D. E.; Sills, D. L.; and Wolf, C. P. (eds.) *Sociopolitical Impacts of Energy Policy and Use.* Washington, DC: National Academy of Sciences, Committee on Nuclear and Alternative Energy Sources (CONAES), 1979.

Yarosh, M. M., and Litka, A. H. *Solar Commercialization: The Consumer Experience.* Cape Canaveral, FL: Florida Solar Energy Center, 1978.

Solar Architects and Engineers

Analysis Methods for Solar Heating and Cooling Applications. Golden, CO: Solar Energy Research Institute, Market Development Branch, 1979. SERI/SP-35-232, This work is also published by SERI in a combined publication entitled: *Analysis Methods for Active and Passive Solar Heating and Cooling Applications.*

Energy. Supplement to the American Institute of Architects' Energy *Notebook.* Washington, DC: American Institute of Architects, 1978.

Fitch, James Marston. *American Building, The Environmental Forces That Shape It.* New York: Schocken Books, Inc., 2nd Ed., 1975.

Franta, G. *Solar Design for Public Housing,*

Seminar Report, July 27 and 28, 1978. Sponsored by the Solar Energy Research Institute, Golden, CO. for HUD, Region VIII, October 2, 1978. SERI/SP-62-131.

Franta, G. *Solar Architectural Planning Seminar Proceedings.* July 10 and 11, 1978. Golden, CO: Solar Energy Research Institute, Market Development Branch, March, 1979. SERI-TP-62-113.

Johnson, D., and Tonelli, A. *A Program to Provide Energy Efficient Solar Assisted Dwelling for Low Income Public Housing.* Denver, CO: U.S. Department of Housing and Urban Development, Region VIII, February 8, 1978, p. 2.

Judkoff, R. *Status of Passive Design Tools.* Golden, CO: Solar Energy Research Institute, Passive Technology Branch, 1979. This work is also to be published by SERI in a combined publication entitled: *Analysis Methods for Active and Passive Solar Heating and Cooling Applications.*

Judkoff, R. *A National Plan for Passive Design and Analysis Tool Development.* Golden, CO: Solar Energy Research Institute, Passive Technology Branch, 1979. Available from R. Judkoff, Solar Energy Research Institute, Golden, CO.

Little, A. D. *Comparative Analysis of Building Energy Analysis Computer Codes for EPRI.* Cambridge, MA: Electric Power Research Institute, 1979.

Mazeria, E. *The Passive Solar Energy Book.* Emmaus, PA: Rodale Press, Inc., 1979.

Morse, F. H. *Status of the United States' Solar Heating and Cooling Program.* Washington, DC: U.S. Department of Energy, May, 1979.

Versteegen, D. L. *Survey of Currently Used Simulation Methods.* McLean, VA: Science Applications, Inc., 1978.

Solar Barriers and Government Actions

Battelle Pacific Northwest Laboratory. *An Analysis of Federal Incentives Used to Stimulate Energy Production.* Richland, WA: Battelle Pacific Northwest Laboratory, December, 1978. PNL-2410. Prepared for the U.S. Department of Energy.

Bezdek, R., et al. *Analysis of Policy Options for Accelerating Commercialization of Solar Heating and Cooling Systems.* Washington, DC: George Washington University, Behavioral Studies Group, Program of Policy Studies in Science and Technology, April, 1977.

Citizen's Solar Program. *State Reports on Barriers and Strategies to Renewable Energy Development.* Washington, DC: U.S. Department of Energy, October, 1978. HCP/U6308-01.

Easterling, J., and Greenberg, J. *An Analysis of the Domestic Policy Review.* Golden, CO: Solar Energy Research Institute, 1979. SERI/RR-55-394.

Federal Action List. White House Press Release, June 20, 1979. Released with Department of Public Resources public announcement.

Federal Operations Panel. *Domestic Policy Review of Solar Energy: Final Report.* Washington, DC: U.S. Department of Energy, October, 1978. TID-28832.

"Non-technical Obstacles to the Use of Solar Energy." *Proceedings of the Joint Organizers' Conference, International Symposium,* June 22, 1979, Brussels, Belgium. Golden, CO: Solar Energy Research Institute, August, 1979. SERI/TP-43-355.

Solar Heating and Cooling of Buildings (SHACOB) Commercialization Report. Part A. Options and Strategies. Vol. 1. Kansas City, MO: Midwest Research Institute, May, 1978. Prepared for REA and DOE Task Force on Solar Energy Commercialization. HCP/M70065-01/2.

Solar Lobby. *Blueprint for Solar America.* Washington, DC: Solar Lobby, January, 1979.

U.S. Congress, Office of Technology Assessment. *Applications of Solar Technology to Today's Energy Needs.* Washington, DC: June, 1978. GPO No. 052-003-00539-5.

U.S. Department of Energy. *Commercialization Task Force Reports.* Vol IV. Washington, DC: U.S. Department of Energy, September, 1978.

U.S. Department of Energy. *Domestic Policy Review of Solar Energy. A Response Memorandum to the President of the United States.* Washington, DC: U.S. Department of Energy, February, 1979. TID/Z8834.

U.S. Department of Energy, Financing and Incentives Panel. *Domestic Policy Review of Solar Energy: Final Report.* Washington, DC: U.S. Department of Energy, October, 1978. TID-28831.

U.S. Department of Energy, Institutional Incentives and Barriers Panel. *Domestic Policy Review of Solar Energy: Final Report.* Washington, DC: U.S. Department of Energy, October, 1978. TID-28836.

U.S. Department of Energy, International Panel. *Domestic Policy Review of Solar Energy: Final Report.* Washington, DC: U.S. Department of Energy, October, 1978. TID-28837.

U.S. Department of Energy, Research, Development and Demonstration Panel. *Domestic Policy Review of Solar Energy: Final Report.* Washington, DC: U.S. Department of Energy, October, 1978. TID-28837.

U.S. Department of Energy. *Solar Energy Research and Development Program Balance. A Review of the Solar Working Group.* Washington, DC: U.S. Department of Energy, February, 1978.

Solar Cooling

deWinter, Frank. (ed.) *Proceedings of the Third Workshop on the Use of Solar Energy for the Cooling of Buildings,* February 15-17, 1978, San Francisco, CA. American Section of the International Solar Energy Society, August, 1978. Conference 780249.

Domestic Policy Review of Solar Energy: A Response Memorandum to the President of the United States. Washington, DC: U.S. Department of Energy, February, 1979. TID-22834.

Franks, L., and Fousel, D. "The Structure of the Solar Cooling Market." *Proceedings of the Third Workshop on the Use of Solar Energy for the Cooling of Buildings:* February 15-17, 1978, San Francisco, CA. Neward, DE: American Section of International Solar Energy Society, August, 1978. pp. 283-287.

LeChevalier, R. "Overview of the National Solar Cooling Program." Frank deWinter, (ed.) *Proceedings of the Third Workshop on the Use of Solar Energy for the Cooling of Buildings,* February 15-17, 1978, San Francisco, CA. Newark, DE: American Section of International Solar Energy Society, August, 1978. pp. 3-13. Conference 780249.

Scholten, W., and Curren, H. "Action Cooling Update." *Solar Age,* 37-41 (June, 1979).

"Solar Cooling." *Solar Engineering Magazine.* **3** (5) (May, 1978).

Solar Energy Research Institute. *Annual Review of Solar Energy.* Golden, CO: Solar Energy Research Institute, November, 1978. SERI/TR-54-066.

U.S. Department of Energy. *Final Report of the Research, Development and Demonstration Panel of the Domestic Policy Review of Solar Energy.* Washington, DC: U.S. Department of Energy, October, 1978. TID-28837.

U.S. Department of Energy. *National Program for Solar Heating and Cooling of Buildings. Annual Report.* Washington, DC: U.S. Department of Energy, July, 1978. DOE/CS-0007.

U.S. Department of Energy: *Solar Heating and Cooling Demonstration Project Summaries.* Washington, DC: U.S. Department of Energy, May, 1978. DOE/CS-0010.

U.S. Department of Energy. *Solar Update, Revised Proceedings.* (Four Regional Conferences on DOE's National Commercial Demonstration Program). Washington, DC: U.S. Department of Energy, July, 1978. Conference 780701.

Wilbur, P. J., and Karak, S. *Solar Cooling.* Philadelphia, PA: The Franklin Institute Press, 1977.

Solar Developers, Builders, Contractors

Bloodgood, J. D. "Four Plans for Solar Houses." *Professional Builder,* **43** (10): 125-32 (October, 1978).

California Building Industries Association. *Solar Energy Handbook for Builders.* Sacramento, CA: California Building Industries Association, August, 1978.

Franklin Institute, National Solar Heating and Cooling Information Center. *Computer Printout of Solar Manufacturers, Distributors, and Other Professionals.* Totals compiled by D. Buchanan, SERI, in a May 1, 1978, Memorandum.

Miller, A. S.; Hayes, G. B.; and Thompson, G. P. *Solar Access and Land Use: State of the Law, 1977.* Rockville, MD: National Solar Heating and Cooling Information Center, 1977.

National Solar Heating and Cooling Information Center. *Solar Heating and Cooling 1979 Solar Buyers Guide.* Rockville, MD: National Solar Heating and Cooling Information Center, 1979.

Phelps, D. L., and Yoxall, R. R. "Solar Energy: An Analysis of the Implementation of Solar Zoning." *Washburn Law Journal.* **17:** 147-62, 1977.

Pollack, P. *The Implementation of State Solar Incentives: Land-Use Planning to Ensure Solar Access.* Golden, CO: Solar Energy Research Institute, March, 1979. SERI/TR-51-163.

Residential Energy from the Sun. Rockville, MD: National Solar Heating and Cooling Information Center, November, 1978.

Sheet Metal and Air Conditioning Contractors National Association. *Fundamentals of Solar Heating.* Vienna, VA: Sheet Metal and Air Conditioning Contractors National Association, January, 1978.

Sheet Metal Industry National Training Fund. *Solar Air Systems; Workbook 1.* Sheet Metal and Air Conditioning Industry, Washington, DC, May, 1979.

Talbott, D. *Solar Energy—Proceed with Caution.* Washington, DC: National Association of Home Builders, February, 1976. Available from: Donald L. Carr, Assistant Director of Technical Services, National Association of Home Builders, 15th and M Street, NW, Washington, DC 20005.

Thomas, W. A.; Miller, A. S.; and Robbins, R. L. *Overcoming Legal Uncertainties About Use of Solar Energy Systems.* Chicago, IL: American Bar Foundation, 1978.

Wisconsin Office of State Planning and Energy. *Protecting Solar Access. Report of the Governor's Special Study Committee on Solar Rights.* Madison, WI, April, 1978.

Solar Economics and Market Projections

Bennington, G., et al. *Solar Energy: A Comparative Analysis to the Year 2020.* McLean, VA: MITRE Corporation, METREK Division, March, 1978. MITRE Technical Report MTR-7579, Report No. HCP/T2322-01.

Butt, S. *Solar Market Capture and Market Penetration.* Washington, DC: Solar Energy Industries Association, October, 1976.

Committee on Nuclear and Alternative Energy Systems (CONAES), Solar Resource Group. *Report of the Solar Resource Group,* February, 1977.

Cone, B. W., et al. *An Analysis of Federal Incentives Used to Stimulate Energy Production, an Executive Summary.* Richland, Washington: Battelle Pacific Northwest Laboratory, March, 1978. PNL-2410.

Costello, D., et al. *Photovoltaic Venture Analysis, Final Report,* Vol 1. Golden, CO: Solar Energy Research Institute, July, 1978. SERI/TR-52-040.

Doane, J. W., et al. *The Cost of Energy from Utility-Owned Solar Electric Systems.* Pasadena, CA: Jet Propulsion Laboratory, June, 1976. ERDA/JPL-1012-76/3.

Energy Research and Development Administration. *The Need for Deployment of Inexhaustible Energy Resource Technologies: Report of Technology Study Panel, Inexhaustible Energy Resources Study.* Washington, DC: Energy Research and Development Administration, September, 1977.

Executive Office of the President, Council on Environmental Quality. *Solar Energy: Progress and Promise.* Washington, DC: Council on Environmental Quality, April, 1978.

Flaim, S. J., et al. *Economic Feasibility and Market Readiness of Ten Solar Technologies.* Golden, CO: Solar Energy Research Institute, October, 1978. SERI/TR-52-055d.

Flaim, T., and Krawiec, F. *Solar Cost Reduction Through Technical Improvements: The Concepts of Learning and Experience.* Golden, CO: Solar Energy Research Institute, Progress Report, February 1979. SERI/PR-52-173d.

Office of Management and Budget. *Discount Rates to be Used in Evaluating Time Distributed Costs and Benefits.* Washington, DC: Office of Management and Budget, March 27, 1972. OMB Circular A-94.

Schiffel, D., et al. *The Market Penetration of Solar Energy: A Model Review Workshop Summary.* Golden, CO: Solar Energy Research Institute, January, 1978. SERI/16.

Stanford Research International. *Solar Energy Research and Development: Program Balance: Annex.* Vols. 1 and 2, February, 1978. NCP/M2693-01 and HCP/M2693-02. Prepared for U.S. Department of Energy, Solar Working Group. Available from: NTIS, 5285 Port Royal Road, Springfield, VA 22161.

U.S. Congress, Office of Technology Assessment. *Application of Solar Technology to Today's Energy Needs.* Washington, DC: Office of Technology Assessment, June, 1978. Washington, DC: Government Printing Office. Stock No. 052-003-00539-5.

U.S. Department of Energy, Deputy Under Secretary for Commercialization. *Overview of Technology Commercialization Assessment.* Washington, DC: U.S. Department of Energy, November, 1978.

U.S. Department of Energy. *Domestic Policy Review of Solar Energy: A Response Memorandum to the President of the United States.* Washington, DC: U.S. Department of Energy, February, 1979. TID-22834.

U.S. Department of Energy, *Market Oriented Program Planning Study (MOPPS).* Washington, DC: U.S. Department of Energy, March, 1978. Final Report.

U.S. Department of Energy. *Solar Energy: A Status Report.* Washington, DC: U.S. Department of Energy, June, 1978. DOE/ET-0062.

U.S. Department of Housing and Urban

Development (HUD). *Working Papers on Marketing and Market Acceptance.* Washington, DC: U.S. Department of Housing and Urban Development, Spring, 1978. Chapter VI.

Waddell, T. D. *The Economic Damages of Air Pollution.* Washington, DC: Environmental Research Center, U.S. Environmental Protection Agency, 1974, GPO No. PB-235701.

Solar Educational Institutions

An Assessment of Need for Developing and Implementing Technical and Skilled Worker Training for the Solar Energy Industry. Report of Navarro College work performed for the U.S. Department of Energy (DOE) under contract number EG-77-S-04-3869.

Coreoleotes, G., et al. *1978-79 National Solar Energy Education Directory.* Golden, CO: Solar Energy Research Institute, January, 1979. SERI/SP-42-141. GPO No. 061-000-00210-3.

Hunt, William D. *Solar Technology: A Task Analysis.* Chico, CA: California State University, June, 1978.

Sheet Metal and Air Conditioning Contractors National Association. *Fundamentals of Solar Heating Correspondence Course.* Washington, DC: Sheet Metal and Air Conditioning Contractors National Association, January, 1978. HCP M4038-01.

Sheet Metal and Air Conditioning Contractors National Association. *Study Guide for Fundamentals of Solar Heating: A Correspondence Course for the Air Conditioning Industry.* Sheet Metal and Air Conditioning Contractors National Association, Washington, DC: January, 1978. HCP M4038-02.

Solar Energy Research Institute. *Organized Labor and Solar Energy:* February, 1979, Golden, CO: Solar Energy Research Institute. SERI/TR-62-148.

U.S. Department of Energy, Assistant Secretary for Intergovernmental and Institutional Relations, Office of Education, Business, and Labor Affairs. *Activities of the Department of Energy in Energy Education.* Washington, DC: U.S. Department of Energy. DOE/IR-0008.

U.S. Department of Energy, Office of Education Business and Labor Affairs. *Solar Energy Task Force Report on Education and Training.* Washington, DC: U.S. Department of Energy, September, 1978. 20585.

U.S. Department of Housing and Urban Development. *HUD Solar Status: Installing Solar: Training Expands to Meet the Need.* Washington, DC: U.S. Government Printing Office. S/N0023-000-00464-0.

Solar Energy

American Institute of Aeronautics and Astronautics. *Solar Energy for Earth.* New York: American Institute of Aeronautics and Astronautics, 1975.

Anderson, Bruce. *Solar Energy—Fundamentals in Building Design.* New York: McGraw-Hill Book Company, 1977.

Beckman, William, et al. *Solar Heating Design—By the F-Chart Method.* New York: John Wiley & Sons, 1977.

Brinkworth, Brian Joseph. *Solar Energy for Man.* New York: John Wiley & Sons, 1973.

Chalmers, Bruce. "The Photovoltaic Generation of Electricity." *Scientific American,* **235** (4): 34-44 (October 1976).

Clark, Wilson. *Energy for Survival: The Alternative to Extinction.* Garden City, NY: Anchor Press/Doubleday, 1977.

Corliss, William R. *Direct Conversion of Energy.* Oak Ridge, TN: U.S. Atomic Energy Commission, Office of Information Services, 1964.

Daniels, Farrington. *Direct Use of the Sun's Energy.* New York: Ballantine Books, 1964.

Davis, Albert J., and Schubert, Robert P. *Alternative Natural Energy Sources in Building Design.* Blacksburg, VA: Passive Energy Systems, 1974.

Duffie, John A., and Beckman, William A. *Solar Energy Thermal Process.* New York: John Wiley & Sons, 1974.

Duffie, John A., and Beckman, William A. "Solar Heating and Cooling." *Science,* **191:** 143-149 (January 16, 1976).

Federal Energy Administration. *Buying Solar.* Washington, DC: U.S. Government Printing Office, June 1976.

Gay, Larry. *The Complete Book of Heating with Wood.* Charlotte, VT: Garden Way, 1974.

Grey, J. "Solar Heating and Cooling." *Astronautics and Aeronautics,* 33-37 (November 1975).

Halacy, D. S. *Earth, Water, Wind & Sun.* New York: Harper & Row, 1977.

Hoke, John. *Solar Energy.* New York: Franklin Watts, 1968.

Hunt, V. Daniel. *Handbook of*

Conservation and Solar Energy. New York: Van Nostrand Reinhold, 1982.

Hunt, V. Daniel. *Handbook of Energy Technology—Trends and Perspectives.* New York: Van Nostrand Reinhold, 1981.

Hunt, V. Daniel. *The Gasohol Handbook.* New York: Industrial Press, 1981.

Hunt, V. Daniel. *Windpower—A Handbook on Wind Energy Conversion Systems.* New York: Van Nostrand Reinhold, 1981.

Keyes, John. *Harvesting the Sun to Heat Your House.* New York: Morgan & Morgan, 1975.

Kreith, Frank, and Kreider, Jan. *Principles of Solar Engineering.* Washington, DC: Hemisphere Publishing Corporation, 1978.

Lucas, Ted. *How to Build a Solar Heater.* 3d Ed., Pasadena, CA: Ward Ritchie Press, 1975.

Mazria, Edward. *The Passive Solar Energy Book.* Emmaus, PA: Rodale Press, 1979.

Meinel, Aden B., and Meinel, Marjorie P. "Physics Looks at Solar Energy." *Physics Today,* 44-50 (February 1972).

Michels, Tim. *Solar Energy Utilization.* New York: Van Nostrand Reinhold, 1979.

Rankins II, William H., and Wilson, David A. *Practical Sun Power.* New York: Lorien House, 1974.

Rau, Hans. *Solar Energy.* Translated by Maxim Schur. Edited and revised by D. J. Duffin. New York: Macmillan, 1964.

Russell, Charles R. *Solar Energy: Elements of Energy Conversion.* Oxford, U.K., New York: Pergamon Press, 1967.

Skurka, Norma, and Naar, John. *Living with Natural Energy. Design for a Limited Planet.* 1st Ed. New York: Ballantine Books, 1976.

Solar Energy for Space Heating and Hot Water. Washington, DC: Energy Research and Development Administration, Division of Solar Energy, 1976, GPO 126239.

Solar Energy Institute of America. *Sun Language.* Washington, DC: Solar Energy Institute of America, 1980.

Stoner, Carol (ed.). *Producing Your Own Power: How to Make Nature's Energy Sources Work for You.* Emmaus, PA: Rodale Press, Book Division, 1974.

Williams, James Richard. *Solar Energy: Technology and Applications.* Ann Arbor, MI: Ann Arbor Science Publishers, 1974.

Yanda, Bill, and Fisher, Rick. *Solar Greenhouse,* Sante Fe, NM: John Muir Publications, 1976.

Yergin, Daniel, et al. (eds.) *Energy Future—Report of the Energy Project at the Harvard Business School.* New York: Random House, 1979.

Solar Energy For Utilities

Aerospace Corporation. *Performance, Value, and Cost of Solar Thermal Electric Central Receiver Plants Outside of the Southwest.* The Aerospace Corporation, May, 1978. Prepared for the U.S. Department of Energy, Division of Solar Technology. DOE Contract No. EY-76-C-03-1101 PA #2.

Aerospace Corporation, Solar Heating and Cooling of Buildings. *Requirements Definition and Impact Analysis.* Palo Alto, CA: Electric Power Research Institute, June, 1978. EPRI-ER-808.

American Gas Association. *Solar Energy Utilization.* January, 1978.

Anderson, S. "Economics of Solar Space Conditioning Systems: Electric Rates and Other Institutional Barriers." In: Hillhouse, K. (ed.) *Legal and Institutional Perspectives on Solar Energy in Colorado.* App. B: November, 1977. Prepared for the National Science Foundation, Grant No. APR-75-18247.

Battelle, Columbus Laboratories. *Annotated Bibliography on Supplemental Firing of Municipal Solid Waste in Electric Utility Boilers.* Palo Alto, CA: Electric Power Research Institute, February, 1978. Report FP-678, TSA 76-46.

Bezdek, R.; Margolin, J.; et al. *Analysis of Policy Options for Accelerating Commercialization of Solar Heating and Cooling Systems.* February, 1978. HEP M 2534-027.

Booz-Allen & Hamilton. *The Effectiveness of Solar Energy Incentives at the State and Local Level.* Booz-Allen & Hamilton, Inc., March, 1976. PB 263-371.

Booz-Allen & Hamilton. *Overview of Research and Development Activities in Biomass.* Booz-Allen & Hamilton, Inc., August, 1978. Prepared for the Gas Research Institute. *Overview of Research and Development Activities in Solar Thermal Energy.* Booz-Allen & Hamilton, Inc., November, 1978. Prepared for the Gas Research Institute.

California Public Utilities Commission. Decision No. 88224. *Application of Southern California Gas Company for*

Authority to Engage in a Solar Demonstration Project and to Include in its Rates the Amounts Necessary to Fund a Solar Energy Program.

Conference Report: Public Utility Regulatory Policies Act. Senate Rep. No. 95-1292, 95th Congress, 2d Session (1978).

Consolidated Edison Company of New York. *General Requirements for Parallel Operation of Windmill Generators with Low Tension Service.* Application and Design Manual No. 4. May 4, 1977.

Dean, J., and Miller, A. "Plugging Solar Power Into the Utility Grid." *Environmental Law Reporter,* **7.** 50069 (1977).

Dean, N., and Miller, A. "Utilities at the Dawn of a Solar Age." *North Dakota Law Review,* **53.** 329 (1977).

Dickson, C.; Eichen, M.; and Feldman, S. "Solar Energy and U.S. Public Utilities." *Energy Policy,* (3) (September, 1977).

An Economic Comparison of Gas/Solar House Heating Systems Versus Electric Solar Systems. American Gas Association, 1978.

Electric Power Research Institute. *Electric Utility Solar Energy Activities, 1976 Survey.* Palo Alto, CA: Electric Power Research Institute, January, 1977. EPRI-ER-321-SR.

Electric Power Research Institute. *Electric Utility Solar Energy Activities, 1977 Survey.* Palo Alto, CA: Electric Power Research Institute, May, 1979. EPRI-ER-649-SR.

Electric Power Research Institute. *Electric Utility Solar Energy Activities, 1978 Survey.* Palo Alto, CA: Electric Power Research Institute, May, 1979. EPRI-ER-966-SR.

Electric Utility Solar Energy Activities 1977 Survey. Palo Alto, CA: Electric Power Research Institute, February, 1978. ER-649-SR.

Energy Users Report. September 28, 1978. "Municipal Solid Waste—Problem or Opportunity?" *EPRI Journal,* **2** (9): 6-13 (November, 1977).

Feldman, S., and Anderson, B. *Utility Pricing and Solar Energy Design.* National Science Foundation, September, 1976. APR-75-18006.

Feldman, S., and Anderson, B. *The Public Utility and Solar Energy Interface: An Assessment of Policy Options.* Washington, DC: Energy Research and Development Administration, Division of Solar Energy, December 31, 1976. ERDA Contract No. E (949-18)-2523.

Feldman, S., and Anderson, B. *The Impact of Active and Passive Solar Building Designs on Utility Peak Loads.* Washington, DC: Energy Research and Development Administration, September 30, 1977. ERDA Contract No. EG-77-G-01-4029. Interim Report.

Feldman, S., and Anderson B., et al. *The Impact of Active and Passive Solar Building Designs on Utility Peak Loads.* Washington, DC: Energy Research and Development Administration, September 30, 1977. Contract No. EG-77-G-01-4029.

Feldman, S.; Wirthshafter, R.; and Wessler, E. "The Effect of Solar Buildings on Peak Load." *Electrical World* (March 15, 1978).

Feuerstein, R. *Utility Rates and Service Policies as Potential Barriers to the Market Penetration of Decentralized Solar Technologies.* Golden, CO: Solar Energy Research Institute, August, 1979. SERI/TR-62-274.

General Electric Company. *Requirements Assessment of Wind Power Plants in Electric Utility Systems.* Palo Alto, CA: Electric Power Research Institute, June, 1978. EPRI-ER-635-SY.

General Electric Company. *Requirements Assessment of Wind Power Plants in Electric Utility Systems.* Palo Alto, CA: Electric Power Research Institute, 1979.

Habicht, C. "Electric Utilities and Solar Energy: Competition, Subsidies, Ownership, and Prices." In: *The Solar Market,* **229:** (June, 1978). *Proceedings of the Symposium on Competition in the Solar Energy Industry. 1977 Solar Energy Symposium,* December 15-16, 1977. Sponsored by the Federal Trade Commission, Bureau of Competition.

JBF Corporation. *A Regional Assessment Study for Solar Electric Systems in the Northeast United States.* JBF Scientific Corporation, Washington, DC, 1978.

Institute of Gas Technology. *Energy From the Sun,* April 3-7, 1978. June, 1978.

JBF Corporation. *A Regional Assessment Study for Solar Electric Systems in the Northeast United States.* JBF Scientific Corporation, Washington, DC, 1978.

Jet Propulsion Laboratories and California Institute of Technology. *Public Policy Issues: A Southern California Gas Company Project SAGE Report,* January, 1978. Report No. NASA-CR-155945, JPL-PUB-77-47. Contract No. NAS 7-100.

Jet Propulsion Laboratories and California Institute of Technology. *Utilization Requirements: A Southern California Gas Co. Project SAGE Report.* January, 1978. Report No. NASA-CR-155948, JPL-PUB-77-40. Contract No. NAS 7-100.

Laitos and Feuerstein. "May Regulated Utilities Monopolize the Sun?" *Denver Law Journal* (1979).

Lorsch, H. *Implications of Residential Solar Space Conditioning on Electric Utilities.* National Science Foundation: December, 1976. AER-75-182701. Prepared by the Franklin Institute Research Laboratories, Contract No. NSF-C1033.

MITRE Corporation, *Solar Thermal Repowering; Utility/Industry Market Potential in the Southwest.* McLean, VA. The MITRE Corporation, December, 1978. MTR-7919.

Noll, R. "Maintaining Competition in Solar Energy Technology." *Proceedings of the Symposium on Competition in the Solar Energy Industry.* Solar Energy Symposium, December 15-16, 1977. Washington, DC: Federal Trade Commission, Bureau of Competition, June, 1978.

Robinson, S. "Solar and Utilities: An Example in Cooperation." *Solar Age,* **3** (12) (December, 1978).

"Solar Homes: The Winning Combinations." *EPRI Journal,* **3** (2): 6-13 (March, 1978).

Solar Hybrid Repowering Project, Public Service Company of New Mexico. 1978. Executive Summary submitted to the U.S. Department of Energy.

Swetman, G.; Elridge, F.; and Jardine, D. *Energy Rate Initiatives Study of the Interface Between Solar and Wind Energy Systems and Electric Utilities.* McLean, VA: The MITRE Corporation, March 31, 1977. PB 265-607.

"Thomas Village." *Solar Age,* **3** (12): 24-25 (December).

U.S. Congress. *National Energy Conservation Policy Act,* Public Law. 95-619, 1978. See also: Conference Report: *National Energy Conservation Policy Act.* S. Rep. No. 95-1294.

U.S. Congress. *The National Energy Act of 1978, Public Utility Regulatory Policies Act of 1978,* 95-617; *Energy Tax Act of 1978,* 95-618; *National Energy Conservation Policy Act,* 95-619; *Power Plant and Industrial Fuel Use Act of 1978,* 95-620; *Natural Gas Policy Act of 1978,* 95-621. For a general summary of the five acts see: Glazer, S. *National Energy Act Fact Sheet.* Prepared by the Environmental Study Conference. Washington, DC: U.S. Department of Energy, Office of Public Affairs, November, 1978.

U.S. Department of Energy. *National Power Grid Study.* Washington, DC: U.S. Department of Energy. Available from: DOE Public Affairs Division, Economic Regulatory Administration, 2000 M Street, N.W., Washington, DC 20461.

U.S. Department of Energy, Solar Energy/Utility Conference, March 28-29, 1978, George Washington University, Washington, DC.

U.S. Department of Housing and Urban Development. *Working Papers on Marketing and Market Acceptance.* Washington, DC: U.S. Department of Housing and Urban Development, Spring, 1978. Chapter VI.

"Utility Involvement: A Roundtable Discussion." *Solar Age,* **3** (12): 12-17 (December, 1978).

Wisconsin Public Service Commission. *Findings of Fact and Order.* Dockets 6630-ER-2 and 6630-ER-5.

Solar Energy Resource Assessment

Bateman, W. *Wind Energy Systems.* Statement Before the Subcommittee on Science and Technology, U.S. House of Representatives. Washington, DC: July 30, 1979.

Changery, M. J. *Initial Wind Energy Data Assessment Study.* Asheville, NC: National Science Foundation/National Oceanic and Atmospheric Administration, 1974. CONF-740780.

Cinquemuni, V.; Owen, J. R.; and Baldwin, R. G. *Input Data for Solar Systems.* Asheville, NC: National Climatic Center; November, 1978.

Coty, U. A. *Wind Energy Mission Analysis.* Washington, DC: Energy Research and Development Administration, September, 1976. SAN/1075-76/1.

Energy Research and Development Administration. *Local Climatological Data:* Annual Summaries for 1977. Asheville, NC: National Climatic Center, 1978.

Elliot, D. L. *Synthesis of National Wind Energy Assessments.* Richland, WA: Battelle Northwest Laboratories, July, 1977. BNWL-2220 WIND-5.

Klein, A. S. "Calculation of Monthly Average Insolation on Tilted Surfaces." *Journal of Solar Energy,* **19** (4): 325-29 (1977).

Lunde, P. J. "What's Happened to Solar Research?" *ASHRAE Journal,* **20** (11): 44-46 (November, 1978).

Randall, C. M., and Whitson, M. E. *Final Report—Hourly Insolation and Meteorological Data Bases Including Improved Direct Insolation Estimates.* December 1, 1977, Aerospace Report No. ATR-78

(7592)-December 1, 1977. Available from: Sandia Laboratories, Albuquerque, NM.

SOLMET Vol. 1 and 2: Users Manual. Asheville, NC: National Climatic Center, August, 1978. TD-9724.

U.S. Department of Energy, Division of Solar Technology. *Fuels from Biomass Program. Program Summary January 1978.* Washington, DC: U.S. Department of Energy, 1978. DOE/ET-0022/1.

U.S. Department of Energy Environmental Resource Assessments Branch Contractors Program Review. Manassas, VA: November 13-16, 1978.

Watt, A. D.; Mathews, F. S.; and Hathaway, R. E. *Open Cycle Ocean Thermal Energy Conversion—A Preliminary Engineering Evaluation.* Golden, CO: Colorado School of Mines, Engineering Physics Department, December 31, 1977. ALO/3723-76/3.

West Associates Solar Resource Evaluations Project, Solar Energy Measurements During 1976. Rosemead, CA: Southern California Edison Co., June, 1977.

Solar Environment

Allen, Margaret. *Joining Forces: Cooperative Environmental Action by Urban Groups—Cases and Lessons.* Washington, DC: Urban Environment Foundation, April, 1979.

Brumleve, T. D. *Eye Hazard and Glint Evaluation for the 5 MWe Solar Thermal Test Facility.* Albuquerque, NM: Sandia Laboratories, May, 1977. SAND 76-8022.

Craig, P.; Christensen, M.; Leirne, M. D.; Mukamel, D. B.; and Simmons, M. *Distributed Energy Systems in California's Future.* Interim Report Vols. I and II. Washington, DC: U.S. Department of Energy, 1978. HEP/P7405-03.

Davidson, M.; Grether, D.; and Horowitz, M. *Assessment of the Socio-Economic and Environmental Aspects of the Central Receiver Power Plants.* Berkeley, CA: Lawrence Berkeley Laboratory.

Davidson, M.; Grether, D.; and Wilcox, K. *Ecological Considerations of the Solar Alternative.* Berkeley, CA: University of California, February, 1977. LBL-5927.

Davidson, M., and Grether, D. *The Central Receiver Power Plant: An Environmental Ecological and Socio-Economic Analysis.* Berkeley, CA: Lawrence Berkeley Laboratory, June, 1977.

Domestic Policy Review of Solar Energy Integration Group. *Status Report on Solar Energy Domestic Policy Review.* August 25, 1978. Public Review Copy.

Energy Research and Development Agency, Division of Solar Energy. *Solar Program Assessment: Environmental Factors.* Washington, DC: Energy Research and Development Agency, March, 1977. March, 1977. Includes the following reports: "Solar Heating and Cooling of Buildings," ERDA 77-47/1; "Solar Agricultural and Industrial Process Heat," ERDA 77-47/2; "Photovoltaics," ERDA 77-47/3; "Solar Thermal Electric," ERDA 77-47/4; "Solar Total Energy Systems," ERDA 77-47/5; "Wind Energy Conversion," ERDA 77-47/6; "Fuels from Biomass," ERDA 77-47/7; and "Ocean Thermal Energy Conversion," ERDA 77-47/8.

Executive Office of the President, Council on Environmental Quality. *Solar Energy: Progress and Promise.* Washington, DC: Council on Environmental Quality, April, 1978.

Gregor, J. J. *Intro-Urban Mortality and Air Quality: An Economic Analysis of the Costs of Pollution Induced Mortality.* Corvallis, OR: Environmental Research Laboratory, U.S. Environmental Protection Agency, 1977. EPA-600/5-77-009.

Hall, E. H., et al. *Comparison of Fossil and Wood Fuels.* Battelle-Columbus Laboratories for the U.S. Environmental Protection Agency, Cincinnati, Ohio, March, 1976. EPA-600/2-76-056.

Heintz, H. T., Jr., et al. *National Damages of Air and Water Pollution.* Washington, DC: Environmental Research Center, 1976. Available from U.S. Environmental Protection Agency, Arlington, VA 22202.

Koshal, R. K., et al. "Air Pollution and the Respiratory Disease Mortality in the U.S.—A Quantitative Study." *Social Indicator Research,* **1** (3) (1974).

Lave, L. B., and Seskin, E. P. *Air Pollution and Human Health.* Baltimore, MD, The Johns Hopkins University Press, 1977.

National Academy of Sciences and National Academy of Engineering. *Air Quality and Automobile Emission Control,* 1974.

Ohi, J. *A Method to Assess and Aggregate the Effects of Solar Technology Deployment on Social and Ecological Systems.* Golden, CO: Solar Energy Research Institute, October, 1978. SERI/TR-53-081.

Rogers, S. E., et al. *Evaluation of the Potential Environmental Effects of Wind Energy Systems Development.* Battelle-Columbus Laboratory, August, 1976. ERDA/NSF/07378-75/1. Prepared for the Energy Research and Development Administration, Washington, DC.

Rubinfeld, D. L. *Market Approaches to the*

Measurement of the Benefits of Air Pollution Abatement. The University of Michigan and National Bureau of Economic Research, 1977.

Searcy, J. O. (ed.) *Hazardous Properties and Environmental Effects of Materials Used in Solar Heating and Cooling (SHAC) Technologies: Interim Handbook.* Albuquerque, NM: Sandia Laboratories, August, 1978. SAND 78-0842.

Sears, D. R.; Marrifield, D. V.; and Perry, M. M. *Environmental Impact Statement for a Hypothetical 1000 MWe Photovoltaic Solar-Electric Plant.* Lockheed Missiles and Space Company, Inc., August, 1977. EPA-600-77-085. Prepared for U.S. Environmental Protection Agency, Office of Research and Development, Cincinnati, Ohio.

Sears, D. R., and McCormick, P. O. *Preliminary Environmental Assessment of Solar Energy Systems.* Lockheed Missiles and Space Company, Inc., August, 1977. EPA-600/7-77-086. Prepared for U.S. Environmental Protection Agency, Office of Research and Development, Cincinnati, Ohio.

Unger, S. G., et al. *National Estimates of Water Quality Benefits.* Development Planning and Research Associates, Inc., 1974.

U.S. Department of Energy, Assistant Secretary for Conservation and Solar Applications and Assistant Secretary for Environment. *Environmental Development Plan (EDP), 1977.* Washington, DC: U.S. Department of Energy, March, 1978. Includes the following reports: "Solar Agricultural and Industrial Process Heat," DOE/EDP-0002; "Fuels from Biomass," DOE/EDP-0005; "Ocean Thermal Energy Conversion," DOE/EDP-0006; and "Wind Energy Conversion," DOE/EDP-0007.

U.S. Department of Energy, Assistant Secretary for Environment, Office of Technology Impacts, Division of Technology Assessment. *A Technology Assessment of Solar Energy Systems. Project Summary for Fiscal Years 1978 and 1979.* Washington, DC: U.S. Department of Energy, July, 1978.

U.S. Department of Energy. *Solar Energy: A Status Report.* Washington, DC: U.S. Department of Energy, June, 1978. DOE/ET-0062.

U.S. Environmental Protection Agency. *Health Consequences of Sulfur Oxides: A Report from CHESS, 1970-71.* Washington, DC: U.S. Environmental Protection Agency, 1974.

Waddell, T. E. *The Economic Damages of Air Pollution.* Washington, DC: Environmental Research Center, U.S. Environmental Protection Agency, 1974. PB-235701.

Yahsch, J. A., et al. *Outpatient Medical Costs Related to Air Pollution in the Portland, Oregon Area.* Washington, DC: Washington Environmental Research Center, U.S. EPA, 1974. PB23846.

Yokell, M.; Lawrence, K.; and Shaheen, A., Jr. *Environmental Benefits and Costs of Solar Energy.* Vols. I and II. Golden, CO: Solar Energy Research Institute, October, 1979. SERI/TR-52-074.

Solar Heating and Cooling

Aerospace Corporation, Solar Heating and Cooling of Buildings. *Requirements Definition and Impact Analysis.* Palo Alto, CA: Electric Power Research Institute, June, 1978. EPRI-ER-808.

American Institute of Architects Research Corporation. *Early Use of Solar Energy in Buildings: A Study of Barriers and Incentives to the Widespread Use of Solar Heating and Cooling Systems.* Washington, DC: American Institute of Architects. August, 1976. PB-267-832/4.

Bezdek, R.; Margolin, J.; et al. *Analysis of Policy Options for Accelerating Commercialization of Solar Heating and Cooling Systems.* Department of Energy, February, 1978. HEP M 2534-027.

Blum, Shelia. *The Status of the IEA & CCMS Multilateral Programs in Solar Heating and Cooling.* Denver, CO: International Solar Energy Society, August 28, 1978.

Comptroller General of the United States. *Commercializing Solar Heating: A National Strategy Needed.* Washington, DC: United States General Accounting Office, July 20, 1979. MED-79-19.

Hooper, F. C., et al. *Solar Space Heating Using Annual Heat Storage.* Washington, DC: U.S. Department of Energy, 1978. Report No. COO/2939-5.

Knasel, T. M., et al. *Validation Status of Solar Heating and Cooling Systems Models.* Science Applications, Inc., October, 1978. Final Report.

Lof, G. O., and Tybout, R. A. "Cost of House Heating with Solar Energy." *Solar Energy,* **14** (3): 253, 1973.

Lorsch, H. *Implications of Residential Solar Space Conditioning on Electric Utilities.* Washington, DC: National Science Foundation, December, 1976. SER-75-182701. Prepared by the Franklin Institute Research Laboratories,

Contract No. NSF-C1033.

Morse, Fred. Chief, Solar Heating and Cooling Research and Development Branch, Conservation and Solar Applications, U.S. Department of Energy. *Status of the U.S. Solar Heating and Cooling Program.* Washington, DC: U.S. Department of Energy, May, 1979.

Morse, Frederick H. *Status of the United States Solar Heating and Cooling Program.* Washington, DC: Department of Energy, May, 1979.

Mueller, R. O., et al. "Optimal Design of Seasonal Storage for One Hundred Percent Solar Space Heating in Buildings." *Proceedings of AAAS Symposium on Efficient Comfort Conditions: Heating and Cooling of Buildings,* February, 1978.

Searcy, J. O. (ed.) *Hazardous Properties and Environmental Effects of Materials Used in Solar Heating and Cooling (SHAC) Technologies: Interim Handbook.* Albuquerque, NM: Sandia Laboratories, August, 1978. SAND 78-0842.

Sheet Metal and Air Conditioning Contractors National Association. *Fundamentals of Solar Heating.* Vienna, VA: Sheet Metal and Air Conditioning Contractors National Association, January, 1978.

Shulze, W. D., and Ben-David, S. *The Economics of Solar Home Heating: A Study Prepared for the Use of the Joint Economic Committee.* Washington, DC: U.S. Congress, March 13, 1977.

Solar Energy Research Institute. *Analysis Methods for Solar Heating and Cooling Applications.* Golden, CO: Solar Energy Research Institute, Market Development Branch, 1979. SERI/SP-35-232.

"Solar Takes on Olympic Proportions." *Solar Engineering,* 26-31 (October, 1978).

Spewak, P. C. *Labor Requirements for Solar Heating, Cooling, and Hot Water.* McLean, VA: MITRE Corporation, September, 1977. MITRE Working Paper No. 12569.

Stanford Research International. *Solar Energy Research and Development: Program Balance.* Stanford, CA: Stanford Research International, February, 1978.

Stepler, R. "Solar Heating—How to Pick the Right System for Your Home." *Popular Science,* 72-75 (July, 1978).

U.S. Congress, Senate Committee on Energy and Natural Resources. Hearing: *Energy: Fiscal Year 1980 Budget Request.* Ninety-sixth Congress, February 7, 1979.

U.S. Department of Energy, Division of Solar Applications. *Solar Heating and Cooling Research and Development Project Summaries.* Washington, DC: U.S. Department of Energy, May, 1978. DOE/CS-0010.

U.S. Department of Energy, Division of Solar Applications. *Solar Heating and Cooling Demonstration Project Summaries.* Washington, DC: U.S. Department of Energy, May, 1978. DOE/CS-0009.

U.S. Department of Energy, Division of Solar Applications. *An Analysis of the Current Economic Feasibility of Solar Water and Space Heating.* Washington, DC: U.S. Department of Energy, January, 1978. DOE-CS-0023.

U.S. Department of Energy, Division of Solar Applications. *Solar Technology Transfer Program: FY 1977 Program Summary.* Washington, DC: U.S. Department of Energy, April, 1978. DOE/CS-0027/1.

U.S. Department of Energy Division of Solar Energy, Energy Research Development Agency. *Solar Program Assessment: Environmental Factors.* Washington, DC: Energy Research Drvelopment Agency, March, 1977. Includes the following reports: "Solar Heating and Cooling of Buildings," ERDA 77-47/1; Solar Agricultural and Industrial Process Heat," ERDA 77-47/2; "Photovoltaics," ERDA 77-47/3; "Solar Thermal Electric," ERDA 77-47/4; "Solar Total Energy Systems," ERDA 77-47/5; "Wind Energy Conversion," ERDA 77-47/6; "Fuels from Biomass," ERDA 77-47/7; and "Ocean Thermal Energy Conversion," ERDA 77-47/8.

U.S. Department of Energy. "Minutes: Solar Heating and Cooling Simulaion and Economic Analysis Working Group Meeting." *U.S. DOE Solar Heating and Cooling Research and Development Loose-Leaf Handbook.* Golden, CO: November 14-15, 1977.

U.S. Department of Energy. *National Commercialization Task Force Reports,* Vol. IV. Washington, DC: U.S. Department of Energy, September, 1978.

U.S. Department of Energy, Office of Solar Applications. *Introduction to Solar Heating and Cooling Design and Sizing.* Washington, DC: U.S. Department of Energy, August, 1978. DOE/CS-0011.

U.S. Department of Energy, Office of

the Assistant Secretary for Conservation and Solar Applications. *National Program for Solar Heating and Cooling of Buildings: Annual Report.* Washington, DC: U.S. Department of Energy, July, 1978. DOE/CS-0007.

U.S. Department of Energy, Office of the Assistant Secretary for Conservation and Solar Applications. *National Program Plan for Research and Development in Solar Heating and Cooling for Building, Agricultural, and Industrial Applications.* Washington, DC: U.S. Department of Energy, August, 1978. DOE/CS-0008.

U.S. Department of Energy, Solar Technology Transfer Program. SOLCOST—Solar Hot Water Handbook: A Simplified Design Method for Sizing and Costing Residential and Commercial Solar Service Hot Water Systems. Washington, DC: U.S. Department of Energy, June, 1978. 3d Ed. DOE/CS-0042/2.

U.S. Department of Energy, Division of Solar Applications. *National Solar Heating and Cooling Demonstration Program: Project Experience Handbook.* Washington, DC: U.S. Department of Energy, September, 1978. DOE/CS-0045.

U.S. Department of Energy. *Solar Heating and Cooling Systems Operational Results.* National Solar Heating and Cooling Demonstration Program, November 28—December 1, 1978, Colorado Springs, CO. Washington, DC: U.S. Department of Energy, 1979. DOE/Solar/0500-79-00.

U.S. Department of Energy, Solar Technology Transfer Program. *An Introduction to SOLCOST: A New Simplified Design Method for Residential and Light Commercial Solar Heating and Cooling.* Washington, DC: U.S. Department of Energy, June, 1978. 3d Ed. DOE/CS-0042/1.

U.S. Department of Energy. *Put the Sun to Work Today,* 1978. Available from DOE Technical Information Center, P.O. Box 62, Oak Ridge, TN.

U.S. Department fo Energy. *SOLCOST Space Heating Handbook With Service Hot Water and Heat Load Calculations.* Washington, DC: U.S. Department of Energy, July, 1978. DOE/CS-0042/3.

U.S. Department of Housing and Urban Development. *Intermediate Standards for Solar Domestic Hot Water Systems, HUD Initiative.* Washington, DC: National Bureau of Standards, 1977. NBSIR 77-1272.

Solar Industry

Aerospace Corporation. *Overview of Photovoltaic Market Studies.* Washington, DC: U.S. Department of Energy, May, 1978.

American Wind Energy Association. *An Index of Manufacturers, Researchers and Distributors of Wind Energy Conversion Systems.* Washington, DC: U.S. Department of Energy, February, 1978. DOE/RE/3533-78/1.

BDM Corporation. *Photovoltaic Power Systems Market Identification, Vol. I,* May, 1977; *Vol. II,* November, 1977; *Vol. III,* November, 1978. Washington, DC: U.S. Department of Energy. HCP/M 2533-01/1.

BDM Corporation. *Photovoltaic Incentive Option.* Washington, DC: August, 1978.

Booz, Allen & Hamilton. *Assessment of Photovoltaics Industry, Market and Technologies.* Washington, DC: U.S. Department of Energy, August, 1978.

Butt, S. "The Solar Future." *Solar Engineering Magazine,* **3** (3): 16-18 (March, 1978).

Butt, S. "President's Column—Marketing and Dealer-Installer." *Solar Engineering Magazine,* **3** (2): (December, 1978).

"The Coming Boom in Solar Energy." *Business Week,* (2555): 88 (October 9, 1978).

Costello, D. *Photovoltaic Venture Analysis.* Golden, CO: Solar Energy Research Institute, July, 1978. SERI/TR-52-040.

Energy Communications Press. *Woodstove, Fireplace and Equipment Directory, 1979.* Manchester, NH: Energy Communications Press, 1978.

Federal Trade Commission, Bureau of Competition. *The Solar Market: Proceedings of the Symposium on Competition in the Solar Energy Industry.* Washington, DC: U.S. Government Printing Office, 1978. GPO 723-397/1045 1-3.

Friedlander, A. F. "Editor's Corner—The Cost of Warranties." *Solar Engineering Magazine,* **3** (1): (January, 1978).

Friedlander, A. F. "Editor's Corner—Growing Pains in the Solar Industry." *Solar Engineering Magazine,* **3** (5) (May, 1978).

R. C. Fuller & Associates. *Impact Analysis of the Proposed Warranty and Conservation Provisions of the State Solar Energy Tax Credit Guidelines.* Sacramento, CA: California Energy Resources Conservation and Development Commission, May, 1978.

Intertechnology/Solar Corporation. *Photovoltaic Energy Technology Market*

Analysis. Washington, DC: U.S. Department of Energy, January, 1978.

MITRE Corporation. *Analysis of Federal Optimistic Support Photovoltaic Industry Growth*. McLean, VA: MITRE Corporation, February, 1979. MTR 79-W/00020.

MITRE Corporation. *Preliminary Wind Commercialization Program*. McLean VA: MITRE Corporation, November, 1976.

Morse, F. H. *Status of the United States Solar Heating and Cooling Program*. Washington, DC: U.S. Department of Energy, Office of Solar Applications, Division of Conservation and Solar Applications, May, 1979.

Muller & Company. *Solar Thermal Collector Manufacturing Activity. July through December 1977; January through March 1978; April through July 1978*. Washington, DC: The Solar Energy Research and Education Foundation, July, 1978.

Orsak, C. G., et al. *Assessment of the Need for Developing and Implementing Technical and Skilled Worker Training for the Solar Energy Industry*. Corsicana, TX: Navarro College, January, 1978. DOE No. EG-77-S-04-3869.

Ramsey, E. M. *Biomass Conversion Industry—A Description of the Market Chain*. Golden, CO: Solar Energy Research Institute, August 9, 1978.

Ratten, E. J. "Survey of Near Term Civilian Application in the U.S." In: *Mission Analysis of Photovoltaic Solar Energy Conversion,* Vol. II, March, 1977. Sandia Laboratories, New Mexico. SAN/1101/PAS-1/2.

Rockwell International, Rocky Flats Plant. *Commercially Available Wind Machines: A Checklist of Systems, Manufacturers, Distributors*. Golden, CO: Rockwell International, June 30, 1978.

Solar Energy Industries Association. "Deskbook Directory of Solar System." *Solar Engineering Magazine,* **3** (12): 17 (December, 1978).

Solar Energy Industries Association. *SEM 79, SEIA Master Catalogue and Solar Industry Index*. Dallas, TX: Solar Engineering Publishers, Inc., 1978.

Solar Energy Industries Association. *Solar Industry Index*. Washington, DC: Solar Energy Industries Association.

Solar Energy Institute of North America. *Solar State of the Union Report 1979*. Washington, DC: Solar Energy Institute of North America, 1979.

Solar Energy Research Institute. "Wind Energy Conversion Systems." In: *Economic Feasibility and Market Readiness of Solar Technologies*. Golden, CO: Solar Energy Research Institute, October, 1978. Chapter 7.

Solar Energy Research Institute. *Putting the Sun to Work in Industry*. Golden, CO: Solar Energy Research Institute, February, 1979. SERI/SP-34-175.

State Farm Fire and Casualty Company. *Solar Energy Group Report No. 77-106*. Bloomington, IL: State Farm Fire and Casualty Company, 1978.

U.S. Department of Commerce, Office of Technology Assessment and Forecast, U.S. Patent and Trademark Office. *Solar Energy Patenting (1963-78)*. Washington, DC: U.S. Department of Commerce, June, 1979. Special Report.

U.S. Department of Energy. *Alcohol Fuels Program Plan*. Washington, DC: U.S. Department of Energy, March, 1978. DOE/US-001/2.

U.S. Department of Energy, Assistant Secretary for Conservation and Solar Applications, Division of Solar Applications. *Solar Heating and Cooling; Research and Development; Project Summaries*. Washington, DC: U.S. Department of Energy, May, 1979. DOE/CE-0010, UC-590.

U.S. Department of Energy. *Domestic Policy Review of Solar Energy. A Response Memorandum to the President of the United States*. Washington, DC: U.S. Department of Energy, February, 1979. TID-28834.

U.S. Department of Energy. *Solar Collector Manufacturing Activity. July 1978 through December 1978*. Washington, DC: U.S. Department of Energy, Energy Information Administration, September, 1979. DOE/EIA-0174(78).

U.S. Department of Energy. *Solar Energy: A Status Report*. Washington, DC: U.S. Department of Energy, June, 1978. DOE/ET-0062.

U.S. Department of Energy. *Solar Heating and Cooling Research and Development Project Summaries*. Washington, DC: U.S. Department of Energy, May, 1978. DOE/CS-0010.

U.S. Department of Energy. *Status Report, Energy Wind Turbines*. Golden, CO: Rockwell International, August, 1979.

U.S. Federal Energy Administration. *WECS Manufacturing and Sales Activity, 1975-76*. Washington, DC: U.S. Federal Energy Administration, April, 1977. FEA/B-77-121.

Solar Industry Associations

American Petroleum Institute. *Energy Diversification by Firms in the U.S.* Washington, DC: American Petroleum Institute, February 28, 1978.

"The Coming Boom in Solar Energy." *Business Week,* Special Report (October 9, 1978).

Electric Power Research Institute. *Electric Utility Solar Energy Activities 1977 Survey.* Palo Alto,CA: Electric Power Research Institute, February, 1978. ER-649-SR.

Kittrell, Charles M.; Executive Vice President Phillips Petroleum Company. Statement made at the Free Enterprise Fair, Western Kentucky University. *Oil Daily* (Tuesday, October 3, 1978).

Linde, H. R., and Rosenberg, R. B. "The Gas Option and the Need for New Technology." *57th National Conference of Regulatory Utility Commission Engineers,* June 18-21, Portland, ME. Chicago, IL: Gas Research Institute, 1979.

Smith, G. P. "Giants Harnessing Solar Development." *Washington Post,* A2 (Sunday, October 3, 1978).

Smith, P. *The Information Needs of the Solar Community—A Management Summary.* Golden, CO: Solar Energy Research Institute, October, 1978.

Solar Energy Industries Association. *Solar Industry Index.* Washington, DC: Solar Energy Industries Association.

Solar Energy Research and Education Foundation. *Solar Thermal Collector Manufacturing Activity.* Solar Energy Research and Education Foundation, August, 1978.

U.S. Department of Housing and Urban Development, *Solar Demonstration Program,* September, 1977. Washington, DC. U.S. Department of Housing and Urban Development HUE-PDR-189-4.

Solar Legislation

American Law Institute/American Bar Association. *Energy and the Law: Problems and Challenges of the Late 70's.* July 9-14, 1978. pp. 553-76. Presented at the University of Colorado School of Law.

American Society of Planning Officials. *Planning for Solar Access: A Guidebook for Planners and Site Planning for Solar Access:* and Environmental Law Institute. *Guaranteeing Legal Access for Solar Collectors.* The publications will be available through the National Solar Heating and Cooling Information Center, P.O. Box 1607, Rockville, MD 20850.

Chew, R. *Solar Law Handbook.* Palo Alto, CA: Solar Energy Research Institute, December, 1978.

"Current Developments." *Solar Law Reporter,* **1,** (2) (July/August 1979).

Dean, N. and Miller, A. "Utilities at the Dawn of a Solar Age." *North Dakota Law Review,* **53:** 329 (1977) See also: Dean, N. and Miller, A. "Plugging Solar Power Into the Utility Grid." *Environmental Law Reporter,* **7:** 50069 (1977).

Environmental Law Institute in conjunction with the University of Utah Law School. *Energy and the Public Lands:* The Environmental Law Institute Annual Conference, August, 1978, Park City, Utah. Selected speeches and unedited tapes of the conference are available from: Continuing Legal Education Programs, Division of Continuing Education, 1191 Annex Building, Salt Lake City, UT 84112.

Environmental Law Institute. *Legal Barriers to Solar Heating and Cooling of Buildings,* March, 1978. Prepared for the Department of Energy, Assistant Secretary for Conservation and Solar Applications, Division of Solar Applications.

Environmental Law Institute. *Solar Policy for the 80's.* September 12-22,1978.

Euser, B.; Joyce C.; Morgan G.; et al. *Solar Energy Legal Bibliography.* September, 1978. (Interim Draft Report prepared by the Solar Energy Research Institute for the Department of Energy, Division of Solar Technology, DOE Contract No. EG-77-C-01-4042.) The bibliography is to be made available through NTIS, U.S. Department of Commerce, 5285 Port Royal Road, Springfield, VA 22161.

Feurestein, R. *Utility Rates and Service Policies as Potential Barriers to the Market Penetration of Decentralized Solar Technologies.* Palo Alto, CA: Solar Energy Research Institute, 1978.

Green, M. *Licensing Arrangements and the Development of the Solar Energy Industry,* Palo Alto, CA: Solar Energy Research Institute, 1978.

Gross, J. *Impact of the Antitrust Laws on the Commercialization of Solar Energy.* Palo Alto, CA: Solar Energy Research Institute, 1978.

Hillhouse, R., et al. *Legal and Institutional Perspectives on Solar Energy in Colorado.* November, 1977. Prepared for the National Science Foundation, Grant No. APR-75-18247. See also: Miller, A. "Solar Energy and Land Use in

Colorado." *Environmental Law Reporter,* **6** 50039, 1976.

Hyatt, R. *Legal and Institutional Implications of Providing Financial Incentives to Encourage the Development of Solar Energy Technologies.* Palo Alto, CA: Solar Energy Research Institute, 1978.

Johnson, K. "State Embargo of Solid Waste: Impermissible Isolation or National Solution to a Pressing Problem?" *Dickenson Law Review,* **82:** 325-50. (1978).

Johnson, S. *A Survey of State Legislative Incentives for Solar Energy Development.* Palo Alto, CA: Solar Energy Research Institute., p. 78.

Johnson, Stephen B. "Approaches to Solar Legislation." *Solar Law Reporter,* **1** (1) (May/June 1979).

Knight, et al. *Ocean Thermal Energy Conversion.* Lexington, MA: Lexington Books, 1977.

Kraemer, S. *Solar Law, Present and Future with Proposed Forms.* Colorado Springs, CO: Shepards/McGraw-Hill Book Company, 1978.

Mayo, L. et al. *Legal-Institutional Implications of Wind Energy Conversion Systems (WECS),* September, 1977. Prepared by the Program of Policy Studies in Science and Technology, George Washington University, for the National Science Foundation, NSF Grant No. APR-75-19137. Taubenfeld, R., and Taubenfeld, H. "Wind Energy: Legal Issues and Legal Barriers." *Southwestern Law Journal.* **31** 1053-03 (1977).

Miller, "Solar Energy and Public Lands." *Journal of Contemporary Law,* **4:** 78. (Winter, 1977). See also: Harwood, C. *Using Land to Save Energy.* Cambridge, MA: Ballinger, 1977.

Moore, G. J. *Solar Energy Lesiglation in the 95th Congress, Second Session, With a Summary of the Solar Provisions of the National Energy Act.* Washington, DC: Library of Congress, Congressional Research Service, January, 1979.

Moore, G. J. *Solar Energy Legislation in the 96th Congress, First Session January through September 5, 1979.* Washington, DC: Science Policy Research Division, The Library of Congress, Congressional Research Service, August 27, 1979.

Nanda, V. *Selected Legal and Institutional Issues Related to Ocean Thermal Energy Conversion (OTEC).* Palo Alto, CA: Solar Energy Research Institute, December, 1978.

National Bureau of Standards. *Intermediate Minimum Property Standards for Solar Heating and Domestic Hot Water Systems,* April, 1976. NBSIR 76-1059. Prepared for Department of Housing and Urban Development, Office of Policy Development and Research, Division of Energy, Building Technology and Standards, Interim Report.

National Bureau of Standards. *Plan for the Development of Standards for Solar Heating and Cooling Systems.* Washington, DC: National Bureau of Standards, June, 1978. NBSIR 78-1143 A.

The National Energy Act passed in 1978 and comprises five collectively grouped acts which are: Public Utility Regulatory Policies Act of 1978, PL. No. 95-617; Energy Tax Act of 1978, PL. No. 95-618; National Energy Conservation Policy Act., PL. No. 95-619; Power Plant and Industrial Fuel Use Act of 1978, PL. No. 95-620; Natural Gas Policy Act of 1978, PL. No. 95-621. For a general summary consult the Office of Public Affairs, Department of Energy, Information: The National Energy Act (November, 1978); S. Glazer, National Energy Act Fact Sheet (1978) (prepared by the Environmental Study Conference).

Riley, J.; Odland, B.; Barker, H.; et al. *Solar Energy: Standards, Building Codes, and Certification Procedures.* Palo Alto, CA: Solar Energy Research Institute, 1978, Ch VI, pp. 29-30.

Schwab, S. *Legal Considerations in the Development and Implementation of Biomass Energy Technologies.* Palo Alto, CA: Solar Energy Research Institute, 1978.

Solar Energy Research Institute and the National Conference of State Legislatures. *Solar Warranties Workshop:* Executive Summary, p. 2, March 23-24, 1978. DOE Contract No. EG-77-C-01-4042.

Solar Energy Research Institute. *Law and Solar Energy: A Meeting of SERI and the ABA's Special Committee on Energy Law.* February 17, 1978.

Spivak, P. *Summary of Land Use Barriers and Incentives to the Use of Solar Energy.* Palo Alto, CA: Solar Energy Research Institute.

Thomas, W. A.; Miller, A. S.; and Robbins, R. L. *Overcoming Legal Uncertainties About Use of Solar Energy Systems.* Chicago, IL: American Bar Foundation, 1978.

U.S. Congress, House of Representatives. Hearings Before the Subcommittee on Advanced Energy Technologies and Energy Conservation Research, Development and Demonstration of the Committee on Science and Technology.

95th Congress, 2d sessions, January 26, 27, 30; February 1, 2, 3, and 6, 1978. Vol. 54 (no. 54).

U.S. Congress. *Solar Energy Research.* Development and Demonstration Act of 1974, Public Law 93-473. October 26, 1974, 88 Stat. 1431; 42 U.S.C. 5551 et. seq.

U.S. Department of Energy. *Environmental Development Plan for Solar Agricultural and Industrial Process Heat 17.* Washington, DC: Department of Energy, 1978.

Wallenstein, A. *Barriers and Incentives to Solar Energy Development.* Northeast Solar Energy Center, June, 1978.

Warren, M. "State Solar Legislation: Avoiding the Administrative Morass." *State Solar Legislation 1974-44:* in *Energy Report to the States,* Vol. 4, April 28, 1978. Palo Alto, CA: Solar Energy Research Institute.

White, S. *Municipal Bond Financing of Solar Energy Facilities.* Palo Alto, CA: Solar Energy Research Institute, December, 1978.

Wisconsin Office of State Planning and Energy. *Protecting Solar Access: Report of the Governor's Special Study Committee on Solar Rights.* State of Wisconsin, April, 1978.

Solar Policy

Bezdek, R.; Mogolin, J.; et al. *Analysis of Policy Options for Accelerating Commercialization of Solar Heating and Cooling Systems,* February, 1978. HEP M 2534-027.

"Carter Sets 20% Goal for Solar, Calls for Solar Bank." *Solar Energy Intelligence Report,* **5** (26): 251-53, June 25, 1979.

Community Services Administratiion. *Solar Energy Policy.* Washington, DC: Community Services Administration, July, 1978.

Domestic Policy Review of Solar Energy Integration Group. *Status Report on Solar Energy Domestic Policy Review.* August 25, 1978.

Domestic Policy Review of Solar Energy. Eizenstat, Stuart. Memorandum May 16, 1978, to the Secretary of Energy, et al., Published in *Solar Energy: A Status Report.* U.S. Department of Energy; June, 1978, Washington, DC. DOE/ET-0062. GPO No. 061-000-00090-9.

Environmental Law Institute. *Solar Policy for the 80's.* September 12-22, 1978.

Executive Office of the President, Council on Environmental Quality. *Solar Energy: Progress and Promise.* April, 1978. GPO No. 041-011-00036-0.

Farhar, B. C.; Unseld, C. T.; Caputo, R.; and Easterling, J. *Citizen Participation in the Domestic Policy Review of Solar Energy.* Golden, CO: Solar Energy Research Institute, January 1979. SERI/TR-53-126D.

Federal Operations Panel. *Domestic Policy Review of Solar Energy: Final Report.* Washington, DC: U.S. Department of Energy, October, 1978. TID-28832.

Feldman, S., and Anderson, B. *The Public Utility and Solar Energy Interface: An Assessment of Policy Options.* Washington, DC: Energy Research and Development Administration, Division of Solar Energy, December 31, 1976. ERDA Contract No. E (949-18)-2523.

Franklin Institute Research Laboratories and Institute for Local Self-Reliance. *The Great Adventure: A Report on the 10 Regional Public Hearings on Solar Energy for the Domestic Policy Review.* Washington, DC: U.S. Department of Energy, October, 1978. HCP/U6354-01.

Jet Propulsion Laboratories and California Institute of Technology. *Public Policy Issues: A Southern California Gas Company Project SAGE Report.* January, 1978. Report No. NASA-CR-155945, JPL-PUB-77-47. Contract No. NAS 7-100.

Jones, B., et al. *Social Research on the 1978 Solar Domestic Policy Review.* Denver, CO: Social Change Systems, Inc., August 18-19, 1977.

Jones, R. G.; Sramek, H. M.; and Pelster, J. M. *Analysis of State Solar Energy Policy Options,* June, 1976. PB 254-730.

Moore, J. G. *Solar Energy Legislation in the 95th Congress, Second Session,* Washington, DC: Library of Congress, Congressional Research Service, November, 1978.

The National Energy Act consists of 5 acts, 4 of which include solar-related provisions: The Energy Tax Act of 1978, P.L. 95-613; the Public Utility Regulatory Policies Act of 1978, P.L. 95-617; the National Energy Conservation Policy Act of 1978, P.L. 95-619; the Powerplant and Industrial Fuel Use Act of 1978, P.L. 95-620.

"Solar Industry Just a Little Nervous About Government 'Help.'" *Energy Daily.* **7** (119): 4 (June 22, 1979).

U.S. Department of Energy, Financing and Incentives Panel. *Domestic Policy Review of Solar Energy: Final Report.* Washington, DC: U.S. Department of Energy, October, 1978. TID-28831.

U.S. Department of Energy. Impacts Panel. *Domestic Policy Review of Solar Energy:*

Final Report. Washington, DC: U.S. Department of Energy, October, 1978. TID-28835.

U.S. Department of Energy, International Panel. *Domestic Policy Review of Solar Energy: Final Report.* Washington, DC: U.S. Department of Energy, October, 1978. TID-28830-1.

U.S. Department of Energy. *Domestic Policy Review of Solar Energy: A Response Memorandum to The President of the United States.* Washington, DC: U.S. Department of Energy, February, 1979. TID-22834.

U.S. Department of Energy. Research, Development and Demonstration Panel. *Domestic Policy Review of Solar Energy: Final Report.* Washington, DC: U.S. Department of Energy, October, 1978. TID-28837.

White House Press Secretary, Office of the. *Fact Sheet: The President's Message on Solar Energy,* June 20, 1979.

Solar Power Satellite

Argonne National Laboratory. *Preliminary Environmental Assessment for the Satellite Power System.* Argonne National Laboratory, September, 1978. SPS-EA-001.

Bain, Claud N. *Potential of Laser for SPS Power Transmission.* PRC Energy Analysis Company, Washington, DC, September, 1978.

Fishlock, D. "NASA Chief 'Skeptical' on Solar Satellites." *Energy Daily,* **6** (203) (October 20, 1978).

Flaim, S., et al. *Economic Feasibility and Market Readiness of Solar Technologies.* Golden, CO: Solar Energy Research Institute, October, 1978. SERI/TR-52-055.

Glaser, P. E. "Power from the Sun: Its Future." *Science,* **162:** 857-61 (November 22, 1968).

Gordon, W. D. "Microwave/Ionosphere Interaction—Recent Arecibo Results." Presented at Annual Meeting of the Sunsat Energy Council, October 5, 1978, Rice Universtiy.

National Aeronautics and Space Administration, Lyndon B. Johnson Space Center. *Satellite Power Systems (SPS) Concept Evaluation Program, Systems Definition Study.* Commended Preliminary Baseline Concept. Houston, TX: January 25, 1978.

National Aeronautics and Space Administration. *Preliminary Baseline SPS Concept Recommendations to DOE/NASA.* Houston: TX, Marshall Space Flight Center, January 24, 1978.

National Aeronautics and Space Administration. *Solar Power Satellite Baseline Review.* Houston, TX: Marshall Space Flight Center—Johnson Space Center, July 13, 1978.

Rockwell International, Space Division. *Satellite Power Systems (SPS) Concept Definition Study.* Final Report. Downey, CA: Rockwell International, April, 1978. SD 78-AP-0023-1.

"Supporters of Solar Power Satellites Form Sunsat Energy Council." *Space Solar Power,* 1 (May 3, 1978).

U.S. Congress, House of Representatives. *Solar Power Satellite.* Hearings before the Subcommittee on Space Science and Applications, and the Subcommittee on Advanced Energy Technologies and Energy Conservation Research, Development, and Demonstration, of the Committee on Science and Technology, 95th Congress, April 12-14, 1978.

Solar Standards, Building Codes, Certification

American Institute of Architects Research Corporation. *Early Use of Solar Energy in Buildings: A Study of Barriers and Incentives to the Widespread Use of Solar Heating and Cooling Systems.* Washington, DC: American Institute of Architects, August, 1976. PB-267-832/4.

American Institute of Architects Research Corporation. *Solar Regulatory Provision Study.* Washington, DC: American Institute of Architects, February, 1978.

Bottaro, Drew J. *The Purpose and Role of Product Standards in the Commercialization of New Energy Technologies: A Preliminary Analysis.* Cambridge, MA: Massachusetts Institute of Technology, October, 1978. MIT-EL 78-023WP.

California State Contractors License Board. *Solar Complaints—Statewide.* July 1, 1976 to September 30, 1977. Sacramento, CA: California State Contractors License Board, 1978.

Federal Trade Commission, Bureau of Competition. *The Solar Market: Proceedings of the Symposium on Competition in the Solar Energy Industry.* Washington, DC: U.S. Government Printing Office, 1978. GPO 723-397/1045 1-3.

Friedlander, F. "Editor's Corner—A Rainy Day Story." *Solar Engineering Magazine,* **3** (3): 5 (March, 1978).

General Electric Company. *Survey to Define Impact of Statewide Building Codes on Solar HVAC Systems, Commercial Buildings.* Philadelphia, PA, July, 1976. C00/2682-76/11.

Hittman Associates, Inc. *Barriers Connected with Certifying or Listing of Energy-Conserving Products Used in Buildings.* Springfield, VA: National Technical Information Service (NTIS), May, 1977. CONS/1211-1.

Johnson, Stephen B. *A Survey of State Legislative Incentives for Solar Energy Development.* Golden, CO: Solar Energy Research Institute, October, 1978.

Massachusetts Institute of Technology, Energy Laboratory. *Government Support for the Commercialization of New Energy Technologies.* Cambridge, MA: Massachusetts Institute of Technology, November, 1976. MIT-EL 76-009.

Riley, J.; Odland, F.; and Barker, H. *Solar Energy Standards, Building Codes, and Certification Programs.* Golden, CO: Solar Energy Research Institute, May, 1979. SERI/TR-53-095.

Schoen, R.; Hirshberg, A. S.; and Weingart, J. M. *New Energy Technologies for Buildings: Institutional Problems and Solutions.* Cambridge, MA: Ballinger Press, 1975.

Sheet Metal and Air Conditioning Contractors' National Association, Inc. *Heating and Air Conditioning Systems Installation Standards for One and Two Family Dwellings and Multifamily Housing Including Solar.* Vienna, VA: Sheet Metal and Air Conditioning Contractors' National Association, Inc., 1977.

Smith, O., and Meeker, J. "New England Electric's Solar Project." *Solar Age,* **3** (3): 16 (March, 1978).

Solar Energy Industries Association. "Commercialization Efforts Continue." *Solar Engineering Magazine,* **3** (9): 9 (September, 1978).

U.S. Department of Energy. *Legal Barriers to Solar Heating and Cooling of Buildings.* Washington, DC: U.S. Department of Energy, March, 1978. HCP/M2528-01.

U.S. Department of Energy, Office of Conservation and Solar Applications. *Plan for the Development and Implementation of Standards for Solar Heating and Cooling Applications.* Washington, DC: Department of Energy, 1978. NBSIR 78-1143A.

U.S. Department of Housing and Urban Development. *Intermediate Standards for Solar Domestic Hot Water Systems, HUD Initiative.* National Bureau of Standards, 1977. NBSIR 77-1272.

Solar Technology and End Uses

Beller, M. (ed.) *Sourcebook for Energy Assessment.* Brookhaven National Laboratory, 1979.

Bockris, J. O. *Energy: The Solar-Hydrogen Alternative.* New York: Halsted Press, 1976.

Caudela, B. J., et al. *Life Style and Energy Consumption Scenarios for the Years 1985, 2000 and 2025.* New York: Hudson Institute, 1975.
Report HI-2268-RR; 1975.

CONAES. *Report of the Solar Resource Group. Committee on Nuclear and Alternative Energy Systems.* Washington, DC: U.S. National Academy of Sciences, 1977.

Craig, P., et al. *Distributed Technologies in California's Energy Future.* Vol. I and II. Berkeley, California: University of California, Lawrence Berkeley Laboratory, 1978. HCP/P7405-01

Davis, K. "The Evolution of Western Industrial Cities." *Cities: Their Origin, Growth and Human Impact.* San Francisco, CA: W. H. Freeman, 1973.

"Esquisse d'un Regime a Long Terme Tout Solaire; Etude d'un Avenir Energique pour la France Axe sur le Potentiel Renouvelable." Le Groupe de Bellevue, 85 Boulevard de Port Royal, 75013 Paris.

Executive Office of the President, Council on Environmental Quality. *A National Energy Conservation Program—The Half and Half Plan.* Washington, DC: 1974.

Executive Office of the President, Council on Environmental Quality. *Solar Energy—Progress and Promise.* Washington, DC: 1978.

Ford Foundation. *A Time to Choose.* Cambridge, MA: Ballinger Press, 1974.

Gregory, D. P. "The Hydrogen Economy." *Scientific American,* **228:** 1 (1973).

Hayes, D. *Rays of Hope.* New York: W. W. Norton, 1977.

Hubbert, M. K. "The Energy Resources of the Earth." *Scientific American.* (September, 1971).

The IEA Energy Simulation Model. Institute for Energy Analysis, Oak Ridge, TN, 1975.

Johansson, T. B., and Steen, P. *Solar Sweden* as summarized in *Ambio,* **7** (2): (1978).

Kelly, J. H., et al. *Hydrogen Tomorrow—Demands and Technology Requirements.* Pasadena, CA: California Institute of Technology, Jet Propulsion Laboratory, 1975.

Leach, G. *A Low Energy Strategy*

for the United Kingdom. London: The International Institute for the Environment and Development, 1979.

Lovins, A. B. "Energy Strategy: The Road Not Taken?" *Foreign Affairs,* **55** (1): 65-96 (October, 1977).

Lovins, A. G. "Scale, Centralization and Electrification in Energy Systems." *Conference on Future Strategies for Energy Development: A Question of Scale,* October 1976, Oak Ridge, TN. Oak Ridge, TN: Oak Ridge Associated Universities, 1977.

Marchetti, C. "Primary Energy Substitution Models: On the Interaction Between Energy and Society." *Technological Forecasting and Social Change,* **10:** 345-56 (1977).

Mesarovic, M., and Pestel, E. *Mankind at the Turning Point.* New York: E. P. Dutton & Co., 1974, p. 138.

MITRE Corporation. *System Analysis of Solar Energy Programs.* McLean, VA: MITRE Corporation, December 1973. MTR-6513.

Morrow, W. E. "Solar Energy—Its Time Is Near." *Technology Review,* **76:** 30-42, 1973.

National Aeronautics and Space Administration, American Society for Engineering Education. *TERRASTAR (Terrestrial Applications of Solar Technology and Research).* NASA/ASEE, 1973. NASA Report CR-129012.

National Science Foundation, National Aeronautics and Space Administration, NSF/NASA Solar Energy Panel. *An Assessment of Solar Energy as a National Energy Resource.* University of Maryland, 1972. Report PB-221659.

Project Independence Report. *Federal Energy Administration,* 1974, GPO No. 4118-00029.

Quinn, J., and Ohi, J. *A Compendium of Decentralized Studies, Programs and Projects.* Golden, CO: Solar Energy Research Institute, 1979. RR-53-316.

Robinson, N. *Solar Radiation.* Amsterdam, The Netherlands: Elsevier, 1966.

Standard Research International. *A Preliminary Social and Environmental Assessment of the ERDA Solar Energy Program 1975-2020.* July 1976. Draft Report.

Starr, C. "Energy R&D Planning." *Science and Technology Policies.* Cambridge, MA: Ballinger Press, 1971.

Teller, E. *Energy—A Plan for Action.* Report to the Energy Panel of the Commission on Critical Choices for America, 1975.

U.S. Congress, House of Representatives, Subpanel IX. *Solar Energy Program. Published in the Hearings on the Solar Energy RD&D Act of 1974. Subcommittee on Energy, Committee on Science and Astronautics, Subpanel IX. July 30-August 2, 1974.*

U.S. Energy Research and Development Administration. National Solar Energy Research, Development, and Demonstration Program—Definition Report. Washington, DC: ERDA, 1975. ERDA-40.

U.S. Energy Research and Development Administration. *A National Plan for Energy Research, Development and Demonstration.* Washington, DC: U.S. Energy Research and Development Administration, June 1975. ERDA-48.

Walton, J. D., Jr. "High Temperature, Solar Energy." In:*Energy Technology Handbook,* Considine, Douglas M. (ed.) New York: McGraw-Hill Book Company; 1977. p. 6.

Weingart, Jerome. *Systems Aspects of Large-Scale Solar Energy Conversion.* International Institute for Applied Systems Analysis. Luxenburg, Austria, 1977. RM-77-23.

Weingart, J. "The Helios Strategy." *Technological Forecasting and Social Change.* **12** (6) (December, 1978).

Weingart, Jerome, and Nakicenouic, N. "Market Penetration Dynamics and the Large Scale Use of Solar Energy," NASA, 1978.

Wolfe, M. "Solar Energy Utilization by Physical Means." *Science.* 1972.

Solar Thermal Energy System

Aerospace Corporation. *Performance, Value, and Cost of Solar Thermal Electric Central Receiver Plants Outside of the Southwest,* May, 1978. Prepared for the U.S. Department of Energy, Division of Solar Technology. DOE Contract No. EY-76-C-03-1101 PA #2.

Aerospace Corporation. *Performance, Value and Cost of Solar Thermal Electric Central Receiver Plants Outside of the Southwest.* March 31, 1978. Report No. ATL-78 (7689-04)-1, Contract No. EY-76-C-03-1101.

Alvis, R. L., and Vandevender, S. G. *Solar Irrigation Program Plan.* Albuquerque, NM: Sandia Laboratories, June, 1977. SAND/77-0730.

Battelle Northwest Laboratories. *Prelimi-*

nary Cost Estimates for Heliostat Production Facilities. Richland, WA: Battelle Northwest Laboratories, March, 1978.

Booz-Allen & Hamilton. *Overview of Research and Development Activities in Solar Thermal Energy,* November, 1978. Prepared for the Gas Research Institute.

Bowmar, S. H.; Brown, C. T.; and Teague, H. L. "Advanced Component Test Facility." *Review of Advanced Solar Thermal Power Systems,* November 15-17, 1978, Denver, CO. DOE/JPL-1060-78/5.

Brown, G. W., and Ballance, J. W. *Integration of Solar Thermal Power Plant into Electric Utility Systems.* Vol. 2. Southern California Edison Company, September, 1976. Report 76-RD-63.

Brown, K. C., *Industrial Process End-Use Matching.* Golden, CO: Solar Energy Research Institute, June, 1978.

Brown, K. C. "Notes on the Potential Small Power System Market and Suggestions Regarding Attribute 6: Research, Development and Industrialization Required." *Small Power Systems Study.* Available upon request from K. C. Brown, Solar Energy Research Institute, Golden, CO 80401.

Brumieve, T. D. *Eye Hazard and Glint Evaluation for the 5 MWe Solar Thermal Test Facility.* Albuquerque, NM: Sandia Laboratories, May, 1977. SAND 76-8022.

Burns & McDonnell, Jet Propulsion Laboratory. *Assessment of the Potential of Solar Thermal Small Power Systems in Small Utilities,* August, 1978.

Butler, B. L. *Solar Materials R&D Overview for Solar Thermal Power Systems.* 11th National SAMPE Technical Conference; November 13-15, 1979, Boston, MA. Available from Society for the Advancement of Material and Process Engineering, P. O. Box 613, Azusa, CA 91702.

Curto, P. A., and Nilodem, Z. D. *Solar Thermal Repowering.* McLean, VA: MITRE Corporation, METREK Division, May, 1978. R-1860.

Dickerson, W. C. "Solar Energy for Industrial Process Heat." *Solar Age,* August, 1977.

Doane, J. W., et al. *A Government Role in Solar Thermal Repowering: Draft Report of the Repowering Strategy Analysis.* Golden, CO: Solar Energy Research Institute, August, 1979. SERI/TP-51-340.

Duffie, J. A., and Beckman, W. A. *Solar Energy Thermal Processes.* New York: John Wiley & Sons, 1974.

Executive Office of the President, Council on Environmental Quality. *Solar Energy: Progress and Promise.* Washington, DC: Council on Environmental Quality, April, 1978.

Fujita, T.; El Gabalawi, N.; Geregra, G.; and Turner, R. N. *Projection of Distributed Collector Solar Thermal Electric Power Plant Economics to Year 1990-2000,* December 1977. DOE/JPL-1060-77/1.

Gupta, B. P., and Kreith, F. (eds.) *Proceedings of the Solar Thermal Concentrating Collector Technology Symposium.* Golden, CO: Solar Energy Research Institute, August, 1978. SERI/TP-34-048.

Jet Propulsion Laboratory, Thermal Power Systems Advanced Solar Thermal Technology Project. *Techno-Economic Projections for Advanced Dispersed Solar-Thermal Electric Power Plants to Years 1990-2000,* November 1978. Preliminary Report. DOE/-JPL-1060-78/4.

Kochman, Neal. *Assessment of Solar Thermal Hybrid Repowering,* PRC Energy Analysis Co., May, 1978. R-1860.

Muller & Company. *Solar Thermal Collector Manufacturing Activity.* July through December 1977; January through March 1978; April through July 1978. Washington, DC: The Solar Energy Research and Education Foundation, July, 1978.

"Plastic Mirrors Feature of Heliostat Developed at Brookhaven Lab." *Energy Insider.* **1** (17): 7 (September 4, 1978).

Powell, J. C., and Grosskreutz, J. C. *Dynamic Conversion of Solar-Generated Heat to Electricity.* National Aeronautics and Space Administration, August, 1974. CR134724. NTIS N-75-16079.

Public Service Company of New Mexico. *Technical and Economic Feasibility of Solar Hybrid Repowering,* June, 1977. DOE Contract No. CST 2499.

Rogan, J. E. *Industrial Applications of Solar Total Energy.* Vol. 3. Huntington Beach, CA: McDonnell Douglas Astronautics Company, April, 1977. SAN/1132-2/3.

Sandia Laboratories. *Solar Energy at Sandia Laboratories.* Albuquerque, NM: Sandia Laboratories, December, 1977.

Skinrood, A. C. *Status Report on a High-Temperature Solar Energy System.* Livermore, CA: Sandia Laboratories, September, 1974. SAND 74-8017.

Solar Thermal Test Facilities Users Association, Proceedings of Annual Meeting: Technical Sessions. April 11-12, 1978, Golden, CO.

Steity, P.; May, L. G.; and Perkins, S. P., Jr.

Assessment of the Potential of Solar Thermal Small Power Systems in Small Utilities. Kansas City, MO: Burns & McDonnell Engineering Company, August, 1978. JPL Contract 954971.

Thornton, J. P.; Brown, K. C.; Edgecombe, A. L.; Finegold, J. G.; Herlevich, F. A.; and Kriz, T. A. *Comparative Ranking of 1-10 MWe Solar Thermal Electric Power Systems. An Executive Overview.* Golden, CO: Solar Energy Research Institute, July, 1978. SERI/TR-35-238.

U.S. Congress, House of Representatives, Hearing Before the Committee on Science and Technology. Ninety-sixth Congress. *1980 U.S. Department of Energy Authorization.* Vol. 1. Washington, DC, February 8, 1979.

U.S. Department of Energy. *Division of Solar Technology, Program Summary.* Washington, DC: U.S. Department of Energy, January, 1978. DOE/ET-0018/1.

U.S. Department of Energy. *Solar Thermal Power Systems Programs, Program Summary.* Washington, DC: U.S. Department of Energy, October, 1978. DOE/ET-007.

U.S. Department of Energy. *Solar Thermal Power Systems, Annual Technical Progress Report FY 78.* Washington, DC: U.S. Department of Energy, June, 1979. DOE/ET-0078.

U.S. Department of Energy. *Solar Thermal Power Systems, Program Summary.* Washington, DC: U.S. Department of Energy, December, 1979. DOE/ET 0078.

Vant-Hull, L. L., and Hildenbrant, A. F. "Solar Thermal Power System Based on Optical Transmission." *Solar Energy,* **16** (6): 575-79 (November-December, 1977).

Vant-Hull, L. L. "Development of Solar Tower Program in the United States." *Optical Engineering,* **16** (6): 575-79 (November-December, 1977).

Solar Thermochemical

Copeland, R. J. "Rough Cost Estimates of Solar Thermal/Coal or Biomass Derived Fuels." June, 1979. AIAA #79-0988.

Escher, W. J. D. "Solar-Produced Fuels—A Major Programmatic Gap." Solar Energy Policy Forum, June 26, 1978, Chicago, IL. Available upon request from W. J. D. Escher, Escher-Foster Technology Associates, P.O. Box 189, St. Johns, MI 48879.

Hlavka, G. E. "Overview of Solar Thermal Fuels and Chemical Project Activities." *Semi-Annual Review of Advanced Solar Thermal Technologies.* November 15-17, 1978, Denver, CO. Available upon request from G. E. Hlavka, Jet Propulsion Laboratory, 4800 Oak Grove Drive, Pasadena, CA 91040.

Marchetti, C. "Hydrogen and Energy." *Chemical Economy & Engineering Review,* **5** (1): 6-25, 1973.

U.S. Department of Energy. *Monthly Energy Review.* (June, 1979).

Weingart, J. "The Helios Strategy." *Technological Forecasting and Social Change,* **12** (1978).

Solar Users and Consumers

Campbell, V. N., et al. *An Attitudinal Study of the Home Market for Solar Devices.* Decisions and Designs, Inc., for the Federal Energy Administration, September, 1977. TR-77-5-53.

Community Services Administration. *Solar Energy Policy.* Washington, DC: Community Services Administration, July, 1978.

Cook, J., et al. *Solar Consumers, An Investigation Toward Commercialization.* Tempe, AZ: Arizona State University, June, 1977.

Leonard-Barton, Dorothy. *The Diffusion and Adoption of Solar Equipment Among California Homeowners: Report on a Pretest Study.* Institute for Communication Research, Stanford University, December, 1978.

Morse, F., and Maybaum, Michael. *Commercialization Strategy of Passive Solar Heating.* Washington, DC: U.S. Department of Energy, November, 1978.

Real Estate Research Corporation. *Selling of the Solar Home.* Washington, DC: Department of Housing and Urban Development, April, 1978, HUD-PDR-296.

Rogers, E. M., and Shoemaker, F. F. *Communications of Innovations: A Cross-Cultural Approach.* The Free Press, 1971.

Solar Energy Intelligence Report, **5** (24): 175 (June, 12, 1978).

Sparrow, T., et al. *Socioeconomic Factors Affecting the Adoption of Household Solar Technology:* Preliminary Findings. University of Houston, 1977.

U.S. Department of Energy. *The Energy Consumer: A National Solar Strategy and Solar Resource Guide,* **1** (4) (August/September, 1979).

U.S. Department of Energy. *Status Report on Solar Energy Domestic Policy Review.* Public Review Copy, Department of Energy, August 25, 1978.

Unseld, C. T., and Crews, R. *Residential*

Solar Energy Users: A Review of Empirical Studies and Related Literature. Golden, CO: Solar Energy Research Institute, 1979. TR-53-245d.

Vories, R. *Solar Warranties Workshop: A Summary.* Solar Energy Research Institute, SERI/TR-62-046.

Weis, P., and Farhar, B. *Public Opinion About Energy: A Literature Review.* June, 1979. SERI/TP-51-1550.

Yarosh, M. M., and Litka, A. H. *Solar Commercialization: The Consumer Experience.* Cape Canaveral, FL: Florida Solar Energy Center, 1978.

State and Local Governments

Allen, Margaret. *Joining Forces: Cooperative Environmental Action by Urban Groups—Cases and Lessons.* Washington, DC: Urban Environment Foundation, April 1979.

Davis Energy Conservation Report: Practical Use of the Sun. A Comprehensive Energy Conservation Program for the City of Davis, CA, Washington, DC: March, 1977.

deSerio, F., and Spears, J. W. "Sunact." *Proceedings of the 4th National Passive Solar Conference,* October 3-5, 1979, Vol. 4. pp. 77-80. Kansas City, MO: The American Section of the International Solar Energy Society, 1979.

Docter, R. D. *The Solar Domestic Policy Review: A Task Undone.* California Energy Commission, 1979.

Institute for Local Self-Reliance. *Planning for Energy Self-Reliance: A Case Study of the District of Columbia.* Washington, DC: Institute for Local Self-Reliance.

Pollack, P. *The Implementation of State Solar Incentives: Land-Use Planning to Ensure Solar Access.* Golden, CO: Solar Energy Research Institute, March, 1979. SERI/TR-51-163.

Solar Energy Research Institute. *Solar Energy Information Locator.* Golden, CO: Solar Energy Research Institute.

Solar Energy Research Institute. *The Implementation of State, Solar Incentives: A Preliminary Assessment.* Golden, CO: Solar Energy Research Institute, January 1979. SERI/TR-159.

U.S. Department of Energy. *The Energy Consumer: A National Solar Strategy and Solar Resource Guide.* **1** (4) (August/September, 1979).

U.S. Department of Energy. *Evaluation of the Energy Extension Service Pilot Program.* Report to Congress, DOE Assistant Secretary for Conservation and Solar Applications, Office of State and Local Programs. Washington, DC: U.S. Department of Energy, March 1979. DOE/CS-0074.

"While Washington Fiddles, Localities Move on Energy Shortage." *U.S. News and World Report.* 69-70 (September 3, 1979). Report on 1st. Annual Conference on Community-Scale Renewable Energy Systems, Boulder, CO, August 20-21, 1979.

Thermal and Chemical Storage

Ashbury, J. G., and Mueller, R. O. "Solar Energy and Electric Utilities: Should They Be Interfaced?" *Science,* **195:** 445-50 (February 4, 1977).

An Assessment of Energy Storage Systems Suitable for Use by Electric Utilities. Golden,CO: Solar Energy Research Institute, July, 1976. EPRI EM-264.

Baylin, F. *Evaluation of Low-Temperature Thermal Storage Programs.* Golden, CO: Solar Energy Research Institute, January, 1979. SERI/RR-54-164.

Bryant, H. C., and Zaivgrando, F. "A Salt Gradient Solar Pond." *Solar Age,* **3** (4): 21 (April, 1978).

Davis, J. Michael. *Thermal Energy Storage R&D Program Status Report and Plan.* Washington, DC: U.S. Department of Energy, Solar Division, February, 1978.

General Electric Company. *Requirements Assessment of Photovoltaic Power Plants in Electric Utility Systems,* June, 1978. EPRI Project 651-1.

General Electric Company. *Requirements Assessment of Wind Power Plants in Electric Utility Systems,* July, 1978. EPRI Project 740-1.

Hoffman, H. W., et al. (eds.) *Proceedings of 2nd Annual Thermal Energy Storage Contractors' Information Exchange Meeting,* September, 1976, Gatlinburg, TN. CONF-770955.

Hooper, F. C. "Annual Storage." *Solar Age,* **13** (4): 16-17 (April, 1978).

Hooper, F. C., et al. *Solar Space Heating Using Annual Heat Storage.* Washington, DC: U.S. Department of Energy, 1978. Report No. C00/2939-5.

Iannucci, J., and Eaker, P. "Central Solar/Fossil Hybrid Electrical Generation, Storage Impacts." *Proceedings of the 1978 Annual AS/ISES Meeting,* August 28-31, 1978, Denver, CO, pp. 904-12.

Lof, G. O., and Tybout, R. A. "Cost of House Heating with Solar Energy." *Solar Energy,* **14** (3): 253 (1973).

Margen, P. "A Central System for Annual

Solar Heat." *Solar Age,* **3** (10): 22-26 (October, 1978).

Meyer, C. F. "Status Report on Heat Storage Wells." *Water Resources Bulletin,* **12** (2): 237-52, April, 1976.

Mitchell, B. M.; Marring, W. G., Jr.; and Action, J. P. *Electricity Pricing and Load Management: Foreign Experience and California Opportunities,* March, 1977. Rand Report R-2106-CERDC.

Mueller, R. O., et al. "Optimal Design of Seasonal Storage for 100% Solar Space Heating in Buildings." *Proceedings of AAAS Symposium: Efficient Comfort Conditions: Heating and Cooling of Buildings,* February, 1978.

"Optimal Design of Seasonal Storage for 100% Solar Space Heating in Buildings." *Proceedings of AAAS Symposium: Efficient Comfort Conditions: Heating and Cooling of Buildings,* February, 1978.

U.S. Department of Energy, Domestic Policy Review of Solar Energy. Washington, DC: U.S. Department of Energy, February, 1979. TID-22834.

U.S. Department of Energy. *Solar Geothermal Electric and Storage Systems Program Summary Document,* January, 1979.

Weingart, J. "Systems Aspects of Large Scale Solar Energy Conversion." International Institute for Applied Systems Analysis, Luxemburg, Austria, 1977. RM-77-23.

Wilson, W. F. (ed.) *Proceedings of 3rd Annual Energy Storage Contractors' Information Exchange Meeting,* December, 1978, Springfield, VA.

Wyman, C. *Thermal Storage Technologies.* Golden, CO: Solar Energy Research Institute, December, 1978. SERI/TR-34-089.

Wind Energy

American Wind Energy Association. *An Index of Manufacturers, Researchers and Distributors of Wind Energy Conversion Systems.* Washington, DC: U.S. Department of Energy, February, 1978. DOE/RE/3533-78/1.

Bateman, W. *Wind Energy Systems.* Statement before the U.S. House of Representatives Subcommittee on Energy Development and Applications, Committee on Science and Technology, July 30, 1979.

Bennington, G. *Solar Energy: A Comparative Analysis to the Year 2020,* McLean, VA: MITRE Corporation, METREK Division, March, 1978. MITRE Technical Report MTR-7579. Report No. HCP/T2322-01.

Bingham, C. E., and Jayadev, T. S. *Correlation of Solar and Winds Resources for Hybrid System Applications.* International Solar Energy Society (ISES) Conference, 1978.

Blackwell, B. F., et al. "Engineering Development Status of the Darrieus Wind Turbine." *Journal of Energy,* **1** (1): 50-64 (January-February, 1977).

Blackwell, B. F. *The Vertical-Axis Wind Turbine "How it Works."* Albuquerque, NM: Sandia Laboratories, December, 1974.

Blackwell, B. F., and Feltz, L. V. *Wind Energy—A Revitalized Pursuit.* Livermore, CA: Sandia Laboratories, March 1975. SAND-75-0166.

Braasch, R. H. "Power Generation with a Vertical-Axis Wind Turbine." *Proceedings of the Second Workshop on Wind Energy Conversion Systems,* June 9, 1975, Washington, DC, McLean, VA: MITRE Corporation, September, 1975. NSF-RA-N-75-050.

Braasch, R. H. "The Design, Construction, Testing and Manufacturing of Vertical Axis Wind Turbines." The Second International Symposium on Wind Energy Systems, October 3-6, 1978, Amsterdam, Netherlands. Cranfield, Bedford, England: BHRA Fluid Engineering; 1978, pp. E3-23 to E3-38. Two vols., ISBN-0-906085-03-0.

Changery, M. J. *Initial Wind Energy Data Assessment Study.* Asheville, NC: National Science Foundation/National Oceanic and Atmospheric Administration, 1974. CONF-740780.

Clark, Wilson. "Energy from the Winds." *Energy for Survivial.* Garden City, NY: Doubleday, 1974.

Consolidated Edison Company of New York. *General Requirements for Parallel Operation of Windmill Generators with Low Tension Service.* Application and Design Manual No. 4, New York, May 4, 1977.

Coty, U. A. *Wind Energy Mission Analysis.* Washington, DC: Energy Research and Development Administration, September, 1976. SAN/1075-76/1.

Darrieus, G. J. M. *Turbine Having Its Rotating Shaft Transverse to the Flow of Current.* United States Patent No. 1, 835, 018, December 8, 1931.

Divone, L. V. "Recent Developments in Wind Energy." The Second International Symposium on Wind Energy Systems, October 3-6, 1978, Amsterdam, Netherlands. Cranfield, Bedford, England: BHRA Fluid Engineering, 1978. pp. A3-17 to A3-28. Two vols., ISBN-0-906085-03-0.

Elderkin, C. E.; Ramsdell; J. V.; and Tennyson, G. P. "Wind Characteristics Workshop, June 2-4, 1976." Boston, MA: *American Meteorological Society Bulletin,* **58** (1): 45-51 (January, 1977).

Eldridge, Frank R. *Wind Machines.* 2d Revised Ed. New York: Van Nostrand Reinhold, 1980.

Electric Power Research Institute. *Electric Utility Solar Energy Activities, 1978 Survey.* York, W. L. (Compiler). Palo Alto, CA: Electric Power Research Institute, May, 1979. ER-966-SR.

Elliot, D. L. *Synthesis of National Wind Energy Assessment.* Richland, WA: Battelle Pacific Northwest Laboratories, July, 1977. Report No. BNWL/WIND-5.

Energy Research and Development Agency, Division of Solar Energy, *Solar Program Assessment: Environmental Factors.* Washington, DC: Energy Research and Development Agency, March, 1977. Includes the following reports: "Solar Heating and Cooling of Buildings," ERDA 77-47/1; "Solar Agricultural and Industrial Process Heat," ERDA 77-47/2; "Photovoltaics," ERDA 77-47/3; "Solar Thermal Electric," ERDA 77-47/4; "Solar Total Energy Systems," ERDA 77-47/5; "Wind Energy Conversion," ERDA 77-47/6; "Fuels from Biomass," ERDA 77-47/7; and "Ocean Thermal Energy Conversion," ERDA 77-47/8.

Energy Task Force. *Windmill Power for City People: A Documentation of the First Urban Wind Energy System.* New York City: May, 1977.

Flaim, S., et al. *Economic Feasibility and Market Readiness of Eight Solar Technologies.* Golden, CO: Solar Energy Research Institute (SERI), October, 1978. SERI/TR-52-055.

General Electric Company. *Requirements Assessment of Wind Power Plants in Electric Utility Systems.* Palo Alto, CA: Electric Power Research Institute, June, 1978. EPRI-ER-685-SY.

General Electric Company. *System Dynamics of Multi-Unit Wind Energy Conversion Systems Applications,* Final Report, General Electric Co., February 15, 1978. Prepared under DOE Contract No. E (49-18)-2332.

Golding, Edward. *The Generation of Electricity by Wind Power.* London: E & F. N. Spon, 1955.

Gustavson, M. R. *Wind Energy Resource Parameters,* McLean, VA: MITRE Corporation, METREK Division, February, 1977. Report M77-29.

Hackleman, Michael A. *Wind and Windspinners: A Nuts and Bolts Approach to Wind-Electric Systems.* Sangus, CA: Earthmind, 1974.

Hamilton, R. "Can We Harness the Wind?" *National Geographic* **148:** 812-829 (December, 1975).

Handbook of Homemade Power. By the staff of the Mother Earth News, New York: Bantam Books, 1974.

Hunt, V. Daniel. *Windpower—A Handbook on Wind Energy Conversion Systems.* New York: Van Nostrand Reinhold, 1981.

Jayadev, T. S., and Henderson, J. "Industrial Applications of Solar-Wind Hybrid Systems." *Wind Technology Journal,* January, 1979.

Jayadev, T. S., et al. *Assessment of Conversion Technologies.* Vol II: *Solar-Wind Hybrid Systems.* Golden, CO: Solar Energy Research Institute, February, 1979. SERI/TR-35-078.

McConnell, R. D. "Giromill Overview." Wind Energy Innovative Systems Conference, May 23-25, 1979, Colorado Springs, CO. Golden CO: Solar Energy Research Institute, May, 1979. SERI/TP-35-263.

MITRE Corporation. *Preliminary Wind Commercialization Program.* McLean, VA, November, 1976.

"Other Business Technology News." *Solar Energy Intelligence Report.* **5** (39): 388 (September 24, 1979).

Putnam, Palmer Cosslett. *Power from the Wind.* New York: Van Nostrand Reinhold, 1974.

Reynolds, John. *Windmills & Watermills.* New York: Praeger, 1970.

Rockwell International, Rocky Flats Plant. *Guide to Commercially Available Wind Machines.* Golden, CO: Rockwell International, April, 1978. Report #RFP-2831/3533/78/3.

Rogers, S. E., et al. *Evaluation of the Potential Environmental Effects of Wind Energy Systems Development.* Battelle-Columbus Laboratory, August, 1976. ERDA/NSF/07378-75/1. Prepared for the Energy Research and Development Administration, Washington, DC.

Simmons, Daniel M. *Wind Power.* Park Ridge, NJ: Noyes Data Corporation, 1975.

South, P., and Watts, A. "Magdalen Islands VAWT Field Test." Workshop on

Economic and Operational Requirements and Status of Large Scale Wind Machines, March 28-30, 1979, Monterey, CA. Palo Alto, CA: Electric Power Research Institute, July, 1979. ER-1110-SR.

Stokhuyzen, Frederick. *The Dutch Windmill.* Trans. from Dutch by Carry Dikshourn. New York: Universe Books, 1963.

Stoner, Carol. *Producing Your Own Power; How to Make Nature's Energy Source Work For You.* Emmaus, Pa.: Rodale Press, Book Division, 1974.

Swetman, G.; Elridge, F.; and Jardine, D. *Energy Rate Initiatives Study of the Interface Between Solar and Wind Energy Systems and Electric Utilities.* McLean, VA: The MITRE Corporation, March 31, 1977. PB 265-607.

Tompkins, D. M. *Wind Energy Conversion Systems (WECS) for Central Station and Dispersed Power Applications.* Arlington, VA: JBF Scientific Corporation, 1979.

Torrey, Volta. *Wind-Catchers—American Windmills of Yesterday and Tomorrow.* Brattleboro, VT: Stephen Greene Press, 1976.

U.S. Department of Energy, Assistant Secretary for Energy Technology. *Solar, Geothermal, Electric and Storage Systems: Program Summary Document FY 1980.* Washington, DC: U.S. Department of Energy, January, 1979.

U.S. Department of Energy, Assistant Secretary for Conservation and Solar Applications and Assistant Secretary for Environment. *Environmental Development Plan (EDP), 1977.* Washington, DC: U.S. Department of Energy, March, 1978. Includes the following reports: "Solar Agricultural and Industrial Process Heat," DOE/EDP-0002; "Fuels from Biomass," DOE/EDP-0005; "Ocean Thermal Energy Conversion," DOE/EDP-0006; and "Wind Energy Conversion," DOE/EDP-0007.

U.S. Department of Energy. *Commercialization Task Force Reports.* Vol. 1, *Electric Markets.* Washington, DC: U.S. Department of Energy, September, 1978.

U.S. Department of Energy, Division of Solar Technology. *Wind Energy Systems Program Summary.* Washington, DC: U.S. Department of Energy, December, 1978. DOE/ET-0093.

U.S. Department of Energy, Division of Solar Energy. *Federal Wind Energy Program: Summary Report.* Washington, DC: U.S. Department of Energy, January, 1978. DOE/ET-0020/1.

U.S. Department of Energy. *Domestic Policy Review of Solar Energy: A Response Memorandum to the President of the United States.* Washington, DC: U.S. Department of Energy, February, 1979. TID-22834.

U.S. Department of Energy, Office of Energy Data, Energy Information Administration, *Monthly Energy Review.* July 27, 1979. For information contact: Editor, *Monthly Energy Review,* National Energy Information Clearinghouse, U.S. Department of Energy, 1726 M Street N.W., Washington, DC 20461.

U.S. Department of Energy. *Status Report, Energy Wind Turbines.* Golden, CO: Rockwell International, August, 1979.

U.S. Federal Energy Administration. *WECS Manufacturing and Sales Activity, 1975-76.* Washington, DC,: U.S. Federal Energy Administration, April, 1977. FEA/B-77-121.

Vertical-Axis Wind Turbine Technology Workshop, May 17-20, 1976. Albuquerque, NM: Sandia Laboratories, 1976. SAND-76-5586.

Wilson, R. E.; Lissaman, P. B. S.; and Walker, S. N. *Aerodynamic Performance of Wind Turbines.* Corvallis, OR: Oregon State University, June, 1976.

Wolff, Alfred R. *The Windmill as a Prime Mover.* New York: J. Wiley & Sons, 1885.

ACRONYMS AND ABBREVIATIONS

ABC	Associated Builders and Contractors, Inc.
ACES	Annual Cycle Energy System
AIA	American Institute of Architects
AIARC	American Institute of Architects Research Corporation
AID	Agency for International Development
AIPH	Agricultural and Industrial Process Heat
angstrom (Å)	Unit of length, 10^{10} m
ANL	Argonne National Laboratory (a part of the Department of Energy, located in Argonne, IL)
ANSI	American National Standards Institute
ARI	American Refrigeration Institute
ARS	Agricultural Research Service (a part of the Department of Agriculture)
AS	Assistant Secretaries (especially in Department of Energy)
ASHRAE	American Society of Heating, Refrigeration and Air Conditioning Engineers
ASME	American Society of Mechanical Engineers
ASTM	American Society of Testing Materials
AT	Appropriate Technology
AWEA	American Wind Energy Association
BATF	Bureau of Alcohol, Tobacco, and Firearms. Under the U.S. Department of Treasury
bbl	Barrels (of oil)
BCF	Bioconversion to fuels
BCI	Battery Council International
BEST	Battery Energy Storage Test Facility
BHP	Brake horsepower
BLM	Bureau of Land Management
BLS	Bureau of Labor Statistics
BNL	Brookhaven National Laboratory (Stonybrook, Long Island, NY)
BSF cell	Back surface field solar cell
Btu	British thermal unit
BTX	Benzene, and its derivatives, toluene (methyl benzene) and xylene (dimethyl benzene); aromatic hydrocarbons
CABO	Council of American Building Officials
C&SA	Office of the Assistant Secretary for Conservation and Solar Applications within the Department of Energy.
CBO	Congressional Budget Office
ccf	One hundred cubic feet. A unit of measure used for small amounts of natural gas
CEQ	Council on Environmental Quality
cfs	Cubic feet per second
CNRS	Centre Nationale de la Recherche Scientifique; organization running the solar furnace in France

CONAES	Committee on Nuclear and Alternative Energy Sources
CO_2	Carbon dioxide
COP	Coefficient-of-performance
CPC	Compound parabolic collector
cps	Cycles per second
CRR	Center for Renewable Resources (a Washington-based solar public interest organization)
CRS/LOC	Congressional Research Service/Library of Congress
CSA	See C&SA
CSA	Community Services Administration
DOC	Department of Commerce
DOD	Department of Defense
DOE	Department of Energy
DOI	Department of Interior
DOL	Department of Labor
DOT	Department of Transportation
DPR	Domestic Policy Review
DTL	Direct-to-load photovoltaic power generation.
EB	Electrochemical Branch
EDP	Environmental Development Plan
EES	Energy Extension Service (part of DOE)
EFG	Edge-defined film-fed growth
EIA	Energy Information Administration (part of DOE)
ELI	Environmental Law Institute (Washington-based public interest organization)
EPAB	Environmental Planning and Analyses Branch
EPRI	Electric Power Research Institute
ER	Office of Energy Research, DOE
ERA	Economic Regulatory Administration
ERCs	Energy Research Centers, ERDA
ERDA	Energy Research and Development Administration
ET	Office of Assistant Secretary for Energy Technology, DOE
ETS	Office of Solar, Geothermal and Electric Energy Systems
EV	DOE Office of the Assistant Secretary for Environmental
FEA	Federal Energy Administration
FEO	Federal Energy Office, the predecessor of FEA
FERC	Federal Energy Regulatory Commission, part of DOE
FFB	Fuels from biomass
FFRDC	Federally Funded Research and Development Center, research organizations, within any Federal agency
FHA	Federal Housing Administration
FHLBB	Federal Home Loan and Bank Board
FOB	Free on Board
FPC	Federal Power Commission (now FERC)
FPGMA	Federal Power Generation and Marketing Agencies
FY	Fiscal year
GA	General Atomics Corporation
GAC	General Advisory Committee

GAO	General Accounting Office
GNP	Gross National Product
GPO	Government Printing Office
GRI	Gas Research Institute (located in Chicago, IL)
GSA	General Services Administration
GSO	Geostationary orbit (for satellite positioning)
H_2	Hydrogen gas
HAWT	Horizontal-axis wind turbine
HLE cell	High-low emitter cell
HUD	Department of Housing and Urban Development
HVAC	Heating, Ventilating, and Air Conditioning
HVI	Home Ventilating Institute
IAEA	International Atomic Energy Agency, Vienna, Austria
IAPMO	International Association of Plumbing and Mechanical Officials
IBC cell	Interdigitated back contact photovoltaic cell
ICBO	International Conference of Building Officials
IEA	International Energy Agency, Paris, France
IEEE	Institute of Electrical and Electronics Engineers
IGT	Institute of Gas Technology
IHS	Indian Health Service
IPH	Industrial process heat
IPTASE	Interagency Panel on Terrestrial Application of Solar Energy
ISCWG	International Solar Commercialization Working Group
ISES	International Solar Energy Society
ISO	International Standards Organization
ISWG	International Solar Working Group, part of DOE
JPL	Jet Propulsion Laboratory
JSC	Johnson Space Center. A NASA Research Center in Houston, TX
kW	Kilowatt
kWh	Kilowatt-hour
LASL	Los Alamos Scientific Laboratory
LAT	Local apparant time
LBL	Lawrence Berkeley Laboratory. A DOE National Laboratory in Berkeley, CA
LDC	Less developed country. Often referred to as a developing nation
LLL	Lawrence Livermore Laboratory. A DOE research center in CA
LNG	Liquefied natural gas
LRC	Lewis Research Center
MASEC	Mid-America Solar Energy Center
MBtu	One million (10^6) British thermal units (Btu). The predominant unit of energy in the United States. Generally, one million Btu are worth between $1 and $10. Frequently written MMBtu or MBTU
MeV	One million (10^6) electron volts. Also written as Mev
MHD	Magnetohydrodynamic generation
MIS	Metal insulator semiconductor

MITRE/METREK	The Metrek Division of the MITRE Corporation
MJ	Megajoule, one million (10^6) joules
MMBtu	One million (10^6) Btu (British thermal units)
MOPPS	Market Oriented Program Planning Study
MRI	Midwest Research Institute. Contractor to DOE responsible for the operation of SERI
MSFC	Marshall Space Flight Center. A NASA research center located in Huntsville, AL
MSSTF	Mid-Temperature Solar Systems Test Facility
MW	Megawatt, one million (10^6) watts, a unit of power
MWh	Megawatt-hour, a unit of energy
NAHB	National Association of Home Builders
NANRDF	Native American Natural Resource Development Federation
NASA	National Aeronautics and Space Administration, active in several solar technology areas, especially through JPL, LRC, and MSFC
NAS/NRC	National Academy of Sciences (closely associated with NAE and NRC, located in Washington, DC, performing some energy analysis for DOE)
NBS	National Bureau of Standards
NCAR	National Center for Atmospheric Research (active in CO_2 and other pollutant and weather topics; located in Boulder, CO)
NCAT	National Center for Appropriate Technology (active in many low-cost solar areas. Principal funding from Community Services Administration. Located in Butte, MT)
NCSBCS	National Conference of States on Building Codes and Standards
NEA	National Energy Act. Major energy legislation passed in 1978
NEMA	National Electrical Manufacturers Association
NEP	National Energy Plan
NEPA	National Environmental Policy Act
NESCA	National Environmental Systems Contractors Association
NESEC	Northeast Solar Energy Center (one of four TSEC's located in Cambridge, MA)
NESS	National Energy Supply Strategy (a 1978 DOE study emphasizing near term energy production)
NHAW	North American Heating and Air Conditioning Wholesalers Association
NOAA	National Oceanic and Atmospheric Administration
NO_x	Nitrogen oxides (a fossil fuel pollutant)
NPAC	National Plan for the Academic Commercialization of Solar Energy
NPS	National Park Service
NRC	National Research Council (the research organization of the NAE and NAS)
NSF	National Science Foundation
NSPC	National Standard Plumbing Code
NTIS	National Technical Information Service
NWS	National Weather Service
OECD	Organization for Economic Cooperation and Development (an organization of cooperating countries, mostly in Europe)
OER	Office of Energy Research
OERI	Office of Energy-Related Inventions (a part of DOE/IR)
OMB	Office of Management and Budget
OPEC	Organization of Petroleum Exporting Countries

ORNL	Oak Ridge National Laboratory (a DOE Research Center located in Oak Ridge, TN)
OSHA	Occupational Safety and Health Administration
OSTP	Office of Science and Technology Policy (directly reporting to the President)
OTA	Office of Technology Assessment
OTEC	Ocean thermal energy conversion
PCMs	Phase-change materials
PE	DOE Office of the Assistant Secretary for Policy Evaluation
PEPS	Photovoltaic Electric Power System
PL	Public Law
PNL	Pacific Northwest Laboratories (a DOE Research Center located in Richland, WA; sometimes referred to as Battelle Northwest since Battelle is the contractor)
POCE	Proof of concept experiment
ppm	Parts per million
psi	Pounds per square inch, a unit of pressure
psia	Pounds per square inch absolute
psig	Pounds per square inch gage (gauge)
PUC	Public Utilities Commission
PV	Photovoltaic
Q	Generally, quads or quadrillion Btu (10^{15} Btu).
RA	DOE's Office of the Assistant Secretary of Resource Applications
R&D	Research and development
RANN	Research Applied to National Needs (the portion of NSF which was responsible for the solar program up to 1974)
RD&D	Research, development, and demonstration
RFP	Request for Proposal
rpm	Revolutions per minute
RSEC	Regional Solar Energy Centers
R-value	Resistance value
SA	The Office of Solar Applications in DOE/CSA
SAE	Society of Automotive Engineers
SANCST	Saudi Arabian National Center for Science and Technology
SBA	Small Business Administration
scf	Standard cubic feet
SEAS	Strategic Environmental Assessment Simulation
SEIA	Solar Energy Industries Association
SEIDB	Solar Energy Information Data Bank
SEIR	Solar Energy Intelligence Report (a privately published, weekly solar newsletter)
SEREF	Solar Energy Research and Education Foundation (closely affiliated with SEIA)
SERI	Solar Energy Research Institute
SEO	State Energy Office

SHAC	Solar Heating and Cooling
SHACOB	Solar Heating and Cooling of Buildings (now less used than SHAC)
SIC	Standard Industrial Classification
SIMSHAC	Solar heating and cooling design model
SMACNA	Sheet Metal and Air Conditioning Contractors National Association
SMWIA	Sheet Metal Workers International Association
SNG	Synthetic natural gas
SO_2	Sulfur dioxide (a major pollutant principally coming from the combustion of coal)
SO_x	Sulfur oxide, a fossil fuel pollutant
SPS	Solar Power Satellite or space power systems; also used to refer to small power systems
SRI	Stanford Research International
SSEC	Southern Solar Energy Center, one of four RSECs located in Atlanta, GA
SSIE	Smithsonian Science Information Exchange
SSPS	Space Solar Power Systems
SSSC	Solar Standards Steering Committee
STC	Solar Thermal Conversion
STCR	Solar Thermal Central Receiver
STES	Solar Thermal Energy System
STOLAR	A random event solar system simulation model used by architects
SUEDE	Solar Utilization, Economic, Development, and Employment (a solar training program)
SUN	Western Solar Network Regional Center
SWECS	Small Wind Energy Conversion Systems
SWG	Solar Working Group
TES	Total Energy System
TIC	Technical Information Center (the group responsible for DOE information dissemination)
TRNSYS	Transient solar system simulation program which is useful as a research tool; a dynamic solar system simulation
TS	Thermal storage
TVA	Tennessee Valley Authority
TW	Terawatt (10^{12} W)
U	Coefficient of heat transmission
USDA	United States Department of Agriculture, often termed ODA (responsible for several parts of the national solar program)
VAWT	Vertical Axis Wind Turbine
WAES	Workshop on Alternative Energy Strategies (a study performed by MIT)
WECS	Wind-Energy Conversion Systems
WEIS	Wind-Energy Innovative Systems (a portion of the ODE/WECS program handled by SERI)
Wh/kg	Watt-hours per kilogram
W/kg	Watts per kilogram

Wp	Peak watt
WSUM	Western Solar Utilization Network (one of four ESECs located in Portland, OR)
WTG	Wind Turbine Generator